T0305521

Materials for Engineers

A working understanding of materials principles is essential in every area of engineering. However, the materials requirements of different engineering disciplines can vary considerably. Existing introductory textbooks on engineering materials adopt a universalist approach, providing theoretical development and surveying a landscape of topics suitable for introducing materials engineers to *their* field. *Materials for Engineers: Principles and Applications for Non-Majors* has been constructed with the requirements of non-materials engineering students ("non-majors") in mind. The theoretical foundations of material structure and behavior are curated and focused, and the description of the behavior of materials as they pertain to performance, measurement, and design is developed in detail.

The book:

- Places applications and essential measurement methods before detailed theory
- Features a variety of types of end-of-chapter exercises, including forum discussion topics for online course components
- Emphasizes computer-based problem solving and includes numerous examples and exercises for MATLAB®
- Includes optional "topic" chapters for course customization, including structures, transportation, and electronics
- Outlines practical details of how and why knowledge of materials is necessary for engineers, including the various roles that materials engineers play and the impact of materials on cost, lifespan, and safety of components and products

This textbook is aimed at undergraduate engineering students taking their first materials engineering course. It can also be used by professional engineers interested in a ready reference. A solutions manual, lecture slides, and example data sets are available for adopting professors.

Jonathan B. Puthoff is an Associate Professor in the Chemical and Materials Engineering Department at California State Polytechnic University, Pomona (Cal Poly Pomona). He received his Ph.D. in Materials Science from the University of Wisconsin – Madison and has a research background in physical metallurgy, materials characterization, adhesive performance, and materials modeling.

Vilupanur A. Ravi is a Professor in the Chemical and Materials Engineering Department, California State Polytechnic University, Pomona (Cal Poly Pomona). He has also served as the Department Chair and as the Director of Research and Partnerships for the College of Engineering. Prior to joining Cal Poly Pomona, Dr. Ravi worked in the industry on a broad range of materials, solved process engineering problems, and led product development efforts. His research interests are in high-temperature materials, coatings, and corrosion. He received his Ph.D. in Metallurgical Engineering from the Ohio State University, Columbus, OH.

Materials for Engineers
Principles and Applications for Non-Majors

Jonathan B. Puthoff and Vilupanur A. Ravi

CRC Press is an imprint of the
Taylor & Francis Group, an informa business

Designed cover image: Shutterstock

First edition published 2025
by CRC Press
2385 NW Executive Center Drive, Suite 320, Boca Raton FL 33431

and by CRC Press
4 Park Square, Milton Park, Abingdon, Oxon, OX14 4RN

CRC Press is an imprint of Taylor & Francis Group, LLC

Library of Congress Cataloging-in-Publication Data
Names: Puthoff, Jonathan B., author. | Ravi, Vilupanur A., author.
Title: Materials for engineers : principles and applications for non-majors / Jonathan B. Puthoff and Vilupanur A. Ravi.
Description: First edition. | Boca Raton, FL : CRC Press, 2025. | Includes bibliographical references and index.
Identifiers: LCCN 2024007989 (print) | LCCN 2024007990 (ebook) |
ISBN 9781032102535 (hardback) | ISBN 9781032102542 (paperback) |
ISBN 9781003214403 (ebook)
Subjects: LCSH: Materials--Popular works. | Materials--Textbooks.
Classification: LCC TA403.2 .P88 2025 (print) | LCC TA403.2 (ebook) |
DDC 620.1/1--dc23/eng/20240620
LC record available at https://lccn.loc.gov/2024007989
LC ebook record available at https://lccn.loc.gov/2024007990

ISBN: 978-1-032-10253-5 (hbk)
ISBN: 978-1-032-10254-2 (pbk)
ISBN: 978-1-003-21440-3 (ebk)

DOI: 10.1201/9781003214403

Typeset in Times
by KnowledgeWorks Global Ltd.

Access the Support Material: routledge.com/9781032102535

To my family, without whom this journey would not have been possible.
To my students for their steadfast support and commitment to learning.
To my colleagues, near and far, for their friendship and collegiality.

V. Ravi

For Megan Maryrose

J. Puthoff

Contents

Preface

Introduction to the First Edition

A knowledge of materials, indeed a *materials understanding*, is essential in every area of engineering. Materials are the medium from which concepts and designs are realized, as well as a product in and of themselves. However, the particular needs of the various engineering disciplines are tied to particular subsets of material types, properties, and application practices. Existing introductory textbooks on engineering materials adopt a quite general approach, providing a theoretical development and surveying a broad swath of topics that are suitable mainly for introducing *materials engineers* to *their* field. This universalist perspective and exposition might not be ideally suited for the needs of students in aerospace, mechanical, biomedical, electrical, and the associated technology disciplines whose needs are downstream from the comprehensive source.

This text has been designed and crafted with the requirements of non-materials–engineering students ("nonmajors") in mind. The theoretical foundations of material structure and behavior have been curated and focused, and the description of the behavior of materials as they pertain to applications, performance, and design has been expanded. In keeping with this more practical outlook, particular application areas (labeled "topics"), including structures, transportation, and electronics, have been selected for emphasis. These topics occupy the latter part of the book, and the intention is that the instructor selects those topics that best intersect with the goals of their courses and programs for dedicated study. Any topics not strictly necessary then become opportunities for enrichment.

Outside of the main goal of providing materials-related information to future engineers, the text covers the essential topics that are on the National Council of Examiners for Engineering and Surveying (NCEES)[1] Fundamentals of Engineering (FE) exam. This makes the book generally useful as a resource for engineers of all kinds who are pursuing professional licensure. Since the book also dedicates considerable attention to engineering design considerations involving materials, its adoption contributes to the fulfillment of ABET[2] design-related outcomes. In particular, ABET Criterion #3, Student Outcome #2 indicates that graduates of engineering programs should have

> *[a]n ability to apply engineering design to produce solutions that meet specified needs with consideration of public health, safety, and welfare, as well as global, cultural, social, environmental, and economic factors.*

The book has significant content related to economic and environmental factors in materials selection, the ethical application of engineering, and the use of materials data in technical decision-making. For this reason, it can be used to reinforce a program's orientation toward several student outcomes at the lower-division level.

The overall approach to solving materials problems has been developed with modern computational and collaborative tools in mind. Students' access to powerful computational platforms that transcend the merely numerical to incorporate the symbolic and the graphical has expanded considerably in the years since other textbook offerings were developed. This text aims to incorporate these tools in a way that both supplements their occurrence at other locations in the curriculum and extends their application in a materials-design–related direction. In addition, other activities leverage the learning-management environments that have become a part of college courses today (face-to-face, hybrid, or remote). These activities are intended not only to provide additional opportunities for confronting and reflecting on the subject matter but also to model different modes of collaborative activity commonplace in the engineering profession.

Finally, every effort has been made to present the content in simple and engaging terms. Materials represent an unavoidable and concrete component of everyday life, though their inner workings are somewhat mysterious to non-specialists. Even the brief introduction to the essential engineering science of materials contained in this text can be profoundly revealing. There is no reason that it cannot also be accessible, relatable, and enlightening.

TO THE STUDENT

As a future worker in engineering or the applied sciences, a good understanding of the materials used in components and systems (whether in "real life" or "on paper") is crucial for your career. It is also important for a student to recognize that engineering materials cannot be studied in isolation but must be discussed alongside other engineering topics; it is important that the relationship between materials engineering and the other disciplines be established from the start. As with any other scientific or technical area that is new to you, developing an understanding of materials can be a difficult undertaking. In any endeavor, difficult ones especially, your success will depend on your gifts, resources, and motivation. Your gifts are your own, and this book, your instructor, and your institution are resources. But what of the third item on the list: motivation? If you haven't considered it before, its nature might be mysterious.

As it turns out, your motivation M in any situation can be represented mathematically, after a fashion:

$$M = B \times E$$

where B is the perceived *benefit* in your view and E is your perceived *efficacy*. The B factor is achieved through success in this endeavor. This is the "prize" to be won through study. Consider the benefits that come with completing your course in materials engineering and learning its lessons well:

1. This course is likely required for your program; successful completion means satisfaction of a degree requirement.
2. Fluency in the language of engineering materials expands your ability to work in multidisciplinary teams, and your work alongside materials specialists will be more productive.

3. The scope of problems that you can understand and solve in your own area of study increases – materials-related techniques impact every branch of engineering and engineering technology.
4. Your problem-solving skills will improve in general, and your exposure to unfamiliar technical topics will expand your general-engineering toolkit.

We have outlined a few (predominantly academic and professional) benefits above, and you should consider some of your own that go beyond this list. Perhaps you are also a curious person and want to learn about these topics for the enjoyment and confidence that come with understanding something new.

The E factor is your own assessment of your ability to complete the endeavor competently. If you do not think you can do it, then you will not be motivated to try (and you will probably not succeed). However, if you believe in your abilities – and you should; you are not in engineering school by accident! – then efficacy will follow. Finally, it is important to recognize that the motivation equation is organized as a *product* of B and E. This has the important implication that if $B = 0$ or $E = 0$, then $M = 0$. Therefore, you must have both B and E to be motivated in your work. We have indicated several benefits that we expect you will find reasonable above. Careful consideration of these benefits, along with recognition of your own efficacy, provides you with the two factors necessary to be motivated in your studies.

A NOTE ON UNITS

In the natural sciences and engineering (and unlike in mathematics textbooks), quantities come with units. The magnitudes of these units are crucial to the engineering description of any system. Recall the essential "SI" unit magnitudes, their symbolic representations, and the physical quantities they are associated with:

Unit Magnitude	Symbol	Associated Physical Quantity
meter	m	length/distance
kilogram[a]	kg	mass
second	s	time/duration
kelvin	K	temperature
ampere/"amp"	A	electrical current
mole	mol	# of entities
candela	cd	luminous intensity

[a] This unit magnitude has a multiplier prefix ("k") built into it; see below.

Note that there are some unit magnitudes that are encountered frequently that are similar to the above but that differ in important ways. For instance, the temperature unit degree Celsius ("degree centigrade" or °C) has the same overall *magnitude* as the kelvin but has a different "zero" location on its scale. For this reason, some quantity x °C is not the same physically as the quantity x K in an absolute sense. There is, however, a simple, exact interconversion between the two:

$$x\,°\mathrm{C} = [x + 5463/20]\ \mathrm{K} = (x + 273.15)\ \mathrm{K}$$

Also, physical quantities that are some countable numbers of discrete entities are frequently used without reference to the unit magnitude (the mole), since it is very large: 1 mol ≈ 6.022×10^{23} entities. The unit designation for these quantities is sometimes simply written "#." For example, "3 cats per square meter" is "3 #/m^2" or sometimes just "3/m^2". There is also infrequent necessity to refer to other common numerical-unit magnitudes, such as the coulomb (which is ≈ 6.2415×10^{18} fundamental electrical charges), as the basis for counting entities.

We also form fractions or multiples of fundamental magnitudes according to the International System of Units (SI) convention, employing the multiplier prefixes "kilo", "milli", "mega", etc. These are provided for reference below.

Multiplier Name	Prefix	Multiplier Value
pico-	p-	10^{-12}
nano-	n-	10^{-9}
micro-	μ-	10^{-6}
milli-	m-	10^{-3}
centi-	c-	10^{-2}
(base unit magnitude)	(none)	10^{0}
kilo-	k-	10^{3}
mega-	M-	10^{6}
giga-	G-	10^{9}

Throughout the text, we frequently require *composite* unit magnitudes based on the above. For example, there is no *fundamental* unit magnitude of a force F, but rather a unit magnitude composed of other unit magnitudes:

$$[F] = [\text{force}] = [\text{mass} \times \text{acceleration}] = \left[\text{mass} \times \text{length/time}^2\right] = \text{kg m/s}^2 = \text{N}$$

where N is the familiar Newton. Note the convention that we use here and throughout the book: "[x]" = "units of x". You will certainly be familiar with many of the composite unit magnitudes we encounter studying the engineering science of materials, but some may be new to you. When you see unit-based calculations like the Newton example above, pay close attention and ensure that you reflect on the nature of the composite unit magnitude involved.

SIGNIFICANT FIGURES

Numerical values that we encounter in the real world are never exact, outside of the simplest counting experiments. Said another way, *all* physical values obtained from measurements are estimates. To indicate quantitatively "how exact" a particular result is, we write our estimates in a way that explicitly indicates the limit in *precision* that our measurement or calculation was able to attain. Consider the following two numbers:

$$3.14 \text{ and } 3.1415926536$$

We recognize both numbers as representations of π, but they are different in their precision. The number of digits used in each case (three digits vs. 11 digits) is meant to indicate how close the number written is to the actual value of π. As another example, consider how we might describe the values

$$2.0 \text{ and } 2.00000000000$$

informally in English. The first we might say is "around two", while a good description of the second would be "very, very, very close to two".

Through the above two examples, you can appreciate how the way we represent our results on the page can convey meaning. These two examples might give you the impression that "more digits = better knowledge". However, this is not generally the case. There are some digits used in representations of numbers that have no meaning save as *placeholders*. For example,

$$3{,}700{,}000 \text{ vs. } 3{,}745{,}098$$

Both of these values have the same number of digits, but the way that the first is written implies a certain degree of "looseness" unless we are reassured otherwise. In fact, the implication here is that the zeros in the first number are there for placeholding purposes and serve only to indicate the *size* of the number and nothing about its precision. Let's rewrite these values in scientific notation:

$$3.7\cancel{00000}\times 10^6 = 3.7\times 10^6 \text{ and } 3.745098\times 10^6$$

Written this way, we can indicate the size of the numbers using the separate factor 10^6. This eliminates the need for any placeholder zeros, and so we strike them out. The difference in precision between the two numbers is now obvious.

The procedure for eliminating placeholder digits from a written number in order to ascertain its precision is equivalent to determining the number of its *significant digits*. The rules for determining the significant digits in a quantity are as follows:

1. Since the only digit in use as a placeholder is "0", any digit that is **not** a zero is significant.
2. Any zeros that occur in between non-zero digits must be interpreted as contributing to the precision and are thus significant. (This is because it is typically impossible to know the first and last digits of a number with certainty, but not those in between.). For example, in "3,745,098" above, the zero is significant.
3. Any zeros that precede the first non-zero digit ("leading zeros") are **never** significant; e.g., in "0.00000534", none of the zeros are counted as significant digits.

4. If a value has a decimal point, all of the zeros after the last non-zero digit ("trailing zeros") are significant. For example, "2.00000000000" above has 12 significant digits.
5. Sometimes a number may have placeholder zeros and significant zeros mixed together. When this happens, some notation must be employed to mark which zeros are significant. For example, $850{,}00\underline{0}{,}000 = 8.50000 \times 10^8$. Here, the underbar indicates a significant zero. We will use this underbar convention throughout the book to indicate its significance. Also, the presence of a decimal point after a value ending in 0 indicates that the zero is significant. For example, "120." has three significant digits.

The number of significant digits in a value is typically determined by the nature of the physical measurement used to determine it or the mathematical routine used to calculate it. Within a given value containing a certain number of significant digits, the digits further to the left are more significant, and the digits further to the right are less significant.

In any arithmetic operation, the precision of the various input values limits the precision of the result. The essential principle is that the *final* result can be no more precise than the least precise input value. This is important to keep in mind given the ubiquity of digital calculators that are compelled to provide results that typically have many more digits of precision than the input values.

1. For **addition** and **subtraction**, the result must be rounded and truncated so that it contains no digits that are less significant than the least significant digit among all the inputs. For example, $25\underline{0}$, 42.5, and 6.33 all have three significant digits. Out of all of these, the value $25\underline{0}$ has its rightmost (least significant) digit ($\underline{0}$) in the "ones" place. We construct the sum $25\underline{0} + 42.5 + 6.23 = 298.73$. Since this result can contain no digits less precise than those in the "ones" place (i.e., to the right of 8), we round and truncate the result to 299.
2. For **multiplication** and **division**, the result should be rounded and truncated to have the same number of significant digits as input with the fewest significant digits. For example, $1.00/3.0 = 0.33333\ldots$ by inspection, but the denominator 3.0 (with two significant digits) limits the result to the two most significant digits of 0.33333…, i.e., 0.33.

Throughout this book, it should be apparent from the values provided with examples and exercises (typically, 2–4 significant digits) what the expected precision of the result should be, and students are expected to truncate their numerical results accordingly.

PHYSICAL CONSTANTS

There are several physical constants that are frequently used in materials science and engineering. These are summarized below. Some of these constants have exact values based on the definitions of the unit magnitudes, but not all. For most engineering calculations, only a few of the digits of the exact value of a constant are required. The typical approximate "engineering" values of these constants are also provided. These may be used throughout this book without altering the quantitative result of any exercise.

Constant	Value	Engineering Value
Avogadro constant, N_A	$6.02214076 \times 10^{23}$ #/mol[a]	6.022×10^{23} #/mol
Speed of light in vacuum, c	299,792,458 m/s[a]	3.00×10^{8} m/s
Permeability of vacuum, μ_0	$4\pi \times 10^{-7}$ (kg m)/(s^2 A^2)	1.257×10^{-6} N/A^2
Permittivity of vacuum, ϵ_0	$1/(\mu_0 c^2)$	8.854×10^{-12} C^2/N m^2
Electron rest mass, m_e	$9.10938370 \times 10^{-31}$ kg	9.109×10^{-31} kg
Elementary charge, q_e	$1.602176634 \times 10^{-19}$ A s[a]	1.602×10^{-19} C
Boltzmann's constant, k_B	1.380649×10^{-23} (kg m^2)/(s^2 K)[a]	1.381×10^{-23} J/K
Gas constant, R	$k_B \times N_A$[a]	8.31 J/mol K
Planck's constant, h	$6.62607015 \times 10^{-34}$ (kg m^2)/s[a]	6.626×10^{-34} J s
Faraday's constant, $\mathcal{F}$	$q_e \times N_A$[a]	96,485 C/mol
Gravitational acceleration (Earth, sea level), g	9.8067 m/s^2	9.81 m/s^2

[a] Exact value.

NOTE ON MATHEMATICAL SYMBOLS

Characters that are used to represent as-yet unfixed quantities (i.e., variables) are typeset as individual, italicized Roman alphabetical symbols: *a*, *C*, *m*, etc., according to common practice. This practice thereby distinguishes the variable name "*A*" from the article "A" or the unit of current "A" typographically. In materials science, as with many areas of applied physics, symbols from the Greek alphabet are utilized as variable names as well. These are provided below for you to review or familiarize yourself with.

Greek Character	Uppercase Symbol	Lowercase Symbol
alpha	Α	α
beta	Β	β
gamma	Γ	γ
delta	Δ	δ
epsilon	Ε	ε or ϵ
zeta	Ζ	ζ
eta	Η	η
theta	Θ	θ
iota	Ι	ι
kappa	Κ	κ
lambda	Λ	λ
mu	Μ	μ
nu	Ν	ν
xci	Ξ	ξ
omicron	Ο	ο

(Continued)

Greek Character	Uppercase Symbol	Lowercase Symbol
pi	Π	π
rho	Ρ	ρ
sigma	Σ	σ or ς
tau	Τ	τ
upsilon	Υ	υ
phi	Φ	φ
chi	Χ	χ
psi	Ψ	ψ
omega	Ω	ω

NOTES

1. https://ncees.org
2. https://www.abet.org

Introduction: Engineering Materials: History and Outlook

From Past to Future

LEARNING OBJECTIVES

After completing this chapter, you should be able to:

1. Describe and discuss the central role of materials in the major branches of engineering.
2. Reflect on how the development of technology and industry (and society, overall) is connected to the development of materials.
3. List and define the vertices of the materials tetrahedron. Describe the relationships between the various vertices.
4. Classify the different levels of structure in materials according to the length scale they are associated with.
5. Define "materials property" and describe the different categories of materials properties according to the physical phenomena they are associated with.
6. List and describe the different classes of materials. Reflect on the variations in properties among the different classes and how those property differences influence the material's suitability in different applications.

I.1 MATERIALS AND ENGINEERING

It is a profound but generally underappreciated truth that materials strongly influence the course of our lives. The satisfaction of the physical needs of humankind – needs related to health, safety, transportation, enterprise, and leisure – is provided largely by industrial and consumer products that have been engineered to meet these needs. It is quite natural that the ultimate utility of a particular commodity is more important to the typical consumer than the details of its construction or the material of which it is made. However, the overall design and fabrication of the product, including the selection and fabrication of its constituent materials, must be done with care to safeguard the consumer's health, wealth, and time.

If you are reading this book, you are likely training to become one of these select people who undertake these design and production efforts. Your discipline might be one of any of the main branches of engineering: mechanical, industrial/manufacturing, aerospace, biomedical, or an associated area like engineering technology. A common feature of all these technical areas (perhaps *the* most

DOI: 10.1201/9781003214403-1

significant) is the necessity of applying some understanding of materials during the design and development process. Different disciplines deal with this necessity in different ways and consider the material aspects of their work at different levels of detail. Some examples of materials commonly required by different engineering areas (and which you might already be familiar with) are provided in **Table I.1**.

TABLE I.1
Major Engineering Disciplines and Examples of the Role of Materials in Each

Discipline	Applications and Materials Needs	
Mechanical engineering	*design and manufacture machines for industrial and consumer applications* steels, specialty alloys, aluminum alloys, glass- and carbon-fiber composites, engineering plastics	
Manufacturing engineering	*design products and the processes for fabricating them* steels, aluminum alloys, plastics, elastomers, technical ceramics, electronics packaging	
Aerospace engineering	*design and manufacture aircraft and spaceflight vehicles* aluminum alloys, titanium alloys, glass- and carbon-fiber composites, ceramic-matrix composites	

(Continued)

TABLE I.1 ***(Continued)***
Major Engineering Disciplines and Examples of the Role of Materials in Each

Discipline	Applications and Materials Needs	
Biomedical engineering	*design and build equipment and implants to improve health* biocompatible metals (titanium) and alloys (stainless steel, cobalt-chromium), technical ceramics, plastics, electronics packaging	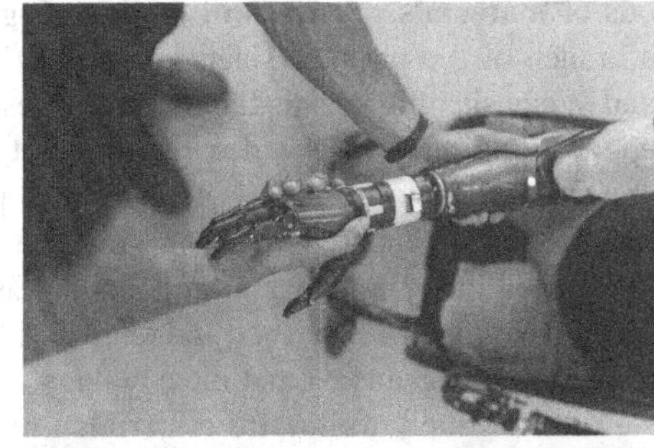
Electrical/ electronics engineering	*design appliances, sensors, and control systems* semiconductors, plastics, optical and magnetic materials	

Note: Most engineers require some understanding of materials to work in their respective areas, and workers in different areas require different materials.

We have established that you will need to know *something* about materials to work in your field, but what? Typical materials-related questions that might arise in the course of your work include

- What material(s) should I select for this product that I have designed?
- Is the material hard or soft, stiff or flexible, strong or weak?
- What conditions will cause the material to break? How does it break?
- Can I heat it to high temperatures? Can I cool it to very low temperatures?
- Will the material be stable in the service environment, or will it degrade?
- Does the material have any negative impacts on health or the environment when in use?
- How much does the material really cost?

As quantitative thinkers and evidence-based decision-makers, engineers should seek to address these questions using numerical values that reflect in some way the characteristics of the materials involved. A curious person recently introduced to such materials-based reasoning might have questions that speak less to the specific application at hand and more to the nature of the materials themselves: "Why should this material possess the characteristics that it does? How can we produce materials whose parameters represent an improvement over the varieties that currently exist? How do you design structures or components using materials information?"

The answers to questions like these are to be found in the theories and methods of **materials science and engineering**.[1] Materials science and engineering is a branch of the physical sciences that has as its objective the description, design, and synthesis of materials themselves. These efforts may be directed at a particular application area, or they may be aimed at resolving fundamental questions about materials behavior without reference to a particular application. The inclusion of both "science" and "engineering" in "materials science and engineering" reflects these combined fundamental and applied aspects. The background knowledge and methods of materials science and engineering overlap with those of chemistry, physics, engineering, and (sometimes) biology.

You are likely to require materials know-how in your career that exceeds the cursory understanding that different materials are suited to different purposes, so you have some responsibility to learn the concepts and methods of materials science and engineering. However, since you have not set out to become a *materials* engineer, only a fraction of the knowledge and practices of materials engineering will be relevant to your work. This book aims to provide the essential materials understanding that you will probably require in your career, as well as presenting detailed examples of applications of that understanding. In this chapter, we familiarize you with the overall extent of the materials field by describing it from two different standpoints. The **historical standpoint** follows the development of materials from prehistory to the modern era. By locating particular materials in the period in which they arose in history, we can establish their relationship to concurrent technologies and their importance to human lifestyles. The **modern standpoint** establishes the basis for materials science and engineering as an independent scientific and technical enterprise. This description of the fundamental concepts and organizing ideas of materials science and engineering will prepare us to study the rest of the topics in this book.

I.2 HISTORICAL DEVELOPMENT OF MATERIALS

Materials represent just one aspect of human technology, but an aspect that in many ways determines the limits on, and future direction of, many other aspects. The capabilities of existing components and devices depend on the materials they are made of, and the development of new materials opens up the possibility of new and better devices and applications. In fact, the different eras of human history and prehistory were originally named after the primary *material* in use at the time, in what is sometimes called the **three-age system**. This naming convention was primarily intended to help categorize archeological artifacts, but with a little imagination, we can extend the concept to fully periodize the historical development of materials. Our interpretation of the history of materials is shown on the approximate timeline in **Figure I.1**.

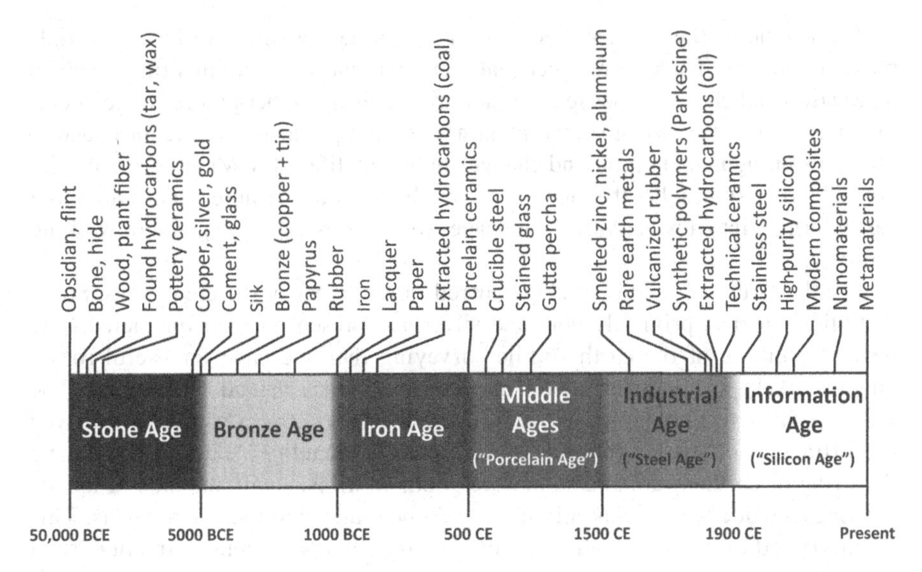

FIGURE I.1 Timeline of the historical development of materials. The central importance of some materials (stone, bronze, iron, porcelain, steel, and silicon) to human living conditions can be used to classify the different eras. Materials of almost all classes – metals, ceramics, and polymers – whether naturally occurring or derived, have been in continuous use since their introduction.

During the **Stone Age**, human technologies were based on found objects and substances harvested from animals, such as sticks, bones, and the like. Modifying these found materials into different and useful forms frequently required the use of tools with significantly different characteristics than the materials to be worked with, i.e., stones. Stones are mixtures of mineral, or **ceramic**, compounds. Because stones are typically quite hard, they can be used to modify the shape of other, softer materials: sticks, bones, hides, etc. Furthermore, because stones are brittle, they themselves can be reshaped by the application of strategically applied percussive forces from other stones. The fracture of some types of brittle stones produces sharp edges, and such reshaped stones are useful in hunting, harvesting plant matter, and food preparation. The adoption of stone tools, which could be manufactured from raw materials found nearly everywhere, presumably helped early humans maintain a prosperous lifestyle as they migrated away from their original habitat and into different climes.

Ceramic materials such as flint and obsidian used in stone-age tools were produced by geologic activity and shaped by hand, but there were other ceramic materials in use at the time that required different sources and strategies to produce. Fired clay pottery for household, agricultural, and artistic purposes became increasingly important as humans adopted sedentary lifestyles. The firing of clay ceramics requires the use of kilns that can reach moderate temperatures. The ability to accumulate heat and apply it in the processing of materials is of significant historical importance and begins in this age.

The people of the **Bronze Age** made an important addition to their materials portfolio: metals. There were other materials introduced during this time, such as more advanced clay ceramics (e.g., stoneware requiring higher processing temperatures) as well as better-engineered plant and animal products, but the introduction of metals brought about profound changes in human lifestyles. Metals are typically not as hard as stones, but they tend to be less brittle. Furthermore, metals are more easily formed into a wider variety of shapes than stones via casting and/or working processes.

The discovery of metals likely occurred in a number of different ways. Some metallic elements, primarily gold and silver, are present in their unreacted state near the surface of the earth. While surveying the landscape for useful rocks, humans likely accumulated metal nuggets and stones veined with metal. The stone-working technology developed during the previous era would have driven the extraction of these noble metals in moderate quantity. Additionally, during the firing of earthenware and stoneware goods in kilns, small amounts of metals existing as oxides in the clay mixtures could be reduced to their pure forms. This discovery led to the introduction of heat-based methods of metal extraction, such as **smelting**.

Bronze is essentially a mixture of two elemental metals: copper and tin, though other formulations exist. Tin melts at a low temperature and is very soft compared to stones. Copper is also soft but melts at a higher temperature. However, a mixture, or **alloy**, of these different metals has two useful properties: the solid alloy is hard but retains a low melting temperature. This means that the copper and tin could be effectively mixed in the molten state, and the liquid mixture could be poured into molds in a process called **casting**. The tools produced in such a fashion were both durable and could possess intricate shapes. The capabilities provided by such tools eclipsed those provided by stones in almost every dimension.

Though copper and tin were the predominant metals in use during the Bronze Age, other metals were obtained and put to limited use, such as arsenic and lead. Another important metal, iron, also became known at this time. Since elemental iron is scarce in Earth's crust, the original sources of iron were strictly meteoric, but advances in the capabilities of smelting **furnaces** eventually enabled the production of metallic iron from ores in quantity. This development inaugurated the **Iron Age**. Since iron is tougher than bronze, tools and weapons made from it are more durable than even the best bronzes. The availability of iron tools leads to tremendous enhancements in agricultural, industrial, and architectural productivity, and thus the overall standard of living. The availability of iron also provided the means to deploy large, well-equipped armies. The weapons and armor (forged swords and such) devised during this period would be standard military equipment for the next several thousand years.

Other materials of note that came into use during this era included several **polymers** of biological origin. The material called lacquer is either the product of the lac insect (deposited on trees as a nesting material) or the sap of the Chinese lacquer tree. When these substances are extracted and redeposited on another surface, they harden and form a waterproof coating. In South and Central America, the natural

elastomer rubber was produced from the sap ("latex") of rubber trees. Rubberized textiles were used in clothing and other domestic applications, in addition to the utilization of rubber balls in sports and games.

What is conventionally called the **Middle Ages** was a time when few new materials were being introduced in the Western world. Elsewhere in the world, however, novel materials continued to be developed. In the Far East, a type of mixed glassy ceramic called porcelain had become commonplace. This material had the qualities of both a very fine finish and significant durability, so it came into widespread use in home furnishings. Its manufacture remained a closely guarded secret for many years and was a significant factor in the development of East-West trade. Since the techniques required to make porcelain were the most advanced examples of materials technology during this time (and it was produced in significant quantities and considered to be quite valuable), we have nicknamed this era the "**Porcelain Age**".

The periods that typically encompass the European Renaissance and **Industrial Revolution** are grouped together on the timeline since the technological developments originating during these periods brought about the mass production of the exceedingly important material called steel. We label this combined era the **Steel Age**. While iron-based materials with similar properties were already known (e.g., "crucible steel" or "Wootz steel"), their use was limited by the amount of effort required to produce them and the small scale of the furnaces available. During this era, however, modern scientific methods were established and applied to the production of metal, establishing the foundation of the modern science of **metallurgy**, and the resulting burst of discoveries in what we now call chemistry and physics drove the development of other technologies. Many new chemical elements, as well as the conditions under which they could be produced, were discovered. The invention of functional **microscopes** expanded our understanding of the previously hidden structures of many substances.

Additional metallurgical advances, such as the introduction of the Bessemer process, established steel as the material whose use defined the era. High-quality steel was cheap and could be fabricated in quantity in many useful forms (e.g., sheets and billets). The application of the steel in machinery and the concentration of productive equipment and personnel in dedicated factories led to rapid industrialization and associated developments in manufacturing, transportation, and architecture. It is also worth noting that the increased energy requirements associated with industrialization necessitated the extraction of more hydrocarbon-based fuels (coal and, later, oil). Other advanced materials synthesis techniques were developed in this era, such as many varieties of synthetic polymers (manufactured from simple chemical precursors) and high-purity (or "technical") ceramics.

Following the expansion of such advanced industrial techniques to every region of the world, a new era, the contemporary era in which we live, began. This era is characterized by rapid technological developments in *all* areas, but the ubiquity of computing resources and high-capacity telecommunications networks have given it the designation of the Information Age. The essential materials that enable modern computing are the **semiconductors** used in the manufacture of compact, low-power, high-speed electronic devices. Though many different varieties of semiconductors exist, most computers utilize silicon processors. For this reason, we call this era the **Silicon Age**. Additionally, developments in photonic technologies, such as the high-purity **glass** used in fiber-optic cables, have come to replace traditional electronic information transmission practices.

There are a number of other important materials that give modern life its qualities, such as advanced **composites** required for aerospace applications and the polymers that constitute a variety of durable or disposable consumer goods. The scientific design of advanced metal alloys and ceramics continues to expand possibilities for applications, such as lightweight terrestrial vehicles and devices designed to operate within the human body. The materials suitable for the latter are sometimes called **biomaterials**. These biomaterials are studied alongside the relevant biology and physiology to understand material-body interactions. (The materials that make up our bodies are also sometimes called biomaterials or "biological materials".) Understanding the bodies of organisms as an assemblage of materials with their own characteristics is an active area of biological and medical research.

It is worthwhile to describe how the development of materials over the centuries has expanded the capabilities of engineers in a quantitative way. As an example, consider the following two **properties** of materials: the material's density and the material's strength. The density ρ (sometimes called "specific mass") is a property you are familiar with from your chemistry courses; it depends on the material's chemical makeup and atomic organization. Density is given in units of $[\rho]$ = [mass/volume] = [mass/length^3] = kg/m^3 or g/cm^3. The "strength" of a material is determined by a complex combination of the material's physical characteristics and has a number of definitions. A common definition of strength is the maximum amount of stress σ that a material can sustain without a qualitative change in its behavior. The units of stress are the same as those of a pressure p: $[\sigma] = [p]$ = [force/area] = [force/length^2] = N/m^2 = Pa (i.e., the pascal).

These properties of the materials available in different eras are compared in **Figure I.2**. **Figure I.2a** shows the materials available prior to the Age of Steel: a collection of metals, ceramics, and various natural materials. The ranges of property values are mapped out as regions in a "space" of density and strength. Engineers of the time would have been constrained to develop devices and structures within this (somewhat limited) space. Applications that required combinations of strength and weight outside of these bounds would not have been possible. At least, they would not be possible until new materials with different properties became available.

Figure I.2b shows the extent of the capabilities of materials currently available in the same property space. The expansion of the materials properties into regions that were previously unattainable has enabled a host of new applications or improvements to existing applications. Consider the properties of aluminum-based alloys in comparison to those of traditional steels; the aluminum alloys provide similar strengths but with a considerably lower weight. As a modern engineer-in-training, you have access to all of these modern materials for your designs and are expected to understand their capabilities.

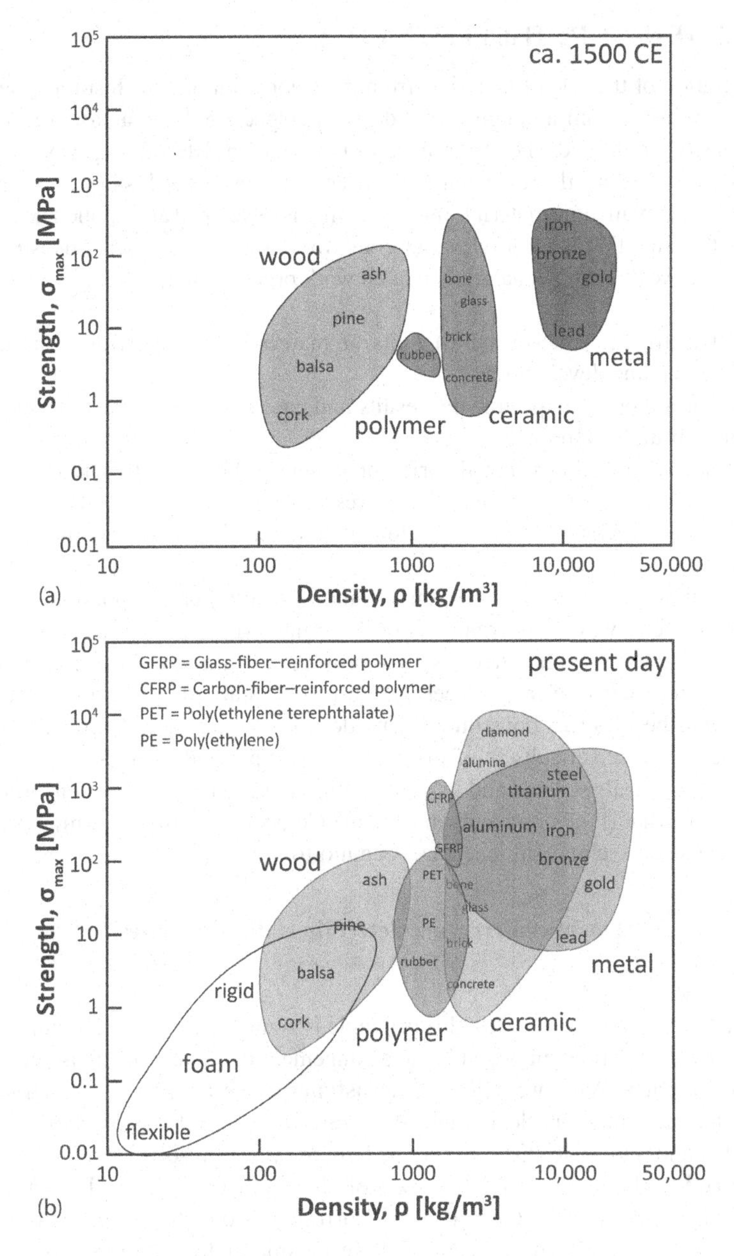

FIGURE I.2 The space of selected materials properties by date. (a) Material strengths and densities available during the early Steel Age. The limited coverage of the space available for the development of applications in earlier eras imposed limitations on the technologies that were possible. (b) Strengths and densities of modern materials. By expanding the capabilities of materials to previously inaccessible regions of the space of properties, many new applications become possible. Note that the ratio of these two properties, strength/density, is sometimes called the "strength-to-weight ratio". (After Fleck et al.[2])

I.3 MATERIALS IN THE PRESENT ERA

Consideration of the role of materials from the standpoint of their historical development is revealing from an engineering-design perspective. Even a casual inspection of this history reinforces the importance of materials in determining why particular technologies were available during a given era. However, the historical perspective does little to explain how materials understanding is developed and applied in the present day, the time that is most relevant to our work as engineers. Some questions that you might have about how materials-related work proceeds in the present day include:

- What are the state-of-the-art tools of materials characterization used in research and development?
- What important experimental results and case studies provide examples to aid our understanding?
- What are the prevailing theories of materials behavior that inform new investigations and help classify new results?
- What strategies exist that can help match materials to applications?

These questions seem quite expansive, and they are, but it is possible to answer them in a concise way if we organize our thinking first.

The tools, results, and theories that provide a framework for understanding materials and their role in engineering are just a small subset of all of the tools, results, and theories that constitute the modern fundamental and applied sciences. In the sciences, select subsets of organizing concepts and ideas are called **paradigms**.[3] If materials science and its applied engineering dimensions constitute such a paradigm, what is included and what is not? Consider first the governing principle of materials science and engineering. As a motto, it is

"The microstructure determines the properties."

Stated in this way, as a motto, this principle is somewhat terse and requires explanation. The structure of an object is the arrangement of and connections between its constitutive parts. What we call the **microstructure** of a material (sometimes just the "structure" when it is clear what we are referring to) includes primarily features situated at a length scale smaller than the human eye can resolve.

Figure I.3 illustrates the relative sizes of objects that interest us. Distances/sizes in the range 10^{-12}–10^{-10} m (or 1–100 pm) correspond to distances in the interior of atoms. Atoms, considered to have a definite volume and no internal structure, are taken to be 100–200 pm (or 0.1–0.2 nm) in diameter. Typical interatomic separations in solid materials are 0.2–0.5 nm. Large compound molecules are a handful of nm across. Above 5–10 nm, materials cease to behave like collections of individual atoms or molecules and begin to behave more like the continuous matter of our common experience.

At this scale, solids may be organized into **crystalline** or glassy arrangements. Objects that possess sizes similar to that of the wavelength of visible light (400+ nm)

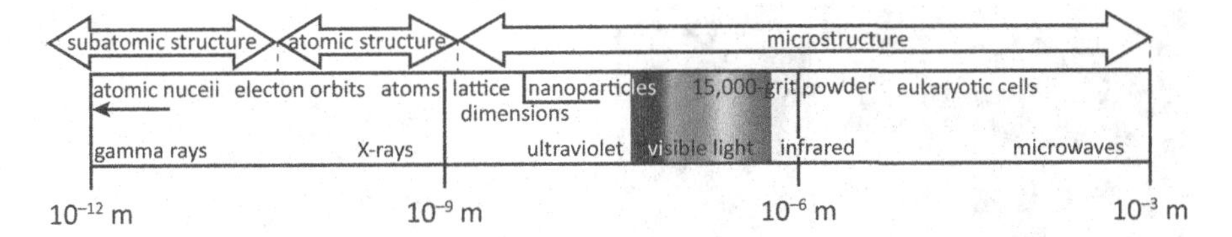

FIGURE I.3 Important length scales in materials science and engineering. The designations for electromagnetic radiation of the specified wavelength are listed along the bottom. Physical objects of the specified size are listed along the top. The lengths/distances typically associated with the microstructure of materials span the range of 10 nm–1 mm.

or greater can be resolved using light **microscopy**, and many important features of materials microstructures exist at or above this size. An example of microstructure, represented by a microscopic image or **micrograph** of a simple carbon steel, is shown in **Figure I.4**. The "grains" of the microstructure are individual crystals of various types, and the irregular grains are collected into a consolidated arrangement.

When we talk about materials properties, we are really talking about parameters that describe some physical behavior of the material and that are amenable to derivation and/or measurement. These property values become important inputs in engineering calculations and assessments. We have already encountered the properties of density and strength. When we think about what a "materials property" means, we consider how the material responds to a particular stimulus, such as the application of a force, the infusion of a certain amount of heat, exposure to an electromagnetic

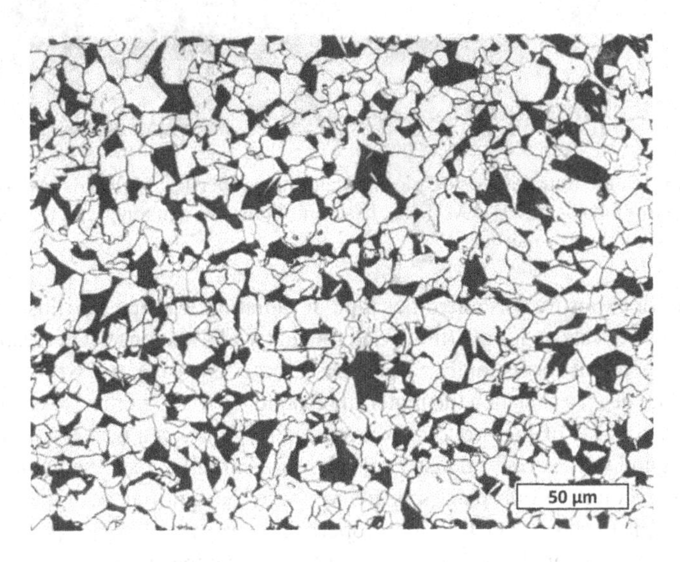

FIGURE I.4 An example of microstructure. This micrograph was collected by focusing a light microscope on the interior surface of a piece of plain-carbon steel. The different crystalline regions or grains (some white, some dark) are separated by boundaries. The scale of the image indicates that the grains are ~30 μm across. (Adapted with permission from Bramfitt and Lawrence.[4])

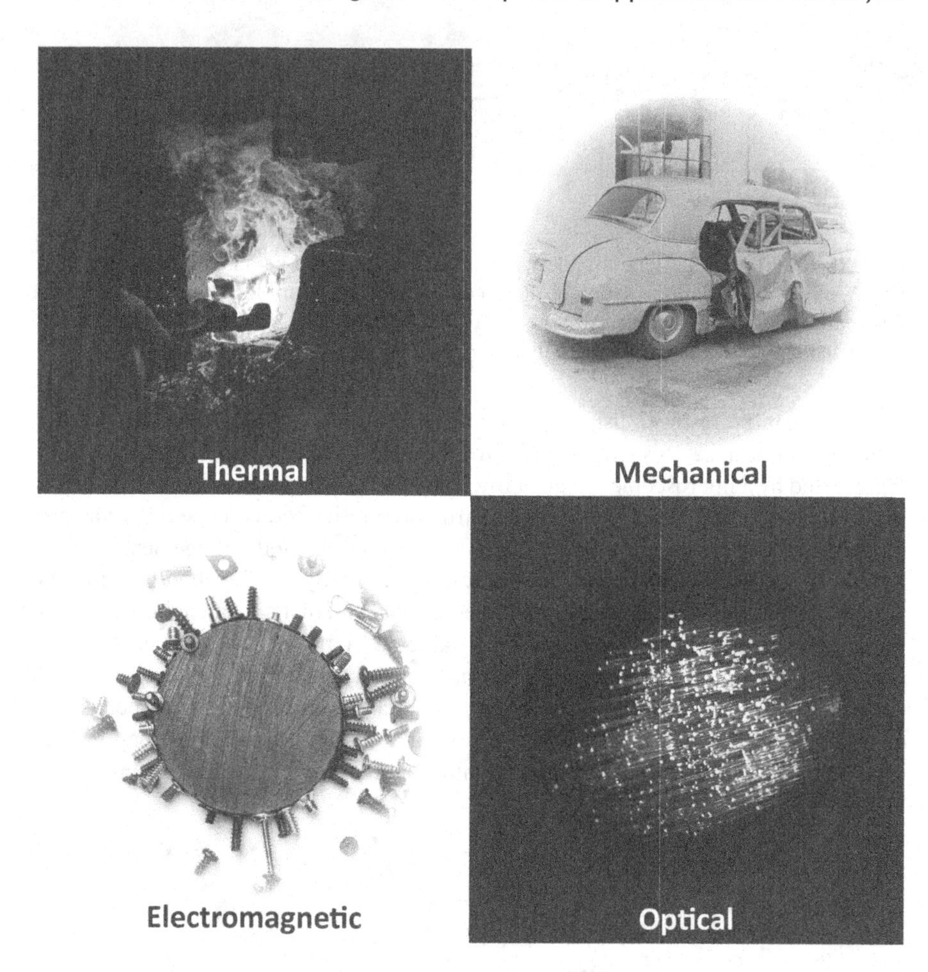

FIGURE I.5 Some important categories of materials properties. Materials respond to their physical environment – forces, EM fields, heat, and light – in a manner and to a degree that is determined by their properties.

(EM) field, or exposure to some other type of radiation. **Figure I.5** illustrates some of these possible categories of stimuli and the material's response. This response can take many different forms, and each of these forms is associated with a particular property. The degree to which the material responds to the stimulus is determined by the numerical value of the property. Each property is given a name, and a physical unit is assigned to it, e.g., for density $[\rho] = kg/m^3$. **Table I.2** lists the names of some commonly encountered properties; some you may be familiar with and some not. It is important to remember that the property is always a physical quantity whose specific value has been determined by an appropriate experiment.

Different materials will have different responses to stimuli, and thus different measured property values should be intuitively obvious. We say that the collection of property values of a material is *characteristic* of that material. What is likely not obvious to you is *why* a particular material should possess the particular values of

TABLE I.2
Selected Materials Properties Names by Category

Mechanical	**Thermal**
How does the material respond to forces?	*How does the material respond to heat?*
elasticity	boiling/melting point
hardness	emissivity
fracture toughness	heat of fusion/vaporization
ductility	thermal conductivity
yield strength	specific heat
tensile strength	glass transition temperature
viscosity	coefficient of thermal expansion
Electromagnetic	**Optical**
How does the material respond to EM fields?	*How does the material respond to light?*
conductivity	absorptivity
permittivity	color
permeability	photosensitivity
dielectric strength	reflectivity
piezoelectricity	refractive index
paramagnetism/ferromagnetism	transmittance

Other
How does the material respond to other stimuli in its environment?
density
environmental stability/corrosion resistance
surface energy
catalytic activity
radioactivity

Note: The properties are organized into broad categories according to the nature of the physical interactions involved.

the properties that it does. Depending on your role, this question may not be of any particular importance, but to a materials engineer, this question is of primary importance. A materials scientist approaches this question by establishing connections between a material's microstructure and its measured property values. Different materials possess different structures, and so they have different property values. Hence: "the microstructure determines the properties".

The determination of fundamental structure-property relationships represents just one axis of activity in the field of materials science and engineering. Observations of a material's structure are made, measurements of the material's properties are taken, and theories are used to draw qualitative and quantitative connections between the two. However, there are other dimensions that require consideration. Consider other activities that materials engineers engage in. For instance, the production of new materials from their precursors: glass from silica, bronze from copper and tin, polyethylene from hydrocarbons, etc. Materials scientists and engineers recognize that the structure a material adopts is the end result of a number of directed and undirected physical and chemical transformations. The design and application of these transformations is called **materials processing**. The net process that a material undergoes in its fabrication results in a particular microstructure. We can therefore identify another axis of activity that strives to relate the processing of materials to their underlying structure.

One more dimension of materials-related activity is worthy of discussion, and that is the identification of relationships between the measured properties of materials and their **performance** in applications. Merely knowing the value of a material property is typically not enough to assess whether or not it is suitable for a given purpose. Typically, part of the **engineering design process** is to identify what overall performance aspects are necessary for a component alongside what performance aspects are merely desirable. These requirements are then translated into requirements on the properties of the material the component is made of via some suitable engineering analysis. Only then is it possible to identify what materials have properties that can provide those characteristics. The phase of the design process that is concerned with identification and selection of the appropriate materials is called **materials selection**. Different engineering areas will typically require different performance aspects in their designs, so workers in these areas will have working familiarity with different types of materials.

All of the different activities related to materials science and engineering are sometimes collected into a schematic called the **materials tetrahedron**, shown in **Figure I.6**. Different materials-related concepts are located at the vertices, and we can identify the structure-properties axis, the processing-structure axis, and the properties-performance axes. The collection of these aspects and activities into a common structure is a recognition that they mutually influence or depend on one another. Frequently, in recognition of the central importance of collecting measurements in all of these activities – recording microstructures, measuring properties, evaluating processes, and quantifying performance – we place another central "vertex" within the tetrahedron for **materials characterization**. As you study materials, pay particular attention to what vertices of the tetrahedron are primarily involved

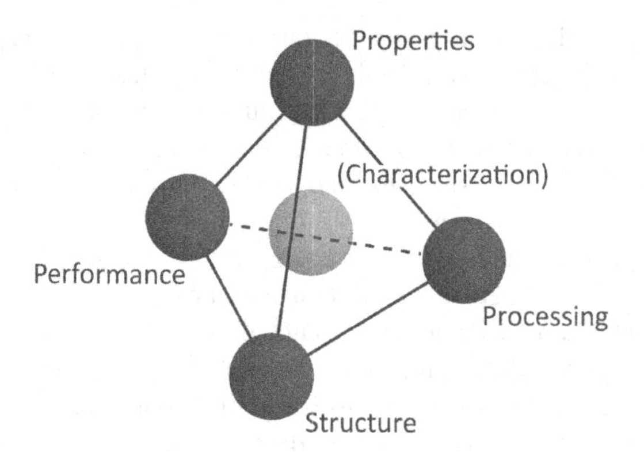

FIGURE I.6 The materials tetrahedron. The activities involved in materials science and engineering are frequently summarized using this representation. We wish to understand how the different vertices are quantified and connected to provide direction to our research and application activities.

in a given topic; this will help you identify the connecting activities and develop a deeper understanding.

Prior to about 75 years ago, materials science and engineering was not considered its own area of investigation, but rather the work was being carried out in physics, chemistry, and other engineering areas. The organization of materials research and development around these structure–properties–processing–performance–characterization principles has produced significant advancements in recent decades, and trends in materials research point toward emerging technologies that could possibly define the next era. Recent (<30 years ago) discoveries in **nanomaterials** can be applied to the design of materials that take advantage of the physical effects that are only accessible at length scales <100 nm. **Metamaterials** are conventional materials that are engineered to produce effects that they do not normally exhibit. Materials-defect engineering has been applied to the storage and retrieval of quantum information for quantum-computing purposes. The technologies that define the next era will likely be based around custom-made matter incorporating structure that spans length scales, from the nanoscopic to the macroscopic, and that have superlative (or conventionally impossible) thermal, optical, tribomechanical, and computational capabilities. If novel applications derived from these discoveries were to become commonplace, human lifestyles would potentially be reshaped in ways as dramatic as those that accompanied the introduction of bronze or iron.

APPLICATION NOTE – MATERIALS CLASSES

At this point in our survey of materials science and engineering, you have probably noticed that the materials we discuss are introduced as members of broad families, **classes**, or "types". Different references/textbooks treat the different classes with different levels of importance, though the classes always have more or less the same members. The primary classes of materials are metals, ceramics, and polymers. We introduce and describe the properties most associated with these classes here, with the understanding that we will discuss them in more detail in later chapters. We also introduce several advanced types of materials – composites and semiconductors – for comparison purposes, along with materials associated with biology and medicine called biomaterials.

Metals

Metals are perhaps the most commonly encountered material in engineering, have been for some time, and will continue to be so for the foreseeable future.

Most metals exist as mineral oxides in the earth's crust, where they (along with oxygen and silicon) are some of the most abundant elements. When extracted from the ground and processed, these ores yield dense, lustrous substances that possess a characteristic set of properties. There are many different metals and a corresponding variation in the properties encountered, but there are some typical behaviors. Metals tend to be heavy (though not always), mechanically durable, and thermally and electrically conductive. They can be employed at moderate temperatures, though the high temperatures typically required to process them tend to increase their overall cost.

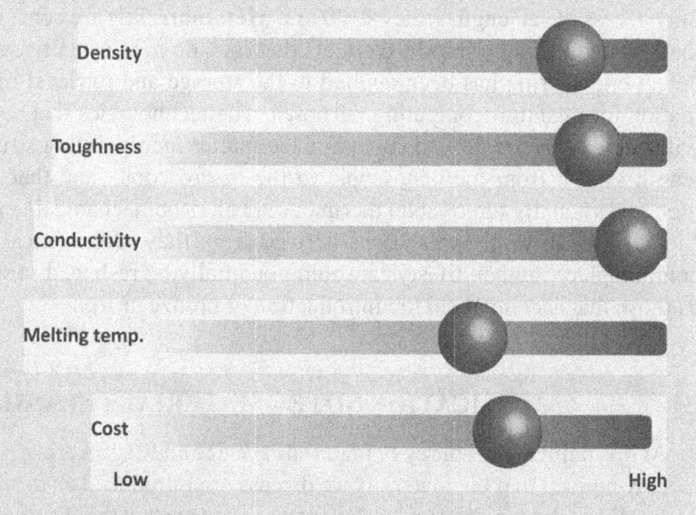

Ceramics

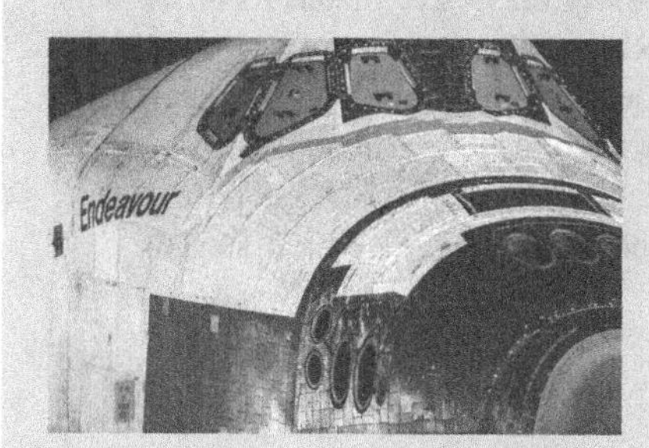

There are many types of ceramics that are used in components of engineered devices. These are sometimes called "technical ceramics" to distinguish them from ceramics used in housewares or as decoration.

Ceramics, consisting chemically of metallic and nonmetallic elements, are akin to rocks and minerals, but they have been deliberately formulated and processed to exhibit a certain set of behaviors. As a class, they tend to be mechanically fragile (in most configurations), though with significant hardness and wear resistance. They are typically less dense than metals and are electrically insulating. They can withstand high temperatures, though this makes them expensive to manufacture.

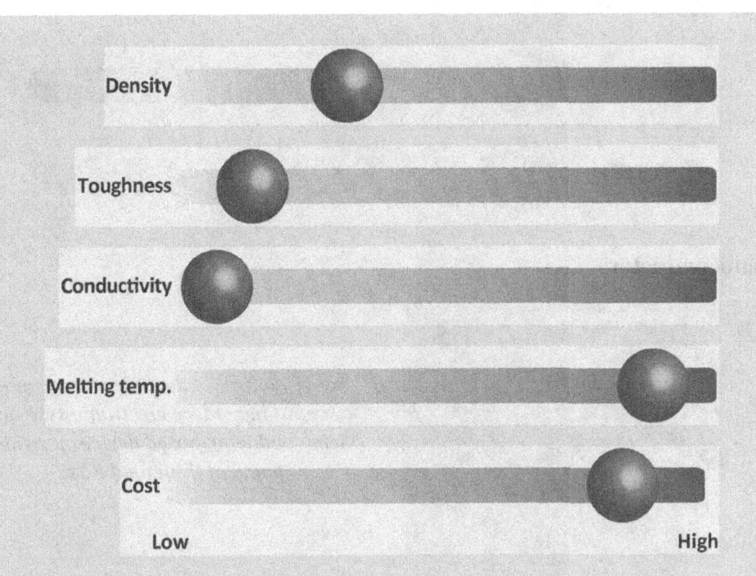

Polymers

From an engineering standpoint, polymers are relatively new materials. They have a reputation as being flimsy and cheap when put in service in place of other materials, but excel in specialized applications where no other materials solution is available.

Polymers are predominantly manufactured from hydrocarbons and can be chemically modified to produce a wide range of properties within the limits of their class. They are relatively tough for their weight and nonconductive, so they frequently appear in the packaging of electrical appliances and electronics. They are restricted to low-temperature applications but can be manufactured in quantity at a low cost.

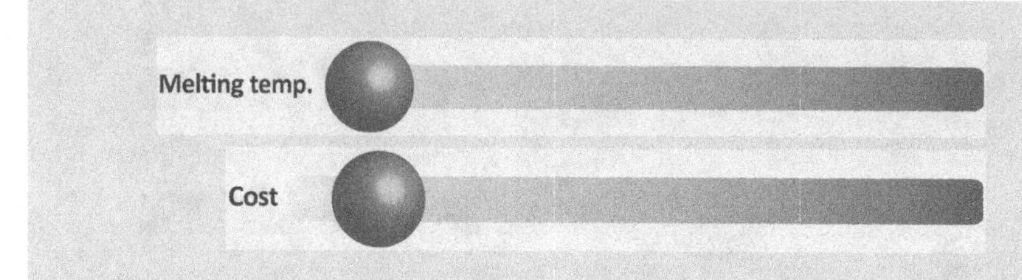

Semiconductors

Active control of the flow of current is essential in modern electronic systems. Semiconductor-based devices provide much of this capability.

The materials that make up electronic devices like microprocessors and various kinds of sensors or transducers are chosen for a different set of properties than most of the materials we typically handle on a day-to-day basis. However, they are no less highly engineered (and, in many cases, even more so) than other kinds of structural materials. In order to attain the fine control over electromagnetic and optical phenomena that is required of modern high-speed, precision electronics in a compact package, materials must be carefully selected and modified to obtain the correct properties. Semiconductors are a class of material that, by virtue of the fact that they are neither fully insulating nor fully conducting, make ideal materials systems when fine-tuning of electrical characteristics is required.

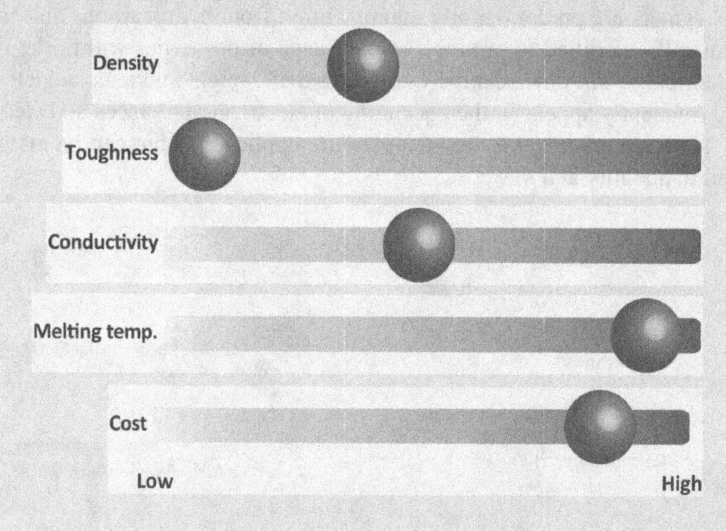

Composites

What happens when you mix materials of two different classes together? The result is a composite that has some of the properties of both.

Metals, ceramics, and polymers are typically formed with all their constituent microstructural components in place. For example, a metal that solidifies from a melt of a particular chemistry attains its ultimate structure during the solidification and heat treatment processes. These materials are sometimes called **monolithic**, meaning that all of the components have (more or less) the *same* origin. Some materials, however, contain microstructural components with *different* origins that have been combined to produce a "mixed" material. Since there are many kinds of metals, ceramics, and polymers, there are many, many kinds of possible ways of combining these monolithic materials to produce composites. Consider a few of these possibilities:

- Metals that contain ceramic reinforcements (e.g., aluminum with Al_2O_3 fibers mixed in).
- Polymers that contain ceramic reinforcements (e.g., "plastics" that contain SiO_2 particles in addition to the base polymer resin).
- Ceramics that contain reinforcements of other ceramics (e.g., glass containing short SiC "whisker" fibers).

As a combination of two different types of material, composites have properties that reflect that mixed identity. For example, a metal/ceramic composite will have an overall density somewhere in between that of the metal constituent and the ceramic constituent. This ability to "tune" the properties of a composite by careful selection of the constituents, their proportions, and their form conveys significant advantages in engineering design.

Biomaterials

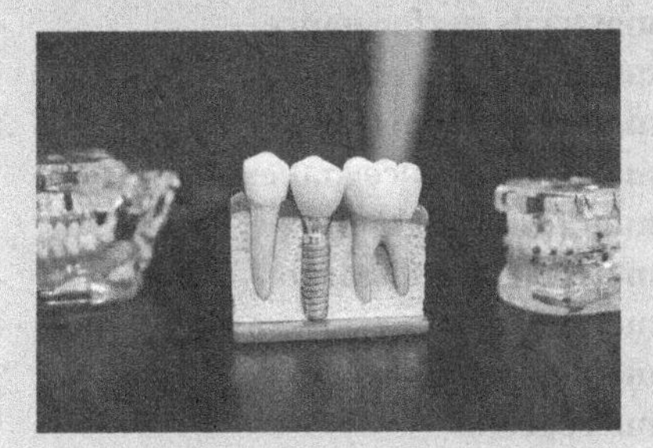

Living things can be understood as assemblies of material components. Devices that interact with these biological systems are made of specially designed biomaterials.

Biomaterials are the constituents of devices that are intended to be used alongside or within living organisms. They can be made of any of the other materials classes – metals, polymers, ceramics, semiconductors, and composites – and are typically formulated so that they are capable of performing their functions in the unique environment of the body. Examples include medical/dental implants, diagnostic test devices, surgical instruments, and replacement tissues. Biomaterials development and biological-materials research are interdisciplinary efforts involving materials science, biophysics, and biology/biochemistry among others.

1.4 CLOSING

By now, you should be convinced of the importance of studying materials (if you weren't before opening this book) and have a good perspective on what that study entails. Even if you are not studying to be a materials engineer, your work will likely be enhanced by an improved understanding of materials principles. There are different but complementary standpoints (the historical and the modern) for approaching the development of materials that are useful to consider as an introduction to the subject. From a modern perspective, we identify 4 (or 5) distinct aspects of materials understanding: processing, performance, properties, and structure (plus characterization). These distinct aspects are unified by their interdependence, as summarized in the materials tetrahedron. An entry point into the application of materials in engineering design comes via a comparison of their recorded properties (which in turn reflect their suitability for a given device or structure). Different classes of materials are associated with different value ranges of these properties, and this classification scheme is important in developing materials understanding for engineering.

1.5 CHAPTER SUMMARY

Key Terms

alloy
biomaterials
Bronze Age
casting
ceramic
composites
crystal
elastomer
engineering design process
furnace
glass
historical standpoint
Industrial Revolution
Iron Age
materials characterization
materials class
materials performance
materials processing
materials properties
materials science and engineering
materials selection
materials tetrahedron
metallurgy
metamaterials
micrograph
microscope

microscopy
microstructure
Middle Ages
modern standpoint
monolithic
nanomaterials
paradigm
polymers
Porcelain Age
semiconductors
Silicon Age
smelting
Steel Age
Stone Age
three-age system

Important Relationships

(None in this chapter.)

I.6 QUESTIONS AND EXERCISES

Concept Review

CI.1 A rapid expansion of the materials available for engineering began with the discovery of modern principles of chemistry (including the principle of the conservation of matter and the nature of gaseous states) in the years 1500–1700 CE. What would you say is the reason for this connection?

CI.2 During the Stone Age, rocks were processed using various strategies (i.e., "treatments") to make them more useful. What are some other examples of treatments that are applied to natural materials (in any era) that you can think of?

CI.3 Another student tells you that one of the properties of a material is its cost (with units [price/mass] = $/kg) because the price is a numerical quantity that describes something about the material. Do you agree? Why or why not?

CI.4 Indicate which label (metal, ceramic, polymer, semiconductor, biomaterial, biological material, or composite) best describes the material in each of these objects: concrete, Pyrex® laboratory glassware, stainless steel ball bearing, silk scarf, nylon jacket, glass-fiber–reinforced epoxy kayak, germanium chip, and artificial heart valve.

CI.5 In upcoming chapters, we will have much to say about the connection between a material's structure and its properties, as well as how materials may be processed to exhibit a particular structure. Describe your interpretation of the relationship between a material's properties and its performance.

CI.6 Nanomaterials, or materials with structural features <100 nm, are difficult to investigate and characterize using common experimental tools like light microscopes. Given the information in **Figure I.3**, why are such microscopes insufficient for the task of imaging nanomaterials?

Discussion-forum Prompt

DI.1 Identify a material that is currently widely used in your field. When was this material introduced? What property category(-ies) are most important in applications that utilize this material? What important applications would you say were not possible before this material was introduced?

PROBLEMS

PI.1 Suppose that you wanted to construct an unpowered mobility device for people with a particular disability. The device must support the weight of a human safely and must be durable enough to last for many years of everyday use. Based on the general characteristics of the materials classes introduced, what would you say are the classes of materials best suited to this application?

MATLAB® Exercises

MI.1 The approximate dates for the materials listed in **Figure I.1** are provided in the file "Exercise MI.1.txt". Plot these data along a linear timeline. You can place text at a particular location on a plot by using the `text()` function. For example, `text(`*xPosition*`,` *yPosition*`,` *string*`, 'FontSize',` *fontSize*`)`. (For information on how to import data sets using MATLAB, see **Appendix A**.)

NOTES

1. Note that here, and frequently throughout the rest of this book, "materials" is used as an adjective, modifying nouns like "science," "development," "property," etc.
2. N. A. Fleck, V. S. Deshpande, and M. F. Ashby. "Micro-architectured Materials: Past, Present and Future". *Proceedings of the Royal Society of London Series A* **466(2121)** (Sept. 2010), pp. 2495–2516.
3. See T. Kuhn. *The Structure of Scientific Revolutions*. Chicago, IL, USA: University of Chicago Press, 1962.
4. Bruce L. Bramfitt and Samuel J. Lawrence. "Metallography and Microstructures of Carbon and Low-Alloy Steels". In: *ASM Handbook, Volume 9: Metallography and Microstructures*. Ed. by G. F. Vander Voort. Materials Park, OH, USA: ASM International, Dec. 2004, pp. 608–626.

1 Fundamental Principles *From Atoms to Materials*

LEARNING OBJECTIVES

After completing this chapter, you should be able to:

1. Describe the structure of atoms, including the organization of electrons into energy levels and orbitals, using the fundamental physical concepts of electrostatics and quantum-mechanical principles.
2. Identify elements with similar chemical behavior based on their membership in a periodic group. Utilize chemical data taken from the periodic table and other sources in calculations.
3. Differentiate intermolecular forces in terms of their relative strengths.
4. Compute the forces between atoms or molecules subject to a particular interaction potential, including a determination of the equilibrium separation.
5. List and describe the important types of bonds in materials.
6. List and describe the primary structural motifs in materials.
7. Differentiate the types of Bravais lattices using lattice parameters.
8. Quantitatively describe the relationship between structure and density in crystalline materials.

1.1 MATTER, CHEMISTRY, AND MATERIALS

You have learned in your chemistry course that the matter we encounter in our day-to-day lives is composed of **atoms**. Atoms themselves have an internal structure, of course, but in most cases, the atom is taken to be the irreducible unit of matter. Your chemistry instructor described to you how atoms can associate (i.e., react) to form stable **molecules** and how to do the atomic accounting required to interpret these reactions. What your chemistry instructor likely did not discuss is how atoms and molecules are assembled in vast quantities to form the materials of our common experience.

We recall from **Figure I.3** that individual atoms and molecules are situated at a scale that is much smaller than the scale at which we typically observe microstructure in materials. You might wonder, "How does matter organize itself at length scales *in between* those of molecules and microstructures?". Adopting a material-oriented mindset, you then refine your question to ask, "What kinds of *structures* bridge the gap between molecules and materials?". To fully answer this question, we need to introduce a number of important chemical and physical ideas; some of these you may be familiar with and some you may not. As an initial attempt to capture the science relevant to the question in a pithy way, consider this quote from Professor Feynman (1919–1988):

DOI: 10.1201/9781003214403-2

If, in some cataclysm, all of the scientific knowledge were to be destroyed and only one sentence passed on to the next generation of creatures, what statement would contain the most information in the fewest words? I believe it is the *atomic hypothesis* that:

> *"All things are made of atoms-little particles that move around in perpetual motion, attracting each other when they are a little distance apart, but repelling upon being squeezed into one another."*

In that one sentence, you will see that there is an *enormous* amount of information about the world if just a little imagination and thinking are applied.

–Richard Feynman (1963)[1]

This statement is short but contains several important ideas:

1. Understanding atomic matter is the foundation of most pure and applied science,
2. Atomic motion is continual,
3. The interactions between atoms, and how those interactions are reflected in atomic behavior.

We will now revisit these core ideas to make them more rigorous and suitable for our work.

1.2 ATOMIC THEORY

We begin by reviewing the essential chemistry, retaining and highlighting those concepts that help us to address the questions above. An atom is depicted schematically in **Figure 1.1**. The constituents of atoms are the primary subatomic particles: **protons** (p^+), **neutrons** (n^0), and **electrons** (e^-). These particles all have mass (= m_p, m_n, and m_e), and you can recall that the protons and neutrons are quite massive compared to the electrons and are both located in the **nucleus** of the atom. The electrons are found in the space immediately outside the nucleus and are more or less confined to this space by their attraction to the positively charged nucleus. The number of protons in the nucleus, which is connected to the overall attractive **electrostatic force** produced, is an important quantity. The number of protons determines the **atomic number** Z of the atom, and this number alone establishes the atom's identity among the other **elements**.

Another important quantity influencing the behavior of the atom is its **atomic mass** A. Mass is a fundamental physical property, and we are frequently interested in how much mass is contained in atoms of different types. **Table 1.1** provides some atomic-mass data for reference. Atomic and molecular masses are conveniently

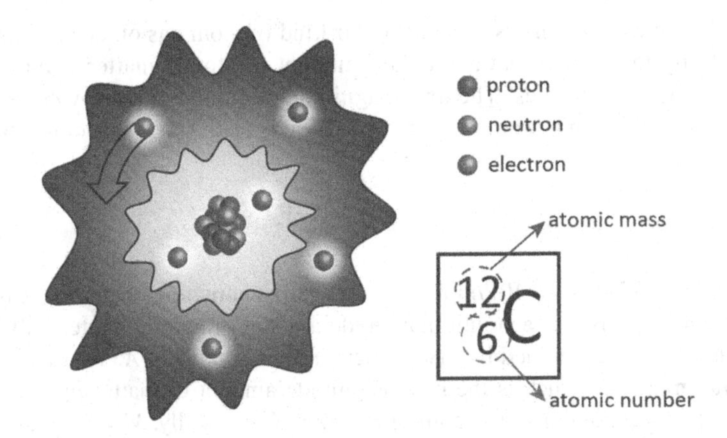

FIGURE 1.1 Schematic diagram of the structure of an atom. The core of the atom is a dense mixture of protons and neutrons called the nucleus. The comparatively light and mobile electrons inhabit the region outside of the nucleus. The electrons' preferred neighborhoods and the nature of their motions are not well defined but are instead *indeterminate*. From the schematic, we count $Z = 6$ protons and thereby identify the atom as carbon (C). Along with the atomic mass ($A = 6m_p + 6m_n \approx 12$ u), we have all the essential atomic data.

TABLE 1.1
Masses of Selected Atoms and Subatomic Particles

Atom or Particle	Symbol	Mass [u or Da]
Electron	e^-	0.0005486
Proton	p^+	1.007
Neutron	n^0	1.008
Hydrogen-1	^1_1H	1.008
Hydrogen-2 ("deuterium")	^2_1H	2.014
Carbon-12	$^{12}_6C$	12[a]
Carbon-13	$^{13}_6C$	13.003

[a] Exact.

represented in units of unified atomic mass units (u) or the equivalent dalton (Da): [A] = [mass] = u = Da. Recall the definition of the u/Da:

$$1\text{ u} = 1\text{ Da} = \frac{\text{mass of } ^{12}_{6}\text{C atom}}{12} \tag{1.1}$$

Within a small error, $m_p \approx m_n \approx 1$ u, and so the atomic masses of atoms are typically rounded off when identifying them. E.g., $^{13}_6C$ is just "carbon-13" and not "carbon-13.003". (The contribution of the electron mass to the overall mass of the atom is typically negligible.)

Knowing the atomic mass of a particular kind of atom has other benefits, such as the ability to measure out a standard amount of atomic matter. Suppose you have N atoms or molecules. The unit magnitude of atomic matter *by count* is the mole (mol), so the number of atoms that you have is also n mol. Since the mole has a size of

$$1 \text{ mol} = N_A \text{ atoms} \tag{1.2}$$

where $N_A = 6.02214076 \times 10^{23}$/mol is the **Avogadro constant**, the numerical value of n is then N/N_A. Now, each unit magnitude amount of atomic matter will have a total mass that is determined by that particular atom's value of A. We then identify the **molar mass** M, which is the unit magnitude amount of matter *by mass*. This has units $[M]$ = [(atomic) mass/number] = g/mol. Numerically, $M = A$ for any given atomic substance (though the two parameters have different units).

Example 1.1:

Suppose you have 8.06×10^{21} atoms of Fe and 5.62×10^{19} atoms of C. How many moles of Fe do you have? How many moles of C? When both samples are combined together, what is the *mole fraction* of Fe in the Fe + C mixture? The mole fraction X_A of a component A is

$$X_A = \frac{\text{Moles of A}}{\text{Total moles of matter}}$$

SOLUTION

Since $N_A = 6.022 \times 10^{23}$/mol, we find

$$n_{Fe} = \frac{8.06 \times 10^{21} \text{ atoms}}{N_A} = \frac{8.06 \times 10^{21} \text{ atoms}}{6.022 \times 10^{23} \text{ atoms/mol}} = \underline{\mathbf{0.0134\ mol}}$$

Similarly

$$n_C = \frac{5.62 \times 10^{19} \text{ atoms}}{N_A} = \underline{\mathbf{0.0000933\ mol}}$$

The mole fraction of Fe in a mixture of these two is

$$X_{Fe} = \frac{n_{Fe}}{n_{Fe} + n_C} = \frac{0.0134}{0.0134 + 0.0000933} = \underline{\mathbf{0.99}}$$

or 99%.

Example 1.2:

How many grams are there in one unified atomic mass unit of material?

SOLUTION

Consider a hypothetical substance with a molar mass M = 1 u/molecule. Since $N_A \approx 6.022 \times 10^{23}$/mol, we find that the mass m of 1.000 u is

$$m = M \times \frac{1}{N_A} = \frac{1.000 \text{ g}}{\text{mol}} \times \frac{\text{mol}}{6.022 \times 10^{23}} = \underline{\mathbf{1.661 \times 10^{-24} \text{ g}}}$$

to four significant digits. (This result will be of some use to us later on.)

We should remind you of an important practical aspect of chemistry here. Though a particular atom will always have a fixed number of protons and a fixed number of neutrons, a **sample** of atoms will always contain atoms of a number of different **isotopes**. Though the atoms may all have the same identity (i.e., the same Z), they could contain different numbers of neutrons and thus have different individual values of A. Compare, for example, the isotope carbon-12 ($^{12}_{6}C$) with A = 12 u and the isotope carbon-13 ($^{13}_{6}C$) with A = 13.003 u. The number of each type of isotope present in a significant sample of atoms (say, one mol worth of them) is determined by that isotope's relative abundance in the environment the sample was taken from. For this reason (unless you are dealing with an isotopically pure sample), values of A for various elements are reported as averages of the isotope masses weighted by their relative abundance.

Example 1.3:

Silicon has three naturally occurring isotopes: 92.23% of the atoms are ^{28}Si (atomic mass 27.9769 u), 4.68% of the atoms are ^{29}Si (atomic mass 28.9738 u), and 3.09% of the atoms are ^{30}Si (atomic mass 29.9738 u). What is the average atomic weight of Si? (Check against the periodic table value.)

SOLUTION

The answer is a weighted average A_{Si} of the atomic weights of all of the isotopes; i.e.,

$$A_{Si} = \sum_i f_i A_i$$

where f_i is the mole fraction of isotope i. We expand the sum to obtain

$$\begin{aligned} A_{Si} &= f_{28Si} A_{28Si} + f_{29Si} A_{29Si} + f_{30Si} A_{30Si} \\ &= 0.9223(27.9769 \text{ u}) + 0.0468(28.9738 \text{ u}) + 0.0309(29.9738 \text{ u}) \\ &= \mathbf{28.1 \text{ u}} \end{aligned}$$

This is basically the same as what the periodic table gives (A_{Si} = 28.085 u), but rounded to reflect the precision of the given values.

Mass is an important physical property that determines much about the behavior of a particle. Another is electric **charge**. It is important to note that the units that we use to describe the charges of protons, neutrons, electrons, etc. are **quantized**. This means that charge Q of an object is divisible only up to integer multiples of a fundamental quantity: $Q = 0, \pm 1\ q_e, \pm 2\ q_e, \ldots$. The quantity q_e is the "elementary" amount of charge. Since each proton has a single positive elementary charge ($= +q_e$) and each neutron is neutral in charge ($= 0$), the total charge of the nucleus is $Q = +Zq_e$. An atom that has as many electrons as protons is, considered altogether, electrically neutral since each electron has a single negative elementary charge ($= -q_e$). Sometimes the excess charge of an object (surplus or deficiency of charges from the neutral state) Q is represented in units of ampere-second, called the coulomb: $[Q]$ = [current × time] = [number × elementary charge] = C. Like the mole standardizes the number of atoms under consideration, the coulomb considers some particular number of charges, and in the SI system, we obtain

$$1\ \mathrm{C} \approx \left(6.241 \times 10^{18}\right) q_e \tag{1.3}$$

Correspondingly, $q_e \approx 1.602 \times 10^{-19}$ C.

It is crucial to our analysis of atomic organization and interaction to consider the Coulomb force F_C that exists between the positively charged nucleus and the e^- that circulate around it. This interaction is essentially electrostatic in nature and therefore obeys **Coulomb's Law**:

$$F_C(r) = -\frac{Q_1 Q_2}{4\pi\epsilon_0} \times \frac{1}{r^2} \tag{1.4}$$

where Q_1 and Q_2 are the charges on the two interacting objects, ϵ_0 is the **permittivity** of vacuum (a physical constant), and r is the separation between the objects. The value of ϵ_0 is approximately $8.854 \times 10^{-12}\ \mathrm{C^2/(Nm^2)}$.

Note the convention for the *sense* of the forces at work in **Equation 1.4**. Positive (+) forces are *attractive*, and negative (–) forces are *repulsive*, as shown in **Figure 1.2**. Indeed, when one charge Q_1 is positive and the other charge Q_2 is negative, the overall Coulomb force F_C is algebraically positive and "opposites attract"

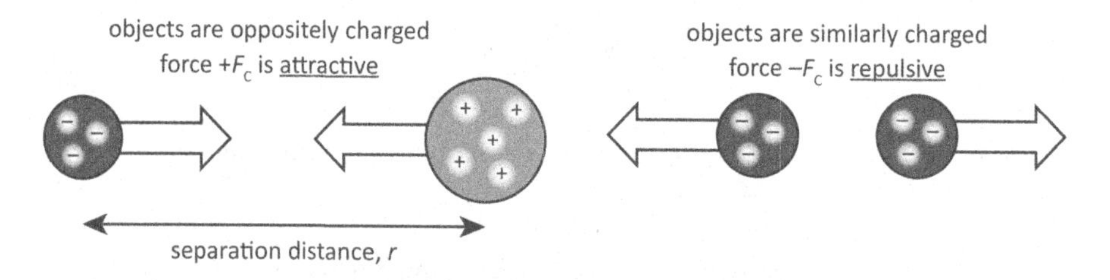

FIGURE 1.2 The sense of forces between charged objects. Attractive forces act to pull oppositely charged objects together; repulsive forces act to push similarly charged objects apart. The magnitude of the Coulomb force F_C depends on the charges of the objects as well as the separation distance r between them.

as we would expect. When the charges Q_1 and Q_2 are alike – either both positive or both negative – the calculated force will be algebraically negative, and repulsion will result. This simple electrostatic interaction, therefore, binds the electrons to the nucleus with the prescribed amount of attractive force while keeping individual electrons apart from one another via repulsion.

Example 1.4:

What is the electrostatic force between two electrons separated by a distance of 0.50 nm? What is the sense of the force?

SOLUTION

We compute the Coulomb force using **Equation 1.4** as

$$F_C(r) = -\frac{Q_1 Q_2}{4\pi\epsilon_0} \times \frac{1}{r^2}$$

where $Q_1 = -q_e$ and $Q_2 = -q_e$ are the electron charges. Since $q_e = 1.602 \times 10^{-19}$ C is the fundamental electron charge, $\epsilon_0 = 8.854 \times 10^{-12}$ C^2/(N $\times$ m^2), and 0.50 nm = 5.0×10^{-10} m, we obtain

$$F_C = -\frac{\left(-1.602\times 10^{19}\ \text{C}\right)^2}{4\pi\left(8.854\times 10^{-12}\ \text{C}^2\text{N}^{-1}\text{m}^{-2}\right)} \times \frac{1}{\left(5.0\times 10^{-10}\ \text{m}\right)^2} = -9.2\times 10^{-10}\ \text{N} = \underline{\mathbf{-0.92\ nN}}$$

Since the sign of the overall force is (–), we recognize this as a repulsive force, as we expect from like charges.

You have probably noticed at this stage that our discussion began as a simple accounting of the atomic components in standard chemical language but has now become blended significantly with some physics related to how these objects behave. These physical interactions are important features of how matter organizes itself.

1.3 ELECTRON CONFIGURATION

Some discussion on the **electron configuration** of atoms is necessary to establish the principles we require to describe materials. We introduced above how the electrons are confined to regions surrounding the nucleus by the electrostatic/Coulombic forces. We sometimes say that the electrons are trapped in a "well" created by the electrostatic interaction and it is necessary to do some physical **work** in order to extract them. This confining **potential** U for a solitary electron inside an atom with atomic number $+Z$ can be represented as

$$U(r) = -\frac{Q_1 Q_2}{4\pi\epsilon_0} \times \frac{1}{r} = -\frac{Zq_e^2}{4\pi\epsilon_0} \times \frac{1}{r} \tag{1.5}$$

This potential is also dependent on separation r, and the connection between this potential and Coulomb's Law is given by $F_C(r) = dU/dr$. We have reframed our picture of the forces in an atom using interaction *potentials* instead of interaction *forces* since it gives us some insight into the way that electron states can be categorized without any definite interpretations of their motions.

This categorization system only makes sense if we appreciate the fact that matter situated at such small scales (i.e., the size of atoms and their constituent parts) follows rules that have no large-scale (or "macroscopic") equivalent. These unfamiliar rules are the principles of **quantum mechanics** (QM). The ones that primarily interest us are:

1. Atoms and their constituent parts are subject to the **uncertainty principle**.
2. Atoms and their constituent parts have wave-based descriptions in addition to their particle-based descriptions; this is **wave/particle duality**.
3. Atoms and their constituent parts possess an important internal property called **spin moment** (or just "spin"), and some particles (electrons in particular) are subject to the **exclusion principle** as a result.

The uncertainty principle (QM principle #1) imposes limitations on how precisely we can *simultaneously* know a particle's location x and its momentum $p = mv$. Our uncertainty is reflected in a pair of quantities, Δx and Δp. For instance, when Δx is small, we know the position with high certainty. Or, when Δp is large, we have little confidence in our knowledge of how fast the particle is going. The complete "uncertainty relation" for this pair of variables is

$$\Delta x \Delta p \geq \frac{h}{4\pi} \tag{1.6}$$

where $h \approx 6.626 \times 10^{-34}$ kgm^2/s is the **Planck's constant**. Note the "simultaneous" aspect; since we cannot make the product $\Delta x \Delta p$ as small as we wish (it must always remain greater than $h/4\pi$), we cannot obtain exact information on *both* x and p. If, for example, we were nearly certain ($\Delta p \approx 0$) that the velocity of a particular particle is zero, we are forced to accept the fact that its location must be exceedingly uncertain ($\Delta x \rightarrow \infty$) so that the uncertainty relation of **Equation 1.6** remains valid. Meaning: we really cannot say for sure whether the particle is anywhere. Fortunately, for everyday macroscopic objects, the uncertainties involved will be moderately large. This means that, since h is so small in numerical value, the uncertainty principle will not confuse our day-to-day activities. However, for electrons in the space interior to atoms, this is not the case, and the uncertainty principle must govern their behavior.

Just because we will always be uncertain about the exact details of an electron's motion does not mean we cannot apprehend and categorize the overall characteristics of that motion. These characteristics are captured in the values of electron **quantum numbers**. The full information on the motion of an electron is contained in a mathematical object called a **wavefunction**, denoted by ψ. In imparting a wavelike identity (QM principle #2) to our electron, we therefore associate it with a wavelike motion, and this wavelike motion ψ must obey an equation of the appropriate form.

This relation is the **wave equation of Schrödinger**, developed for the purpose of describing the wavelike behavior of matter. This equation involves the energy E of the electron (a constant, overall characteristic of its motion) and the forces the electron is subject to, i.e., those derived from the potential U. Solving the Schrödinger wave equation requires advanced mathematical techniques, but the different possible solutions can be cataloged in general terms using the quantum numbers n, ℓ, m, and m_s.

THE SCHRÖDINGER EQUATION AND ITS SOLUTION

In 1D, the time-independent ("standing-wave") Schrödinger equation is

$$-\frac{h^2}{4\pi m_e}\frac{d^2\psi}{dx^2}+U(x)\psi(x)=E\psi(x)$$

We recognize this as a 1D *differential equation* in the rectangular coordinate x. Solving the equation with the known potential function $U(x)$ gives us the unknown function $\psi(x)$. In the space of a 3D/spherical atom, the wavefunction will be a *partial differential equation* and necessarily have a solution $\psi(x, y, z)$.

When this 3D solution to the Schrödinger equation is *quantized*, the wavefunction will also look different according to its quantization parameters n, ℓ, and m: $\psi_{n\ell m}(x, y, z)$. The wavefunction $\psi_{n\ell m}$ completely encodes a particular object's state but does not itself convey anything about a particular object's observable physical properties (position, momentum, etc.). These can be obtained from the wavefunction through a suitable calculation, and then only as *probability densities*. For example, the probability density f that captures a particle's position (x, y, z) is determined by

$$f(x,y,z)=|\psi(x,y,z)|^2$$

This density can be used to calculate how likely a given position is through standard statistical techniques.

First, like the charge a particle possesses, the energy the electron has can only come in discrete, quantized amounts: $E = E_1, E_2, E_3, \ldots$. This quantization is a consequence of the confinement the electron is subject to; the electron is "bound" to its atom. We label these possible energies, now called **energy levels** in the atom, using the **first quantum number** (or "principal quantum number") n. From our solution to Schrödinger's wave equation, we get[2]

$$E_n = E(n) = -\frac{m_e Z^2 q_e^4}{8\epsilon_0^2 h^2}\times\frac{1}{n^2} \tag{1.7}$$

where n = 1, 2, 3, etc. Our electrons can therefore possess one of a increasing set of energies that depends on an electron's value of n. This principle is illustrated in

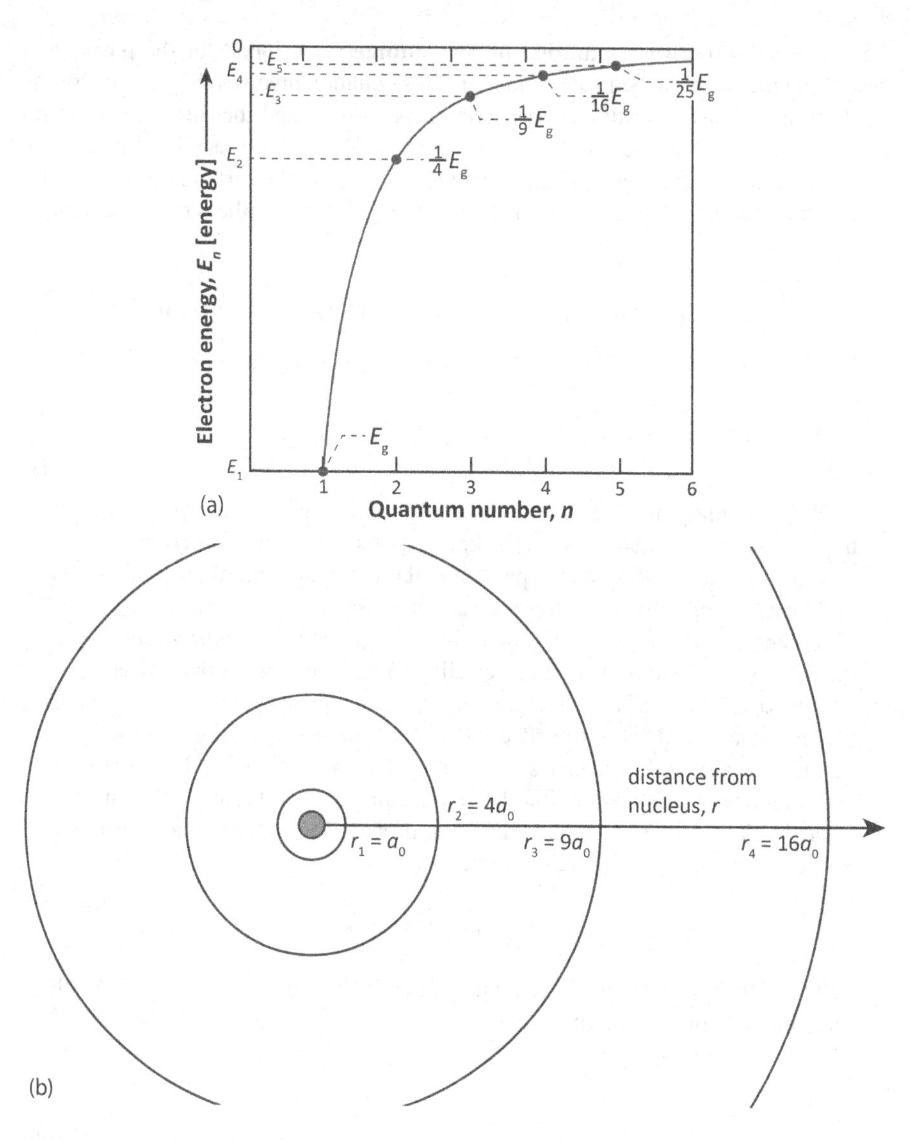

FIGURE 1.3 The values and importance of the first quantum number n. (a) n establishes the energy of the electron E_n according to **Equation 1.7**. This energy follows $E_n \propto -1/n^2$ and has a minimum (= E_g) for $n = 1$. (b) The energy E_n can be associated with the radius of an electron's "Bohr orbit" in an atom. For instance, in hydrogen, we have $r_n = a_0 n^2$, where $a_0 = h^2\epsilon_0/\pi m_e q_e^2$.

Figure 1.3. There is a minimum value of energy possible (called the "ground state") that is obtained when $n = 1$. As an electron's primary quantum number increases, an electron's energy level increases from the ground state $E = E_g = E_1$ toward $E = 0$. This progression of energies is depicted in **Figure 1.3(a)**. Because the progression follows $1/n^2$, the energy levels E_n are increasingly closely spaced as n increases. These energy levels are associated with a rough or "average" orbital radius r_n of the electron about the nucleus, as shown in **Figure 1.3(b)**. As n increases, so does the typical electron distance from the nucleus. (These are sometimes called "Bohr orbits".)

The sizes of the electron orbits (as determined by n) are not the only piece of discriminating information on an electron's configuration. We also wish to know something about the *shapes* of the orbits. The **second quantum number** (or "orbital quantum number") ℓ captures another aspect of the electron's motion: the overall magnitude of its angular momentum L. We understand well the role of angular momentum in determining the trajectory of an everyday object from physics. (Importantly, the angular momentum ensures that the electron, attracted to the nucleus, does not fall into it.) However, in our bound electrons, the magnitude of the angular momentum is also quantized according to

$$L_\ell = L(\ell) = \frac{h}{2\pi}\sqrt{\ell(\ell+1)} \tag{1.8}$$

Furthermore, the possible values of L that the electron can access are limited by the amount of energy E_n the electron has. What this requirement comes down to is that ℓ must be an integer and $0 \le \ell < n$. The differently shaped electron trajectories that result from different values of ℓ are called subshells. **Figure 1.4** shows some simplified subshell shapes (called "Bohr orbits") and their dependence on the value of ℓ.

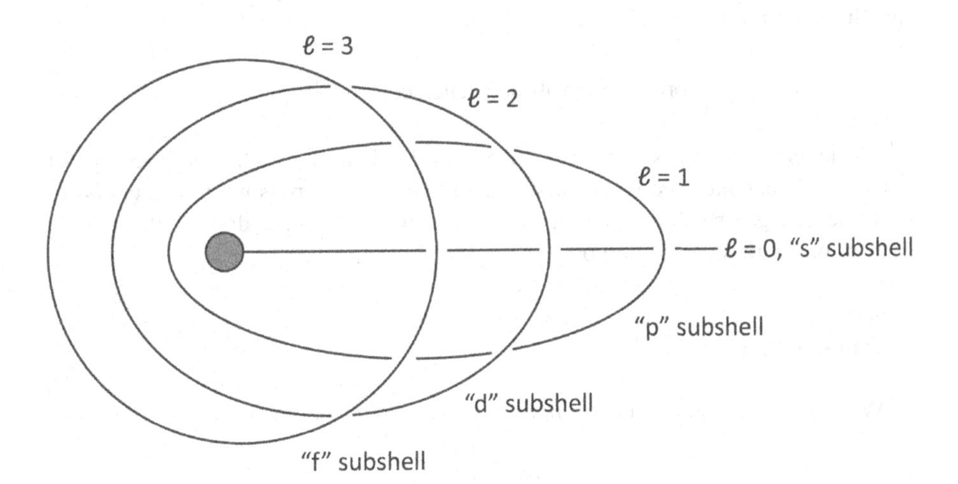

FIGURE 1.4 Effect of angular momentum on electron subshell shape. The angular momentum L depends on the second quantum number ℓ; ℓ thereby plays a role in determining the subshell shape. Representing these shapes is difficult because of the complexity of the Schrödinger equation and the indeterminate nature of the electron's motion, but a simplified picture of modified Bohr orbits demonstrates how different shapes are reflected in the different values of ℓ. The different orbit types are sometimes labeled with letter codes (s, p, d, etc.) rather than their ℓ-values.

The **third quantum number** (also called the "magnetic quantum number") m describes the magnitude of a particular component L_z of the angular momentum. Fixing the value of this component fixes the *direction* of the electron's angular momentum vector. Meaning: two electrons with the same angular momentum magnitude L (i.e., with the same value of ℓ) can have orbits with different *orientations*. These different possibilities are the electron **orbitals**. As you might now expect, the possible values of $L_z = (h/2\pi)m$ are quantized according to the value of m. Since m must be an integer (positive or negative) and $-L \leq L_z \leq +L$, we must have $-\ell \leq m \leq +\ell$. The **fourth quantum number** (called the "spin quantum number") m_s supplements the angular momentum of the electron with an "intrinsic" spin contribution. The value of this contribution is $m_s = \pm½$. The spin property has little to do with the electron's motion but rather establishes limits on the orbital occupancy according to the exclusion principle (QM principle #3).

The motion of an electron with quantum numbers $\{n, \ell, m\}$ in an (x, y, z) coordinate system can be interpreted by considering its "stationary" wavefunction $\psi_{n\ell m} = \psi(x, y, z)$. These wavefunctions are exceedingly complicated functions of the position (x, y, z), but their overall form is fully determined by the quantum numbers; different wavefunctions $\psi_{n,\ell,m,ms}$ correspond to the different orbital/spin combinations that are possible. Though the electron's motion in its assigned orbital is not absolutely determined (unlike the motion of a classical particle whose properties we know), we can still discuss it in terms of probabilities. The probability of finding the electron near (x, y, z) depends on the value of $|\psi|^2$ there. These $|\psi|^2$ calculations are used to generate the electron probability-density plots like those shown in **Figure 1.5**. Orbitals are also frequently labeled by using the shorthand notation

$$\text{orbital} = n(\text{subshell letter code})^{\#\ \text{occupants}}$$

The letter codes are "s" for $\ell = 0$, "p" for $\ell = 1$, "d" for $\ell = 2$, "f" for $\ell = 3$, "g" for $\ell = 4$, etc. This notation does not distinguish between electrons with different m- or m_s-values in a given subshell, but the possible values of m and m_s do limit the number of electrons with this orbital type.

Example 1.5:

What are the quantum numbers n, ℓ, m_ℓ, m_s of a $3d^5$ electron?

SOLUTION

We can ascertain from the labeling scheme that the first quantum number is $n = 3$ (i.e., the first digit). The "d" designation indicates the $\ell = 2$ subshell; this is the second quantum number. The third quantum number could be one of five possibilities: $m_\ell = -2, -1, 0, +1$, or $+2$, and the fourth quantum number could be one of two possibilities: $m_s = -½$ or $+½$. (Since there are five orbitals in the d subshell, the expectation is that the electrons will be evenly distributed among all five and all share the *same* m_s value. See the following filling rules.)

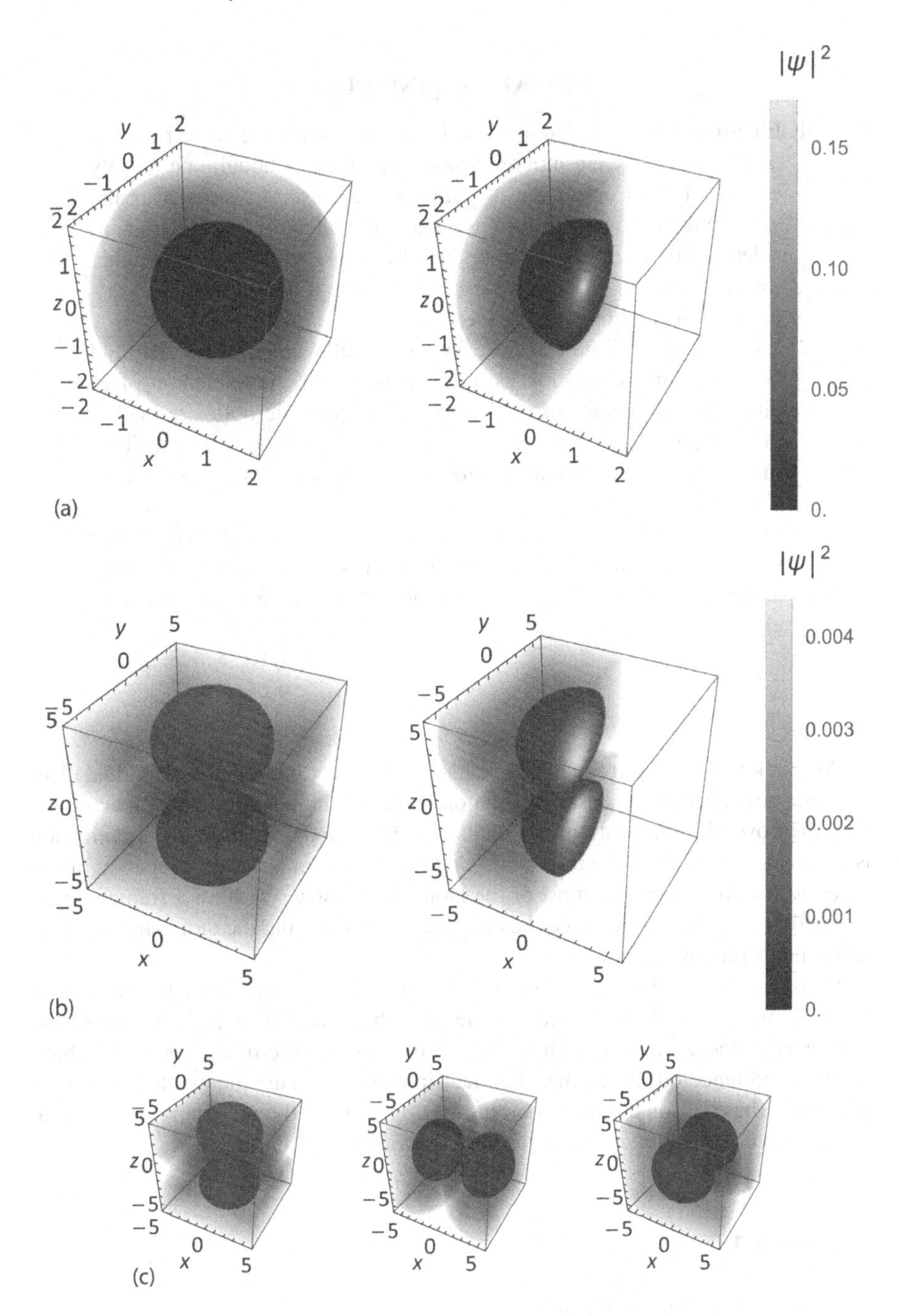

FIGURE 1.5 Electron probability density plots for selected orbitals. (a) s-type orbitals (ℓ = 0) are spherical, and the electron is most likely to be found near the central point. (b) p-type orbitals (ℓ = 1) have a double-lobe structure. The orbital is not symmetric and is aligned along a particular axis. (c) Three p-type orbitals with different values of m_ℓ. (The units of the axes are all multiples of a_0.)

ORBITAL-FILLING RULES

1. Electrons are placed into orbitals, starting with the lowest-energy orbital available and moving upward in energy. Though one naively expects that all orbitals at the same energy level are situated at the same energy, this is not the case when multielectron effects are considered. For instance, the energies of orbitals with $\ell = 1$ are higher than those with $\ell = 0$ for the same n. The energy hierarchy of the various orbitals is shown in **Figure 1.6**.
2. Each orbital orientation (i.e., for a particular value of m) can hold two electrons *maximum* as a consequence of the exclusion principle. The m_s values of the two electrons must be opposite in sign or "antiparallel": one with $m_s = +½$ and one with $m_s = -½$. This means that, for a given orbital type, the number of occupants $\leq 2(2\ell + 1)$.
3. Orbitals are filled uniformly across the orientations. We do not expect to find one orientation of an orbital with two antiparallel-spin electrons while there are orientations of the same orbital with no electrons.

As you know from introductory chemistry, the exact motions of individual electrons are not necessary to describe the atom's chemical interaction with other atoms, only their overall organization into shells, subshells, and orbitals. This organization is an atom's or an ion's electron configuration. In particular, the occupancy of the outermost shell determines most of the atom's chemistry. We have a framework of subshells 1s, 2s, 2p, etc. to place the electrons in, but placing the atoms into the correct orbitals requires a set of rules.

Consider the implications of these rules: orbitals that are closer to the nucleus (lower n) fill first, and these innermost levels, when filled, have little impact on the chemistry of the atom. Rather, it is the electrons in the outermost orbitals that have a strong influence on the atom's chemical interactions. This means that, when we talk about the electron configuration of the atom, we for the most part need only to discuss the organization of the outermost, or "valence", orbitals.

Example 1.6:

What is the electron configuration of Fe?

SOLUTION

Since a neutral Fe atom has $Z = 26$ electrons, we can sort them into orbitals starting at $n = 1$, $\ell = 0$; then $n = 2$, $\ell = 0$; then $n = 2$, $\ell = 1$; etc. The pattern becomes:

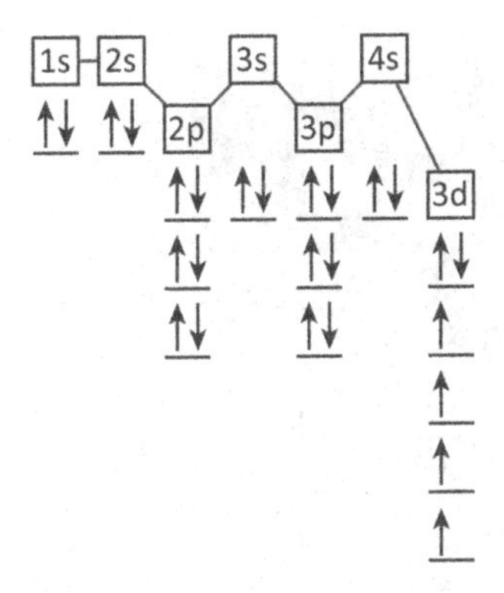

We would label this configuration as **$1s^22s^22p^63s^23p^64s^23d^6$**. Because the inner-core "$1s^22s^22p^63s^23p^6$" configuration is that of the noble gas Ar, we sometimes write **[Ar]$4s^23d^6$**, instead.

These filling rules, combined with the particular ordering of orbital energies illustrated in **Figure 1.6**, give rise to repeating (or "periodic") trends in the electron configurations of the elements. When the different elements are placed in order and spaced according to these filling trends, we obtain the familiar diagram called the **periodic table**, shown in **Figure 1.7**. The periodic table is organized so that elements with similar electron configurations are situated adjacent to one another in *vertical* columns, called "groups". Atoms belonging to the same group will thus have similar properties and tend to interact with other atoms in similar ways. For this reason, the periodic table provides great predictive power when it comes to the behavior of different atoms.

Importantly for our discussion, the periodic table establishes the foundation for understanding the composition of materials. For instance, metal alloys are composed of elements taken almost exclusively from the middlemost transition metals (groups 4–12) in the table. Sometimes, these alloys are further modified

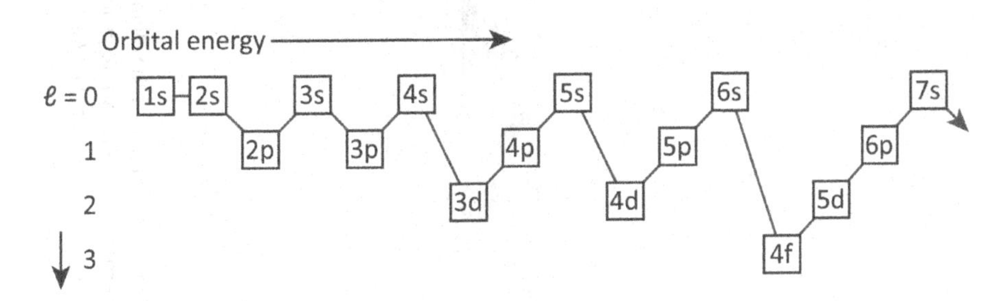

FIGURE 1.6 Ranking of orbital energies. The modification of orbital energy levels by the presence of other electrons in the atomic system produces a staggered ordering. The orbitals are filled in a left-to-right order, and each orbital has its various orientations filled with parallel-spin electrons before antiparallel-spin electrons are introduced. (Some exceptions, like the energies of copper and chromium orbitals, differ slightly from this pattern.)

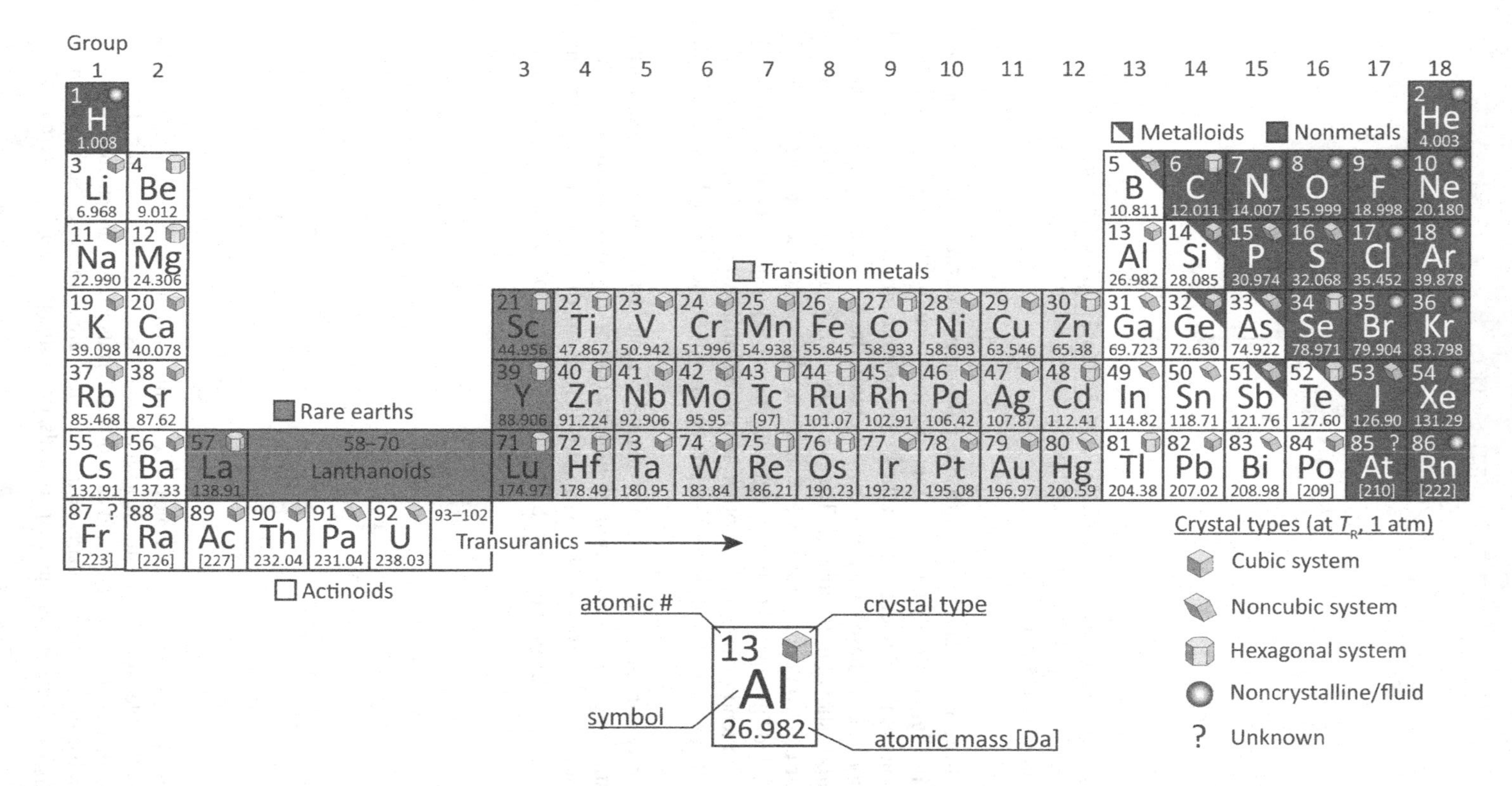

FIGURE 1.7 The periodic table of the elements. When elements are arranged in rows and columns corresponding to the orbital-filling trends, atoms with similar properties fall naturally into groups. Larger categories of elements (spanning multiple groups) include *transition metals*, *rare earths*, *metalloids*, *nonmetals*, and *transuranics*. The *lanthanoids* are the elements {$^{140}_{58}$Ce, $^{141}_{59}$Pr, $^{144}_{60}$Nd, $^{145}_{61}$Pm, $^{150}_{62}$Sm, $^{152}_{63}$Eu, $^{157}_{64}$Gd, $^{159}_{65}$Tb, $^{163}_{66}$Dy, $^{165}_{67}$Ho, $^{167}_{68}$Er, $^{169}_{69}$Tm, $^{169}_{70}$Yb, $^{175}_{71}$Lu}.

by the addition of "rare earths" (the lanthanoids along with elements in group 3). Ceramics are frequently chemical compounds of metals from groups 1–14 and nonmetals from groups 14–17. The combination of these elements forms, for example, carbides, nitrides, oxides, and halides. Semiconductors and their dopants are typically pulled from groups 13–15. Polymers are typically made of carbon, silicon, hydrogen, nitrogen, oxygen, and some halogen elements from group 17.

1.4 INTERMOLECULAR FORCES AND BONDING

Now that we have assembled the essential chemical and physical ideas underlying atomic organization, we return to Professor Feynman's quote from **Section 1.1**. How is it that atoms can feel each other's presence and character, knowing when to be attracted and when to be repelled? What is the net effect of these forces, and what is their importance? The answer lies in an analysis of the **intermolecular forces** that exist between different types of atoms and molecules. **Table 1.2** summarizes these forces, and we review and describe them here.

Recall **Equation 1.4**, which captures the interaction between two charged objects as an attractive or repulsive force F_C. We have applied this interaction to the description of how electrons are bound to a nucleus using **Equations 1.5** (for the electrostatic potential) and the Schrödinger equation. If the attractive force binding the electrons to the nucleus can be overcome (i.e., sufficient work is performed on the particle), then the atom can become *ionized*. This **ion** retains its identity (its Z doesn't change) but attains a nonzero net charge. A neutral atom initially containing Z electrons that has one electron extracted in this way now has an overall charge of $+q_e$. An ion with an overall positive charge (of any magnitude) is called a **cation**. Any other charged species in the vicinity will then interact with the ion according to Coulomb's Law: positively charged objects will be repelled, and negatively charged objects will be attracted. The same goes for an atom that captures an additional electron and becomes a negatively charged **anion**; Coulomb's Law will still apply, but the senses of the forces are reversed.

Example 1.7:

Though atoms and ions do not have well-defined shapes or radii, the atomic radii of ions, taken to be solid spheres, can be estimated from their spacing in an ionic solid (see **Section 1.5**, below). Suppose the radii of Fe^{2+} and Cl^- ions are 0.076 and 0.175 nm, respectively, when they are "touching" (i.e., the spheres of the specified radii are in contact). Calculate the force of attraction between these ions when they "touch".

SOLUTION

The force between charged ions is the electrostatic or Coulomb force described by **Equation 1.4**. Here, $Q_{Fe} = +2q_e$ and $Q_{Cl} = -q_e$ are the ion charges, and the separation

TABLE 1.2
Catalog of Intermolecular Forces

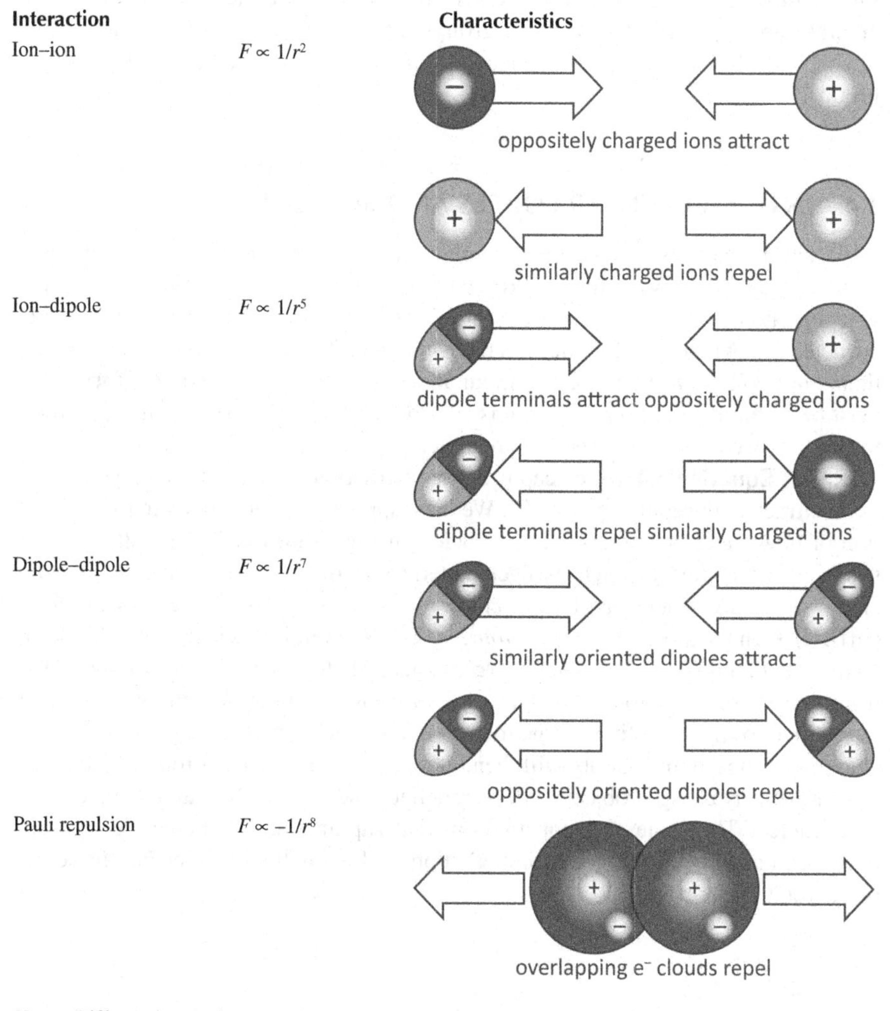

Interaction		Characteristics
Ion–ion	$F \propto 1/r^2$	oppositely charged ions attract similarly charged ions repel
Ion–dipole	$F \propto 1/r^5$	dipole terminals attract oppositely charged ions dipole terminals repel similarly charged ions
Dipole–dipole	$F \propto 1/r^7$	similarly oriented dipoles attract oppositely oriented dipoles repel
Pauli repulsion	$F \propto -1/r^8$	overlapping e⁻ clouds repel

Note: Different interactions are characterized according to whether they are attractive (+) or repulsive (–), as well as by the exponent m on the force law $F \sim r^{-m}$ (note that Pauli Repulsion is never attractive).

of the charges (considered to be concentrated at the center of the spheres) is $r = r_{Fe} + r_{Cl} = 0.076 + 0.175 = 0.251$ nm. Following **Example 1.4**, we have

$$F_C = \frac{(+2)(-1)q_e^2}{4\pi\epsilon_0} \times \frac{1}{\left(0.251 \times 10^{-9}\ \text{m}\right)^2} = +7.32 \times 10^{-9}\ \text{N} = \underline{\mathbf{7.32\ nN}}$$

Note that this force is attractive and relatively large.

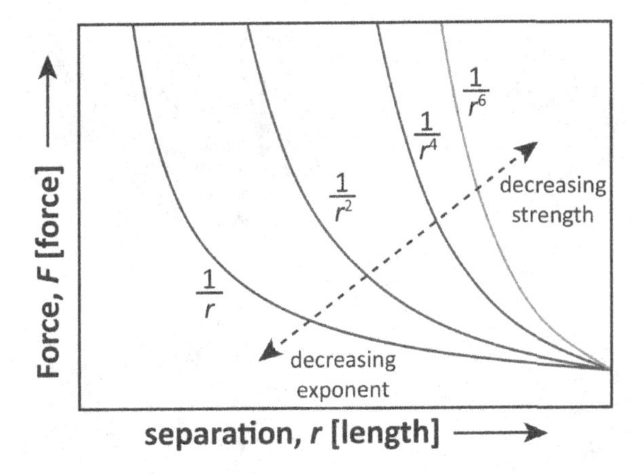

FIGURE 1.8 Example force laws for intermolecular interactions. The curves are generic, so no scale or units are attached. The force laws differ in the value of their force-law exponent: $\alpha = 1, 2, 4, 6$. How rapidly the force decreases with increasing r indicates something about the "strength" of the interaction; higher exponents give steeper fall-off, corresponding to weaker forces.

An important aspect of these intermolecular force laws requires attention. A distinguishing characteristic of these force laws is the force-law exponent α. This exponent determines *how rapidly* the force falls off with increasing r. **Figure 1.8** compares force laws with various values of α. Intermolecular interactions may be classified as "strong" or "weak" according to their value of α. When α is small, force drops off slowly, and the interaction is significant over larger distances. Strong ion–ion interactions ($\alpha = 2$) have this quality. Intermolecular interactions with $\alpha = 7$ are therefore relatively weak; the molecules must be close together or the force between them will be effectively nil. It is also worth recalling at this point that the force laws can be derived from potentials. The general form of such a potential will be

$$U(r) = -\frac{C}{r^{\alpha}} \tag{1.9}$$

where C is a constant that contains the physical parameters relevant to that interaction. E.g., $C = -Q_1Q_2/4\pi\epsilon_0$ and $\alpha = 1$ for electrostatic interactions.

Another type of interaction is that between **polar molecules**. A polar molecule is one that is overall neutral in charge but in which the internal (+) and (–) charges are not uniformly distributed throughout the molecule. The segregation of (+) and (–) charges into different halves of the molecule forms a **dipole**. As examples, consider the molecules shown in **Figure 1.9**. The two halves of the dipole can be treated in many ways as separate charges, giving rise to Coulomb-like interactions between the dipole terminals and other charges/dipoles in the vicinity. The sense of the interaction will vary depending on the relative orientation of the dipole(s), but a given orientation will produce an interaction with a form given by **Equation 1.9**. The strength of dipole–dipole interactions typically corresponds to $\alpha = 7$.

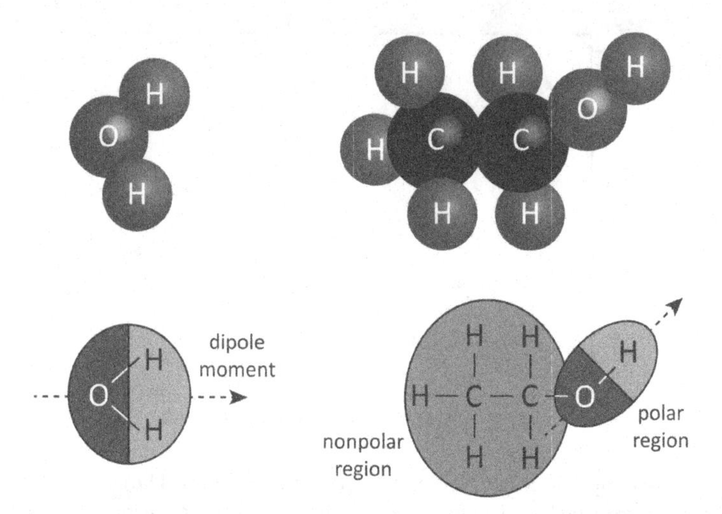

FIGURE 1.9 Some polar molecules. Water (H_2O) and ethanol (C_2H_5OH) molecules have distinct, permanent regions within the molecule that can be locally polarized or non-polar. The magnitude and direction of this polarization are described by the dipole moment.

An important property of many neutral, non-polar species is that they are subject to **polarization**. In isolation, the individual charges in an atom or non-polar molecule will be distributed uniformly throughout it. This situation can be disturbed by the presence of a dipole, as shown in **Figure 1.10**. The action of the dipole forces bound (but still internally mobile) charge in the non-polar species to adopt a configuration

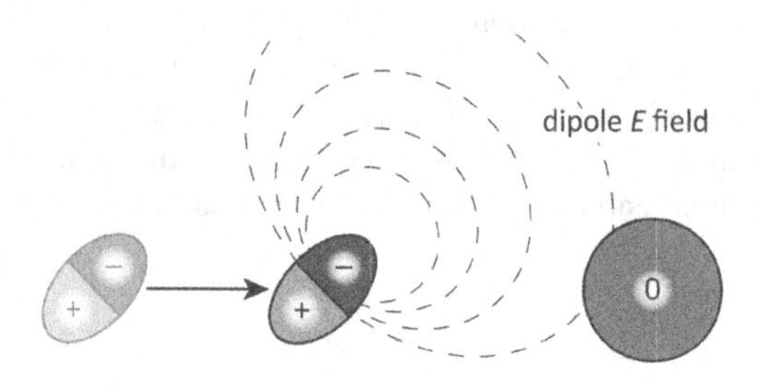

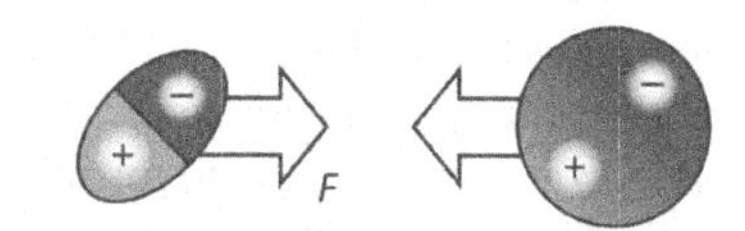

FIGURE 1.10 The polarization of a neutral, non-polar atom/molecule by a dipole via segregation of its interior charges. The electric field of the dipole exerts a force on the charges inside the neutral species, segregating the interior into a positively charged region and a negatively charged region (i.e., a dipole).

that mirrors that of the dipole. This is **induced polarization** and it can produce an interaction between the non-polar and polar species where we would not normally expect to find one. The physics of the dipole–induced dipole interaction are complicated, but the overall force law can still be expressed using an interaction potential like **Equation 1.9** with $\alpha = 7$.

Finally, it is important to recognize that the non-polarity of a neutral species is conditional, even in isolation. An isolated atom or molecule typically has a symmetrical distribution of electrons within it, rendering it non-polar. However, because electrons are constantly in motion, they can adopt arrangements that are not symmetric, i.e., their random motions could distribute them unevenly. This means that the molecule can become **spontaneously polarized** when electrons aggregate on one side of the molecule or the other. This polarized state is typically short-lived. The terminals of the spontaneously polarized species oscillate, and in the absence of other interactions, these oscillations decay away. If, however, another non-polar atom or molecule is nearby, that species can also become polarized by the induction effect described above. The interaction between spontaneously generated dipoles and induced dipoles produces a force law like that of **Equation 1.9** with $\alpha = 7$. It is worth noting that there are numerous intermolecular interactions with high m-values: dipole–dipole, dipole–induced dipole, spontaneous dipole–induced dipole, etc. These interactions are collectively weak but are almost always present in some form. Because these "universal" forces were originally theorized to exist by Johannes Diederik van der Waals (1837–1923), they are called **van der Waals forces**.

We now have a definite answer to the question of what kind of physical effects produce the attraction between atoms and molecules that pulls them closer together. Even molecules that are both neutral and non-polar will still tend to be attracted to one another. But this is still only part of the picture of atomic interaction because they must also "repel upon being squeezed into one another". Meaning: the net interaction between molecules requires the presence of both attractive and repulsive forces simultaneously. These repulsive forces arise as a consequence of the overlapping of electron orbitals that occurs when two molecules are brought close together. Orbitals that are at capacity (i.e., with two antiparallel-spin electrons) cannot accommodate any more electrons, and the energetic cost of reorganizing the orbitals expresses itself as a strong repulsive force. This is called "Pauli repulsion" since it is a direct consequence of Wolfgang Pauli's (1900–1958) exclusion principle. Pauli repulsion is the primary source of repulsive forces between molecules, and it must be the case that these repulsive forces drastically exceed attractive forces at short distances so that the molecules never occupy the same space.

The attractive and repulsive forces between molecules are superimposed to produce a net interaction between them that is attractive at large separations r and repulsive at short ones. The attractive component $U_A(r)$ of the interaction (expressed as a potential) and the repulsive component $U_R(r)$ are

$$U_A(r) = -\frac{C_A}{r^\alpha} \quad \text{and} \quad U_R(r) = +\frac{C_R}{r^\beta} \tag{1.10}$$

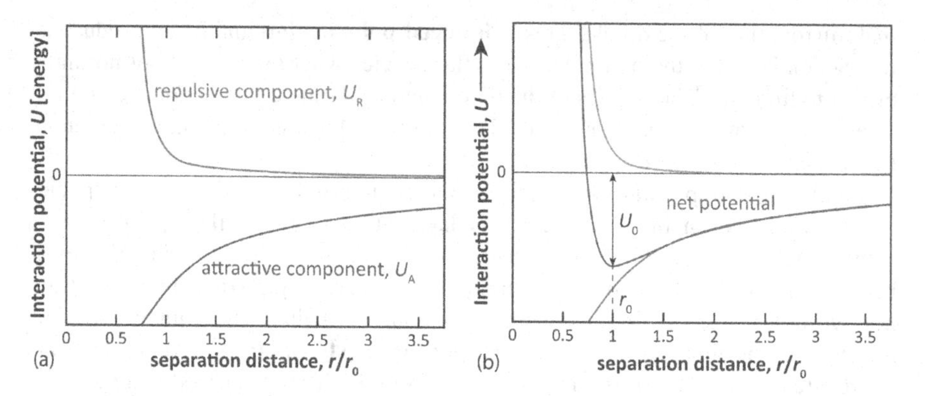

FIGURE 1.11 Components of the interaction potential for a pair of atoms. In (a), the components provide a positive or negative contribution according to their sense, and the value of the exponent determines how the potential changes with separation. (b) shows the sum of the two individual potentials. The net potential has a minimum at a separation of r_0, and this minimum value represents the binding energy of the system.

These potentials are illustrated schematically in **Figure 1.11**. In **Figure 1.11(a)**, the two potentials are depicted with different senses [(+) or (–) according to **Equation 1.10**], and their different exponents ($\beta > \alpha$) give different shapes. The net interaction potential is just the sum of the two components

$$U(r) = U_A(r) + U_R(r) = \frac{C_R}{r^\beta} - \frac{C_A}{r^\alpha} \tag{1.11}$$

This net potential is illustrated in **Figure 1.11(b)**. This potential has an important feature: the curve has a minimum at a special value $r = r_0$. This minimum value U_0 is the binding energy of the two-atom system.

The force $F(r)$ that is associated with this potential can be determined from the normal rule:

$$F(r) = \frac{dU}{dr} = \frac{d}{dr}\left[U_A(r) + U_R(r)\right] = \alpha\frac{C_A}{r^{\alpha-1}} - \beta\frac{C_R}{r^{\beta-1}} \tag{1.12}$$

This force is plotted schematically in **Figure 1.12**. This curve has a profound ability to explain much about the way that matter works. First, the curve shows that force values are large and negative at short separations; hence, the atoms will repel when too close. Second, when the atoms are not so close that they push each other away, they will be attracted to one another. Note that this attraction persists for an unlimited distance, though it does become smaller as r increases. Finally, the "special point" r_0 from **Figure 1.11** takes on a more explicit physical meaning in this force curve. r_0 marks the distance at which the two molecules are in a state of force **equilibrium**, meaning that the attractive and repulsive force components are equal

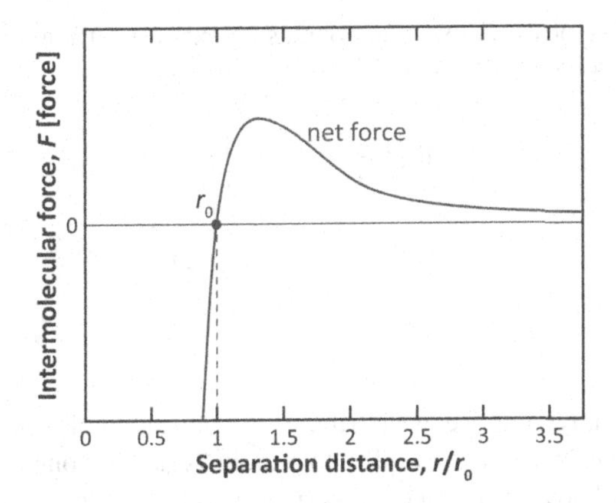

FIGURE 1.12 The net force between the two molecules is shown in **Figure 1.11**. This force is given by $F(r) = dU/dr$. It is large and repulsive (negative) at short separations and small and attractive (positive) at large separations.

in magnitude and opposite in sense or sign. Two molecules separated by this distance will neither attract nor repel but will remain at this separation indefinitely.

Example 1.8:

An important interatomic potential $U(r)$ in the analysis of multiatomic systems is the Lennard-Jones or "6-12" potential, given by

$$U(r) = 4U_0\left[\left(\frac{a}{r}\right)^{12} + \left(\frac{a}{r}\right)^{6}\right]$$

In the 6-12 potential, U_0 is the depth of the potential [see **Figure 1.11(b)**], and a is the root of $U(r)$, i.e., $U(a) = 0$. What is the equilibrium separation r_0 for this potential?

SOLUTION

The condition for force equilibrium between the two atoms can be expressed by the equation

$$F = 0$$

i.e., the atoms are at equilibrium when there is no force pushing them apart or pulling them together. The value of r that satisfies this condition is the equilibrium spacing r_0. Since we have the potential, we can find the force as

$$F(r) = \frac{dU}{dr} = \frac{d}{dr}\left\{4U_0\left[\left(\frac{a}{r}\right)^{12} + \left(\frac{a}{r}\right)^{6}\right]\right\} = 4U_0\left(\frac{6a^6}{r^7} - \frac{12a^{12}}{r^{13}}\right)$$

(Note the senses of these two components.) Next, we set this expression for F equal to 0 and solve for r to get r_0:

$$4U_0\left(\frac{6a^6}{r^7} - \frac{12a^{12}}{r^{13}}\right) = 0$$

$$\downarrow$$

$$r_0^6 = 2a^6$$

or $r_0 = a(\sqrt[6]{2}) \approx$ **1.12a**.

Based on the analysis above, we attribute to atoms and molecules not just an ability to come together but also the ability to remain "stuck" to one another. This is an important ability because it makes materials possible. Atoms positioned in stable arrangements are held in place by **bonds**. You are familiar with the chemical bonds that form stable molecules, but these molecules only incorporate a handful of atoms. These bonds are persistent, rather than temporary, because the complex of atoms is more *stable* than the atoms individually. We review the essential bond types here.

Ionic bond. Ion–ion interactions are rated "strong" in our ranking of intermolecular forces, and atoms that are ionized can form extremely stable arrangements. Ionically bonded compounds are formed via an electron-transfer process between a metal and a nonmetal. The metal species *loses* an electron/electrons and the nonmetal species *acquires* the negative charge(s). This exchange stabilizes the electron structure of one or both of the atoms by emptying or filling their outermost orbitals, but leaves the atoms ionized. The resulting metal cation(s) and nonmetal anion(s) are then bonded by virtue of the powerful attractive Coulomb forces that arise between the ions. A variety of different possible compounds exist depending on how many electrons are transferred and how many atoms participate, but note that the resulting ionic compound must be overall neutral. Some common ionic substances are listed in **Table 1.3**.

TABLE 1.3
Some Materials Based on Ionic Bonding between Components

Groups 1 & 2	Groups 3–14
LiF, LiI	$FeCl_2$, $FeCl_3$
NaF, NaCl	HgO, Hg_2O
KCl, KI	CuO, Cu_2O
CsCl, CsBr	PbO, PbS_2
MgO, $MgCl_2$	Fe_2S
CaO, CaS, $CaCl_2$	$CrCl_2$
BaO, BaS	NiF_2

Note: Metal atoms are electron donors (cations) and nonmetal atoms are electron acceptors (anions).

Covalent bond. Covalent bonds are based on the "sharing" of electrons in a compound and form between atoms that have incomplete outer orbitals and that are not both metals. Two nearby atoms (brought together by, say, van der Waals forces) will find that their outermost orbitals overlap, generating a shared orbital space that electrons can inhabit. Since the occupancy of this overlapping state is limited to two electrons by the exclusion principle, electrons are shared in pairs. This new shared orbital also establishes a particular axis connecting the centers of the bonded atoms, typically represented by a "—" drawn between them. This gives the molecule a particular fixed geometry, as represented by **Lewis structures**. Because both atoms can count both shared electrons as "theirs", the overall stability of the pairing is increased by the bond and they form a stable association as a molecule. Covalent bonds can have a variety of strengths depending on how much the orbitals overlap, and materials based entirely on covalent bonding tend to have a very rigid structure.

Metallic bond. You are likely familiar with ionic and covalent bonding from your chemistry courses, but you might not be familiar with the primary type of bonding between exclusively metallic elements. Neutral metal atoms are attracted to one another by van der Waals forces, of course, and arrange themselves into stable configurations based on the balance of attractive and repulsive forces. This configuration is additionally stabilized by an electron-sharing scheme described by the model of Paul Drude (1863–1906). This type of bonding is not based on the sharing of pairs of electrons between two atoms, but rather *all* electrons are shared among *all* metal atoms that make up the arrangement. Large numbers of electrons (say, 6.022×10^{23} of them) obviously cannot share a common state ψ in the metal, so the electrons occupy closely spaced states throughout the entire metallic material. This means that no atom "owns" a particular electron; rather, the atoms exist as ionic "islands" in a "sea" of electrons that can flow to and fro in the structure. Metallic bonds do not possess the same directionality as covalent bonds, so metallic structures possess some capacity for reorganization.

van der Waals bond. If the bond holding two atoms in place is neither ionic, nor covalent, nor metallic, what else could it be? As we know, van der Waals forces are always present between atoms and molecules. In the absence of other bonding effects, the weak intermolecular forces provide a means for atoms to associate. The resulting bonds are themselves weak, but they can nevertheless be physically significant in contexts where the other types of binding are absent. **Hydrogen bonding** is a type of van der Waals bonding particular to the exposed protons (H^+) present in hydrogen-bearing dipoles. Ice is an example of a solid composed of covalently bonded water molecules held together in a common structure by hydrogen bonds.

The strength and nature of the bonds present in a substance determine many of the overall properties of the material. The binding energy reflects how much work must be invested to disrupt a bond, and the configuration of the bonds gives rise to a particular structure. We will explore these ideas in the following section.

1.5 STRUCTURAL MOTIFS IN MATERIALS

Atoms and molecules will associate with one another under the influence of intermolecular interactions and will persist in their associations through the stabilizing influence of bonding. When the atoms are arranged with an equilibrium spacing r_0

between one another, the network of atoms will fill a certain volume of space, and these large networks constitute the materials we are interested in. You might now ask: what do these networks look like? As you might suspect, there is an uncountable number of ways to assemble atoms into a consolidated mass. Fortunately for our understanding, most arrangements fall into one or more of a small number of **structural motifs** exhibiting similar features. Individual materials reflect one or more of these motifs, though the exact details of the arrangement give materials different structures (and thereby different properties, as alluded to in **Chapter I**).

Perhaps the most common motif in materials is the **crystal** motif, illustrated in **Figure 1.13**. A crystal is an arrangement of atoms that has a very regular geometry; it is so regular that it can be continued indefinitely without any significant disruption of

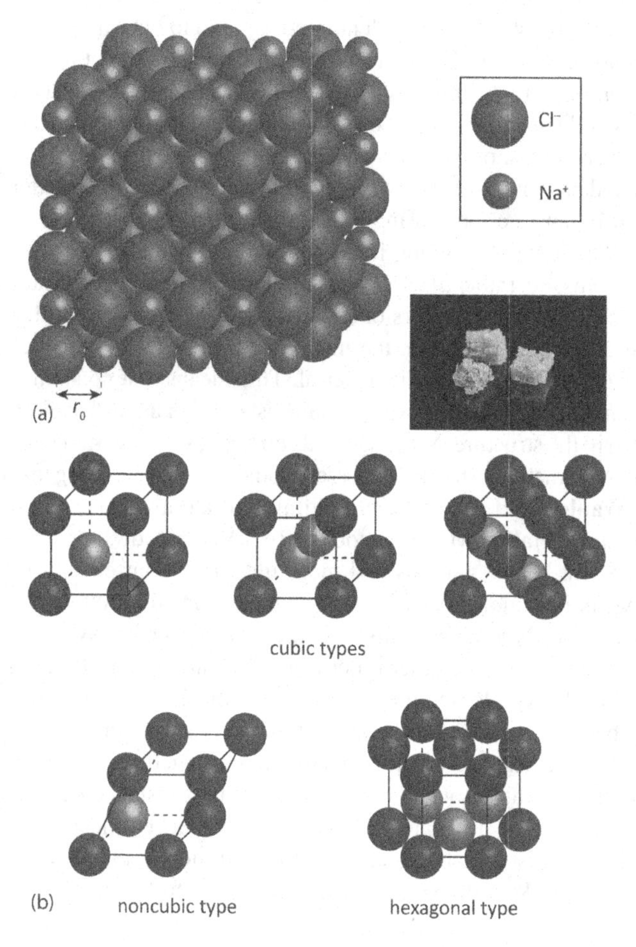

FIGURE 1.13 The crystal structural motif. (a) An example of an ionic NaCl crystal. All of the Na^+ and Cl^- ions are arranged at a fixed separation r_0 from one another in a 1:1 ratio. The arrangement is highly regular, and the cubic symmetry of the structure is reflected in the shape of macroscopic salt crystals. (b) Crystals are based on small units, or "unit cells", that can be stacked indefinitely. These cells are based on cubic, noncubic, and hexagonal arrangements.

the basic arrangement. Consider the crystal depicted in **Figure 1.13(a)**. We recognize the atomic/ionic constituents of this crystal as sodium and chlorine ions. The sodium cations will be attracted to the chlorine anions (forming strong ionic bonds), and the like ions will repel one another. This "rocksalt" crystal is a geometrical arrangement that places anions next to cations and separates the similar charges as much as possible. This arrangement is, therefore, *stable* and *regular.* The diagram shows only a small crystal – 100 ions or so – but the essential pattern may be repeated indefinitely. The crystal motif is based on small arrangements of atoms that act as "building blocks" or "unit cells" from which larger crystals are assembled. Some of the basic types of unit cells are shown in **Figure 1.13(b)**. Most isolated elements from the periodic table take on a crystalline form. (The most common unit-cell type for each of the elements under ambient conditions is illustrated in the periodic table of **Figure 1.7**.)

The crystal motif is based on unit cells organized into *regular* arrangements. The glass motif, shown in **Figure 1.14**, has a different organization. Like crystals,

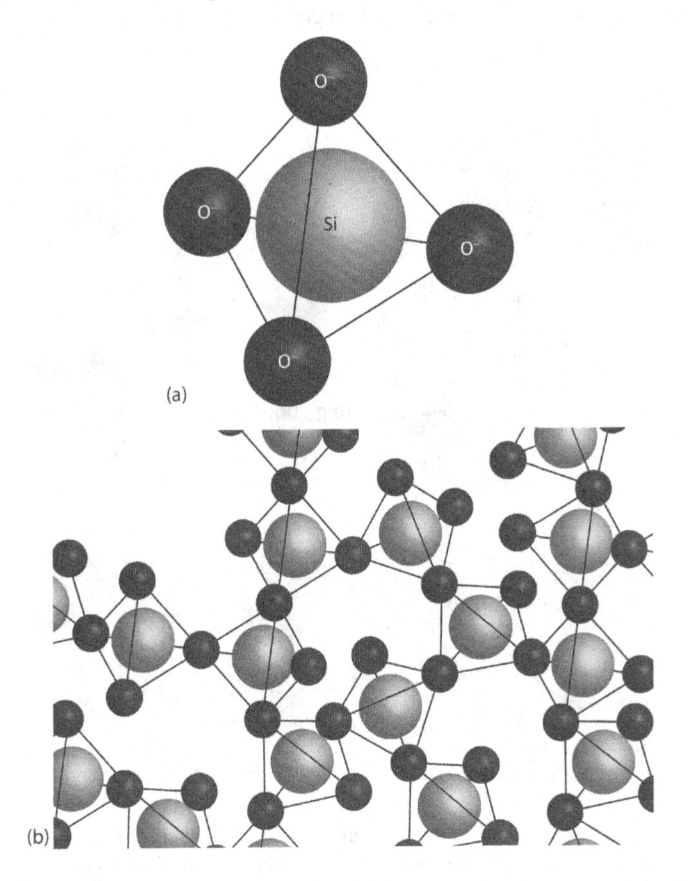

FIGURE 1.14 The glass structural motif. (a) Like crystals, glasses are based on small structural units involving only a few atoms. For example, orthosilicate ions (SiO_4^{4-}) are the repeated units in common silicate glasses. (b) Unlike the crystal motif, the glass units are disordered, forming a random network-type structure. Sometimes, other atoms are added to the network to modify the glassy structure.

glasses typically have structural units consisting of a few atoms, such as the silicate unit in **Figure 1.14(a)**. The four O atoms at the corners of the structure have one electron to share each and can all bond to the same central Si atom. (This kind of tetrahedral bonding, requiring the participation of combined s- and p-type orbitals $\psi_{n(sp3)}$, is called "sp^3 hybridization".) This bonding arrangement also means that the silicates can form an extended network of structural units joined at the corners. If the arrangement of the tetrahedra is regular, you obtain a crystalline mineral called quartz. However, if the formation of a regular network is not possible (such as in the presence of network-modifying atoms like Ca), a disordered arrangement is obtained. This is the structure of a silicate glass and is shown in **Figure 1.14(b)**. The **glass** motif is one where there is no repeating structure beyond the small structural unit.

The macromolecule motif has as its structural unit large, typically organic, molecules called **macromolecules**. This motif is illustrated in **Figure 1.15**. **Figure 1.15(a)** shows an example macromolecule based on covalently bonded C atoms. Each C atom

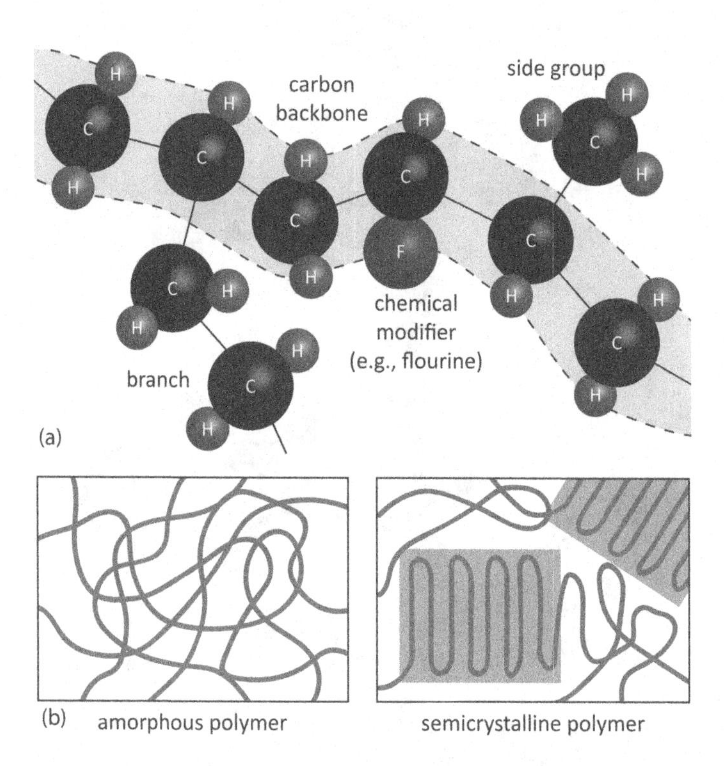

FIGURE 1.15 The macromolecule structural motif. (a) Macromolecules are covalently bonded compounds with high molecular weights. They are typically composed of carbon atoms arranged in a "backbone" chain saturated with hydrogen atoms. The backbone can be branched or chemically modified to replace the hydrogen with other atoms or groups of atoms, modifying the structure of the macromolecule. (b) Possible arrangements of macromolecules into bulk materials include disordered/amorphous/"glassy" styles (left) and ordered/crystalline styles (right).

can form four bonds in an sp^3-type configuration: one bond to the previous atom in the chain, one to the next atom in the chain, and two bonds to any of the following:

- hydrogen atoms (e.g., in a hydrocarbon),
- C-bonded *branches*,
- other C-based groups [e.g., a methyl ($-CH_3$) group], or
- other chemical species (e.g., halogens F or Cl).

The long (linear or branched) macromolecules then aggregate to form a polymer. As with glasses, an ordered arrangement of macromolecules forms a crystalline polymer, and a disordered arrangement of macromolecules forms an amorphous polymer. Therefore, polymers can express multiple motifs at once, and different polymer structures compose materials with different properties. These structures are illustrated in **Figure 1.15(b)**.

Not every material is either a single crystal or an amorphous arrangement. In fact, most materials typically encountered in engineering are crystalline but are composed of many microscopic crystals; i.e., they are **polycrystalline**. The individual crystals are compacted into a bulk material, as shown in **Figure 1.16**. The microstructure of an object is revealed when inspected by a light microscope. At magnifications typically in the range of 10× to 400×, the individual crystals, or "grains", and their collective arrangement are revealed. The grains, possessing a distribution of orientations, sizes, and shapes, adjoin one another to fill the material's volume. The grains are separated by **grain boundaries**: regions of the structure that contain disordered atoms that can't be said to belong to either adjoining grain. When viewed at higher magnification (e.g., using electron microscopy), the underlying crystalline organization of the grains is revealed. Note that, in the case of this steel alloy, the crystalline structure incorporates multiple different elements.

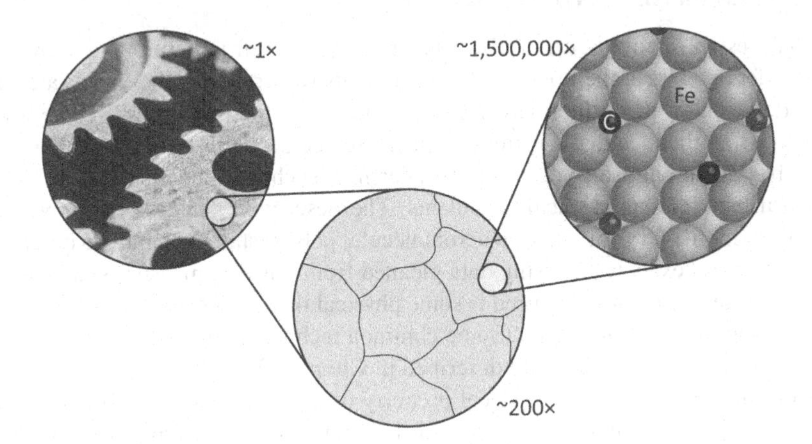

FIGURE 1.16 The polycrystalline structural motif. The bulk solid/steel gear is composed of individual crystals (resolvable via microscopic techniques), separated from one another across grain boundaries. Higher-magnification microscopy reveals that the grains of the microstructure themselves reflect the crystal structural motif.

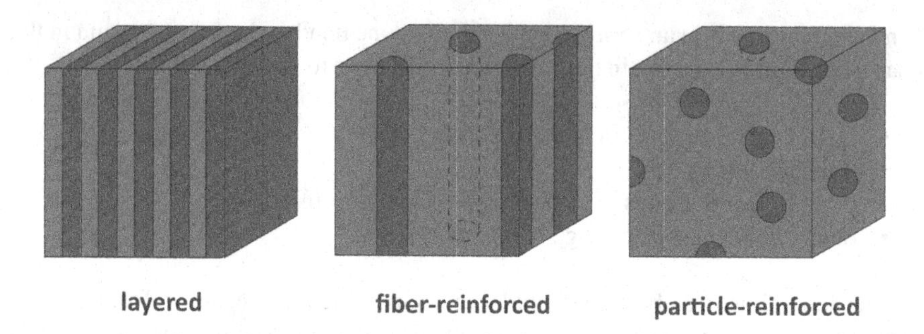

FIGURE 1.17 Examples of composite architectures. The matrix material is the majority of the volume of the composite, overall, and the reinforcing material may come in 0D (particle), 1D (fiber), or 2D (sheet/layer) forms.

Composite materials, as they are built from (typically different) materials from two (or more) different classes, require additional description. In contrast to most monolithic materials, there may be different motifs existing side-by-side in distinct regions of a composite. For example, crystalline ceramic particles (the minority component) reinforce a bulk amorphous polymer (the majority component). The material that is being reinforced is called the **matrix**. The matrix surrounds and supports the reinforcing material, and the reinforcing material enhances the properties of the matrix synergistically. The shape and arrangement of the reinforcement material in the matrix determine the composite's architecture. **Figure 1.17** shows some typical architectures. Layered, fiber-reinforced, and particle-reinforced matrices are the most common.

1.6 PROPERTIES AND STRUCTURE

We have established that materials differ in terms of their structure and know that they differ in terms of their properties. We have also asserted that these facts are connected; we claim that "the microstructure determines the properties". Though you probably find the idea that a material's structure and properties are connected quite plausible, as an engineer you might wonder how such a connection can be established using rigorous physical arguments. The descriptive features of the various structural motifs (crystal, glass, macromolecule, polycrystal, etc.) may be described in more or less exact terms using data gleaned from physical measurements, as can the properties. All that is required is some physical theory that establishes the properties using the structural data. (Some common techniques for obtaining structural information about a material are described in **Chapter 3**.)

Consider the density ρ, a physical property of materials. Recall that density has units $[\rho]$ = [mass/volume] = [mass/length^3] = kg/m^3, so establishing the associated structure/property connection for a given material requires us to determine how much mass is contained in a representative volume of material. A crystalline material that possesses a simple cubic crystal structure is shown in **Figure 1.18**. An individual unit cell of this structure – the cube's bounding faces and its interior – includes

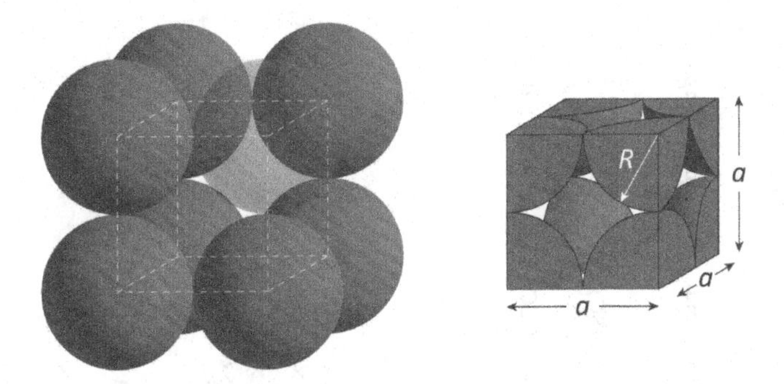

FIGURE 1.18 The packing of atoms/spheres in a simple cubic unit cell. The atom centers occupy the corner positions in the cube. In the cutaway diagram, the contents of the cell are revealed to be eight atomic segments, each with 1/8th of an atomic volume/mass. The cube has side length a, and from the diagram, we can infer that $a = 2R$.

8 "partial" atoms (one centered at each corner). Each partial atom represents 1/8 of an entire atom, so we can compute the mass m inside the confines of the unit cell as

$$m = 8 \times \frac{1}{8} \times A = A \tag{1.13}$$

where A is the atomic mass of the atom. The volume V associated with the unit cell can be inferred from the schematic. If the cube has a side length of a, then the total volume is $V = a \times a \times a = a^3$. Furthermore, from the atomic packing geometry, we can determine that the cube side length $a = 2R$, where R is the radius of the atom. This gives

$$V = a^3 = 8R^3 \tag{1.14}$$

We then obtain

$$\rho = \frac{m}{V} = \frac{A}{8R^3} \tag{1.15}$$

i.e., the density can be inferred from the structure of the crystal and some data taken from the periodic table. This is an important result. It captures to a high-numerical precision the relationship of a material's density to the characteristics of its constituent atoms (size and mass) and its structural organization.

For other crystalline materials, possessing different structures, and constituent atoms, the calculation above requires modification. The fundamental crystalline structures, also called **Bravais lattices,** are given in **Table 1.4**. The Bravais lattices are organized into seven different systems, each system corresponding to a particular kind of polyhedron that can be stacked with others of its kind to fill space according to the requirements of the crystalline motif. These polyhedra have atomic contents: atoms/molecules that lie at fixed positions within the shape. The "simple" variant of each system has its atomic positions at each of the eight corners only (similar to the

TABLE 1.4
Basic Crystal Structures (Bravais Lattices) and Their Properties; All Angles between Edges of the Polyhedra Are 90° unless Otherwise Noted

System/Variant	Unit-cell geometry	Mass [amu]	Volume [length³]
Cubic	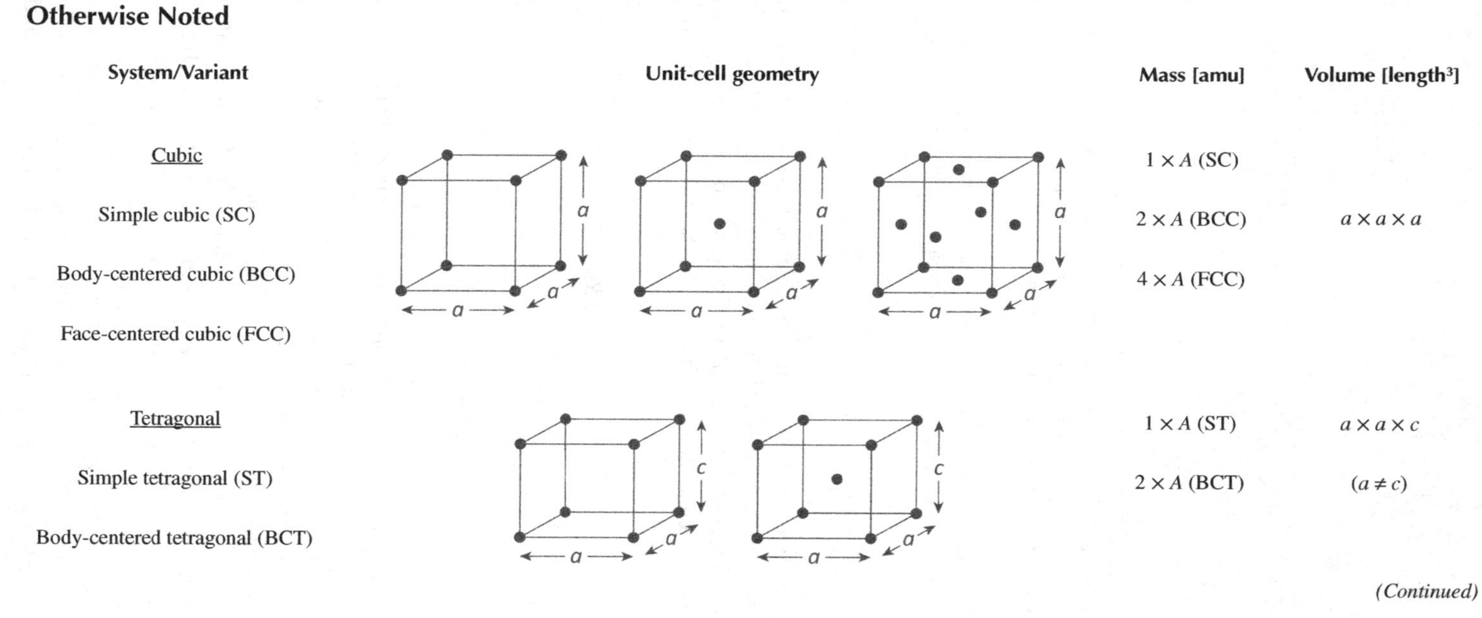	$1 \times A$ (SC)	
Simple cubic (SC)		$2 \times A$ (BCC)	$a \times a \times a$
Body-centered cubic (BCC)		$4 \times A$ (FCC)	
Face-centered cubic (FCC)			
Tetragonal		$1 \times A$ (ST)	$a \times a \times c$
Simple tetragonal (ST)		$2 \times A$ (BCT)	$(a \neq c)$
Body-centered tetragonal (BCT)			

(Continued)

TABLE 1.4 *(Continued)*

Basic Crystal Structures (Bravais Lattices) and Their Properties; All Angles between Edges of the Polyhedra Are 90° unless Otherwise Noted

System/Variant	Unit-cell geometry	Mass [amu]	Volume [length³]
<u>Orthorhombic</u>			
Simple orthorhombic (SO)	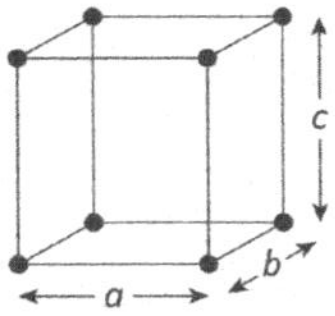 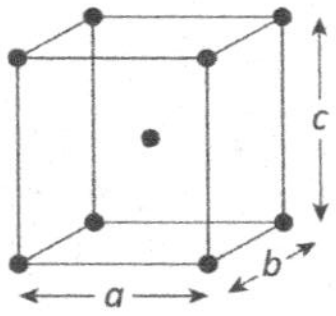 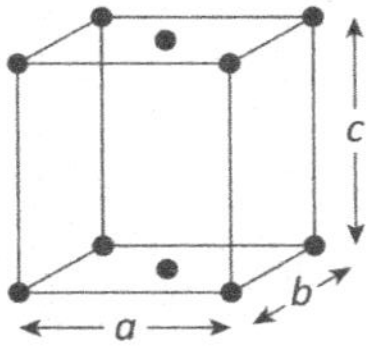	$1 \times A$ (SO)	
Body-centered orthorhombic (BoCO)		$2 \times A$ (BoCO)	$a \times b \times c$
Base-centered orthorhombic (BaCO)	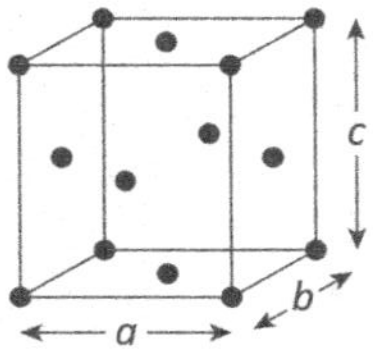	$2 \times A$ (BaCO)	$(a \neq b \neq c)$
Face-centered orthorhombic (FCO)		$4 \times A$ (FCO)	
<u>Rhombohedral</u>	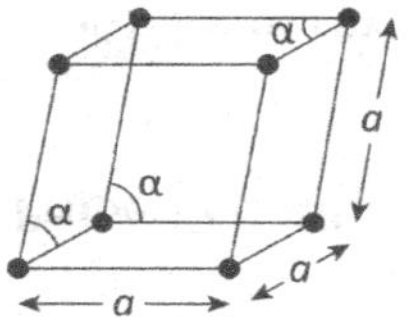		
Rhombohedral (R)		$1 \times A$	$a^3\sqrt{(1 - 3\cos^2\alpha + 2\cos^3\alpha)}$ $(\alpha \neq 90°)$

(Continued)

TABLE 1.4 *(Continued)*
Basic Crystal Structures (Bravais Lattices) and Their Properties; All Angles between Edges of the Polyhedra Are 90° unless Otherwise Noted

System/Variant	Unit-cell geometry	Mass [amu]	Volume [length³]
Hexagonal	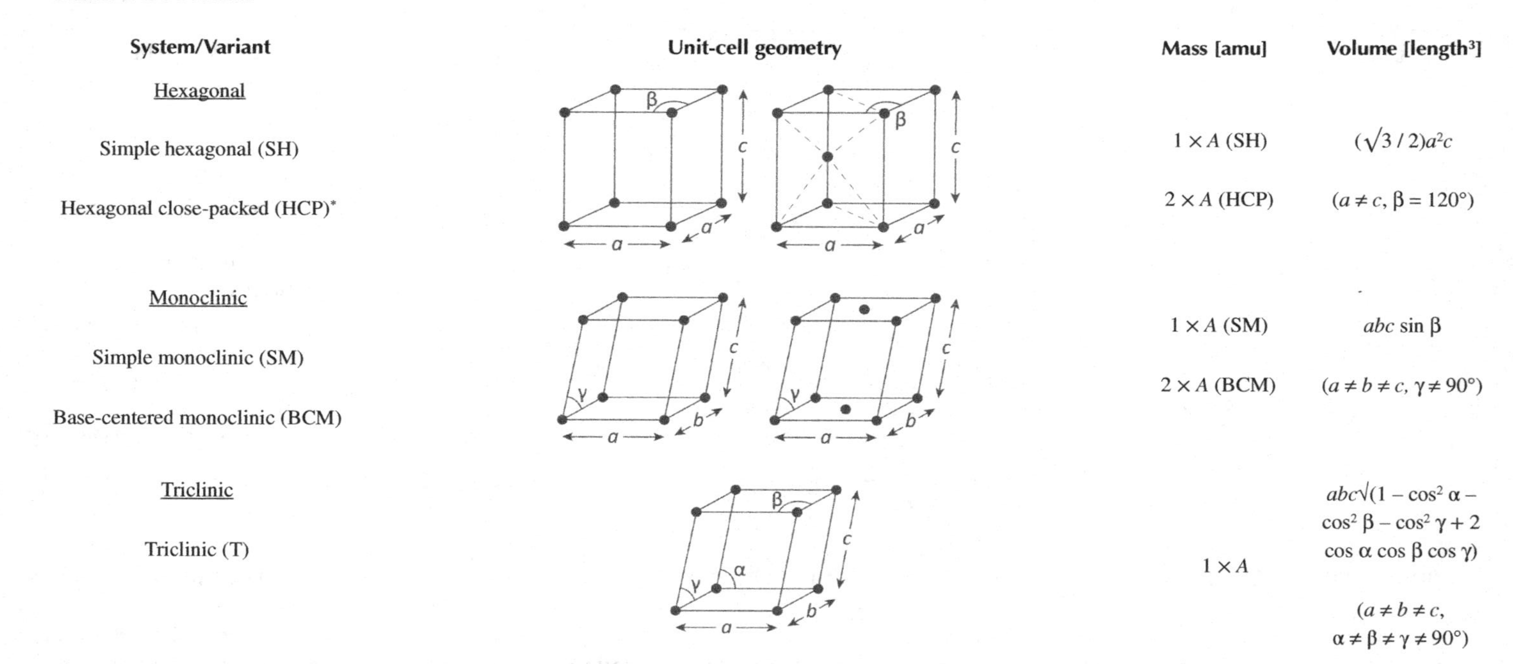		
Simple hexagonal (SH)		$1 \times A$ (SH)	$(\sqrt{3}/2)a^2c$
Hexagonal close-packed (HCP)*		$2 \times A$ (HCP)	$(a \neq c, \beta = 120°)$
Monoclinic			
Simple monoclinic (SM)		$1 \times A$ (SM)	$abc \sin \beta$
Base-centered monoclinic (BCM)		$2 \times A$ (BCM)	$(a \neq b \neq c, \gamma \neq 90°)$
Triclinic			
Triclinic (T)		$1 \times A$	$abc\sqrt{(1 - \cos^2 \alpha - \cos^2 \beta - \cos^2 \gamma + 2 \cos \alpha \cos \beta \cos \gamma)}$ $(a \neq b \neq c, \alpha \neq \beta \neq \gamma \neq 90°)$

* The HCP structure is not strictly a Bravais lattice but is quite common in engineering materials/metals, so it is depicted alongside the simple hexagonal unit cell for convenience.

situation in **Figure 1.17**), while the variants have additional atomic positions. For instance:

Body-centered – an additional position is located at the center of the cell.
Face-centered – additional positions located at the center of all the faces of the cell.
Base-centered – additional positions located at the center of two (top and bottom) faces of the cell.

Though different crystalline materials may have the same Bravais lattices as their structure, unit cells are further distinguished by their **lattice parameters**. The parameters include the cell edge lengths *a*, *b*, and *c*, as well as the interior angles α, β, γ. The atomic contents (in terms of the total unit-cell mass *m*) are provided for each variant, as are the unit-cell volumes in terms of the lattice parameters. For some of the unit cells, there is a simple geometric relationship between the lattice parameters and the atomic size *R*, but in many cases, such relationships are absent, and so density calculations must be supplemented by other pieces of experimentally determined information.

Example 1.9:

Compute the density ρ of a crystal of Cr, which has the BCC structure. Take the radius of a Cr atom as $R = 0.125$ nm. If the experimentally determined value of the density is $\rho' = 7.19$ g/cm^3, what is the *percent error* in your calculation?

SOLUTION

The BCC unit-cell layout is shown below.

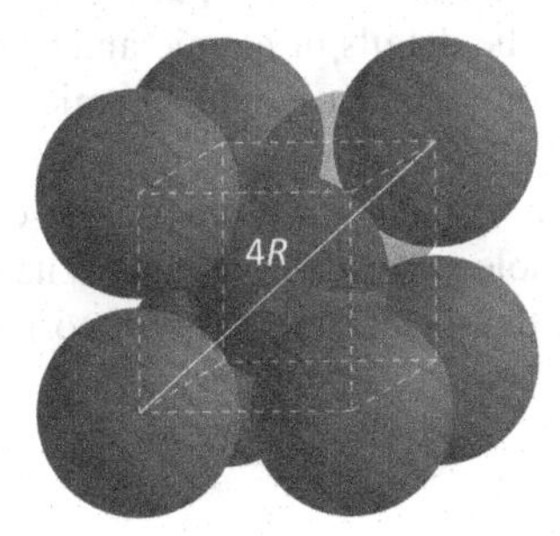

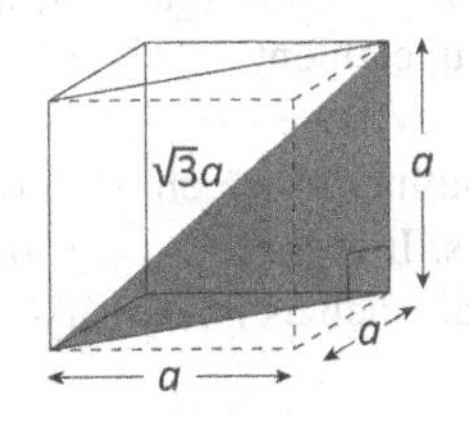

In the BCC arrangement, the Cr atoms are all collinear and touch along the body diagonal of the cube. A geometrical analysis reveals that the length of this body diagonal is $a\sqrt{3}$, so we make the association

$$a\sqrt{3} = 4R$$

Since the volume V of a BCC unit cell is a^3, we obtain

$$V = a^3 = \left(\frac{4R}{\sqrt{3}}\right)^3$$

From **Table 1.4**, the atomic contents of a BCC unit cell is a mass $m = 2A$, where $A = 51.996$ u for Cr. We can now compute

$$\rho = \frac{m}{V} = \frac{2A}{\left(4R/\sqrt{3}\right)^3} = \frac{3A\sqrt{3}}{32R^3} = \frac{3(51.996\text{ u})\sqrt{3}}{32(0.125\text{ nm})^2} = 4320\text{ u/nm}^3$$

The units "u/nm^3" are not particularly useful, so we convert. Recalling the facts that 1 nm = 1 × 10^{-7} cm and 1 u = 1.661 × 10^{-24} g:

$$\rho = 4320\frac{\text{u}}{\text{nm}^3} \times \frac{1\text{ nm}^3}{\left(10^{-7}\text{ cm}\right)^3} \times \frac{1.661\times10^{-24}\text{ g}}{1\text{ u}} = \underline{\mathbf{7.18\ g/cm^3}}$$

This is close to the known/accepted value of $\rho_0 = 7.19$ g/cm^3, and the percentage error is

$$\%\text{err} = \frac{\rho_0 - \rho}{\rho}\times 100 = \frac{7.19-7.18}{7.18}\times 100 = \underline{\mathbf{0.163\%}}$$

Density is one example of a property of materials that can be inferred from the molecular characteristics and bonding structure of the constituents. There are numerous other examples as well; properties like melting temperature, vapor pressure, specific heat, and elasticity can be related to bond strength and configuration. The elastic properties are of particular importance in engineering design, and we will take up a discussion of them in the next chapter.

1.7 CLOSING

The materials of our common experience are composed of, at the very small scale (at the "bottom"), atoms and molecules. The details of atomic and molecular composition and electronic organization not only determine the chemical identity and reactivity of an element or compound but also inform us about the kinds of physical interactions they have with each other. It is these interactions that contribute the most to the manner in which the atomic/molecular matter organizes itself into stable configurations. It is these configurations that determine some important properties of the material, such as its density.

1.8 CHAPTER SUMMARY

Key Terms

anion
atom
atomic mass
atomic number
Avogadro constant
Bond
Bravais lattice
cation
charge (electric)
Coulomb's Law
covalent bond
crystal

dipole
electron
electron configuration
electrostatic force
element
energy/work
energy level
exclusion principle
force equilibrium
first quantum number
fourth quantum number
glass
grain boundary
hydrogen bond
induced polarization
intermolecular forces
ion
ionic bond
isotope
lattice parameters
macromolecule
matrix
metallic bond
molar mass
molecule
neutron
nucleus
orbital
periodic table
permittivity
Planck's constant
polar molecule
potential energy
proton
quantization
quantum mechanics
quantum number
sample
Schrödinger wave equation
second quantum number
spin moment
spontaneous polarization
structural motif
third quantum number
uncertainty principle
van der Waals forces
van der Waals bond
wave/particle duality
wavefunction

Important Relationships

$$X_A = \frac{\text{Moles of A}}{\text{Total moles of matter}} \quad \text{(mole fraction)}$$

$$A_{avg} = \sum_i f_i A_i \quad \text{(average atomic mass)}$$

$$F_C(r) = -\frac{Q_1 Q_2}{4\pi\epsilon_0} \times \frac{1}{r^2} \quad \text{(Coulomb force)}$$

$$U(r) = -\frac{Z q_e^2}{4\pi\epsilon_0} \times \frac{1}{r} \quad \text{(electron Coulomb potential)}$$

$$\Delta x \Delta p \geq \frac{h}{4\pi} \quad \text{(uncertainty relation)}$$

$$-\frac{h^2}{4\pi m_e}\frac{d^2\psi}{dx^2} + U(x)\psi(x) = E\psi(x) \quad \text{(1D Shrödinger eqn.)}$$

$$E_n = -\frac{m_e Z^2 q_e^4}{8\epsilon_0^2 h^2} \times \frac{1}{n^2} \quad \text{(energy quantization)}$$

$$L_\ell = \frac{h}{2\pi}\sqrt{\ell(\ell+1)} \quad \text{(quantized angular momentum)}$$

$$\text{orbital} = n(\text{subshell letter code})^{\#\ \text{occupants}} \quad (\text{e}^- \text{ configuration})$$

$$U(r) = \frac{C_R}{r^\beta} - \frac{C_A}{r^\alpha} \quad \text{(net interaction potential)}$$

$$F(r) = \alpha\frac{C_A}{r^{\alpha-1}} - \beta\frac{C_R}{r^{\beta-1}} \quad \text{(net interaction force)}$$

$$U(r) = 4U_0\left[\left(\frac{a}{r}\right)^{12} + \left(\frac{a}{r}\right)^{6}\right] \quad \text{(6–12 potential)}$$

$$\rho = \frac{m}{V} \quad \text{(density of unit cell)}$$

1.9 QUESTIONS AND EXERCISES

Concept Review

C1.1 The energy-level values derived from **Equation 1.7** apply to a system with one nucleus and one electron, e.g., an H atom or He^+ ion. However, this expression is inaccurate for atomic systems with multiple electrons. Why?

C1.2 We have depicted the atoms in a crystal as sitting at *fixed* positions in a unit cell. However, if the atoms were truly at a standstill, the uncertainty principle indicates they wouldn't have *any* resolvable position. What does this imply about the behavior of atoms in the lattice?

C1.3 Think about what you know about the properties of materials as introduced in the **Introduction Chapter**. What are some properties that you think are primarily dependent on the strength of the intermolecular forces between the particles in a material? What properties are mostly not dependent?

C1.4 What type of bonding would you expect to be predominant in brass (a mixture of Cu and Zn)? What type of bonding would you expect in poly(ethylene) (a polymer made of C and H)? What type of bonding would you expect in the mineral sylvite (with the formula KCl)?

Discussion-forum Prompt

D1.1 Choose a material that you use in your area of engineering and identify the structural motif or motifs that it exhibits. Do some background research on the typical behaviors, properties, and uses of this material and include those in your forum post. After you have done this, look at the posts of the other students in the class and identify one of those that give the same motif (but likely for a different material). What are the similarities and differences between your material and theirs in terms of their behaviors and applications?

PROBLEMS

P1.1 Suppose that you are given a specimen of lead that contains 35 g of ^{208}Pb, 15 g of ^{207}Pb, and 2 g of ^{206}Pb. What is the average atomic weight of your Pb sample?

P1.2 The electronic configuration of aluminum is $[Ne]3s^23p^1$. The most common "oxidation state" of Al is Al^{+3}. What is the electronic configuration of this cation?

P1.3 What is the concentration in wt% of each of the elements in the alloy "Ti90Al10"? The alloy has 90 Ti atoms and 10 Al atoms for every 100 atoms total. The atomic weight of Ti is 47.87 u, and the atomic weight of Al is 26.98 u.

P1.4 Polymer materials are frequently composed of chemical "repeat units" with fixed chemical constituents, and these repeat units can be thought of as a submolecule of the larger macromolecule. The repeat unit of poly(vinyl chloride) (PVC) is $-(C_2H_3Cl)-$. What is the molecular weight of this repeat unit? If a typical macromolecule has 10^5 repeat units, what is the molecular weight of the PVC macromolecule?

P1.5 Suppose you have two particles, an ion and a dipole, that interact according to the relationship of **Equation 1.11**. If the exponents $\alpha = 4$ and $\beta = \alpha + 4$, and the constants $C_A = 1.5$ eV/nm^4 and $C_R = 0.2$ eV/nm^8, what is the equilibrium separation between the particles? (Note that an "electron-volt" is a small unit of energy with 1 eV $\approx 1.602 \times 10^{-19}$ J.)

P1.6 The density of gold is 19.3 g/cm^3 and gold atoms have an atomic radius of $R = 0.144$ nm. Does it have an FCC or BCC structure?

P1.7 The "atomic packing factor" (APF) is a representation of how much space is taken up by the atoms in a unit cell:

$$\text{APF} = V_{\text{atoms}} / V_{\text{cell}} = N V_{\text{atom}} / V_{\text{cell}}$$

where the volume of a (spherical) atom is $V_{\text{atom}} = (4/3)\pi R^3$. Mo is a BCC metal with $R = 0.136$ nm. What is the APF of Mo?

MATLAB® Exercises

M1.1 The data file "Exercise M1-1.txt" contains data on the atomic radii R of the elements according to their atomic number Z. Place these data in a scatter plot: R vs. Z. What are the trends you observe?

M1.2 **Example 1.7** illustrates a calculation of the force between two ions with given charges Q_1 and Q_2 at a given separation r. Write a MATLAB function that computes the coulombic force FC between two ions of *any* charge at *any* separation. The ionic charges should be specified using the built-in `input()` function (as multiples of q_e), and the distance should be specified using `input()` (in units of nm). The resulting force should be given in nN.

M1.3 Plot the Lennard-Jones potential of **Example 1.8** using MATLAB. To do this, take a = one distance unit = 1 and U_0 = one energy unit = 1. How does the graph change when you set $a = 2$ units or $a = ½$ unit? How does it change when you increase or decrease U_0?

NOTES

1. R. P. Feynman, R. B. Leighton, and M. L. Sands. *The Feynman Lectures on Physics, Volume 1.* First edition. Menlo Park, CA, USA: Addison-Wesley, 1965, p. 3.
2. These results really only apply to simple or idealized atoms/ions but provide a good conceptual basis for understanding atomic structure.

2 Materials Properties and Performance

From Structure to Behavior

LEARNING OBJECTIVES

After completing this chapter, you should be able to:

1. Distinguish the study of the mechanics of materials from the other types of mechanical studies.
2. Define the concepts of stress and strain as they relate to materials subject to forces or deformations and compute their values for a given geometry and loading.
3. Define the elastic moduli for materials (tensile, compressive, and shear), name the deformations that they are associated with, and obtain their numerical values from data tables.
4. Utilize the stresses, strains, elastic moduli, and various forms of Hooke's Law for solids in the solution of problems in elastic deformation.
5. Describe the relationship between atomic bonding and the elastic moduli.
6. Relate the deformation of a loaded structural member to the loading/support configuration, materials properties, and beam geometry.
7. Compute beam deflections and stresses using geometry and provided data.

2.1 DEVELOPING MATERIALS UNDERSTANDING

Up to this point, we have discussed materials through the lens of the fundamental sciences of chemistry and physics. These sciences provide us with a description of materials that is detailed, revealing, and worth knowing, but not perhaps in exactly the way we, as engineers and applied scientists, require in our everyday work. When an engineer designs a component made from a certain material, they are unlikely to start their calculations with the atomic or molecular constituents. The design, validation, and application of materials generally require understanding them at a different *level of description*. The fundamental or structural level of description introduced in **Chapter 1** is distinct from the engineering or "property" level of description. Both levels of description have the same overall purpose: the development of a rational approach to resolving materials-related questions. However, the types of problems they are capable of addressing are different.

At the end of **Chapter 1**, we discussed how the atomic properties and structural arrangement of crystalline systems can be used to infer the value of an important property (the density) of the bulk material. This calculation therefore bridges the gap between the structural level of description and the engineering level. Let's investigate further these two levels of description and how they relate to one another. The

DOI: 10.1201/9781003214403-3

focus of our investigation will be on a particular class of properties – the mechanical properties – so we begin by describing the engineering context that makes these properties important and the physical context needed to define them.

2.2 MATERIALS PROPERTIES

Common engineering descriptions of materials revolve around their macroscopic properties, some examples of which you will recall from **Section I.3**. As an entry point into our discussion of properties in general, we introduce and define some of the **mechanical properties** of materials. From your physics classes, you may recall that mechanics is the study of forces and related motions. In the context of materials, we are typically not interested in the overall motion (i.e., the "rigid-body motion") of the material object, but rather how it responds to the stimulus of the force in terms of *deformation.*

APPLICATION NOTE – MECHANICS AND ITS SUBBRANCHES

Mechanics is a branch of physics that addresses the interaction between forces and motions of discrete material objects or continuous substances. Mechanical analyses fall into a number of different internal categories according to the principles involved, typically Newton's Laws and analytic geometry. A mechanical analysis typically begins with a mathematical description of the bodies and forces. As the above image shows, a moving body has some properties associated with it (location of center of mass, velocity of center of mass), and these properties will change (or not) depending on the orientation and magnitude of the forces it is subject to (represented as vectors).

Rigid-body Mechanics describes how an array of forces determines the motion of a body through space (dynamics) or, in the case where the forces are mutually negating, establishes a state of equilibrium (statics). The influence of these forces on the shape of the body itself is not considered.

Kinematics is the description of motion through space without reference to any forces that produce or influence it, a.k.a. "the geometry of motion".

Continuum mechanics captures the role of forces (both internal and external) in determining the flow of a fluid (fluid mechanics) or the deformation of a solid (solid mechanics). Solid mechanics includes the physical description of elastic, plastic, and fracture behavior of solids.

The organization of these subbranches is represented schematically in the following diagram.

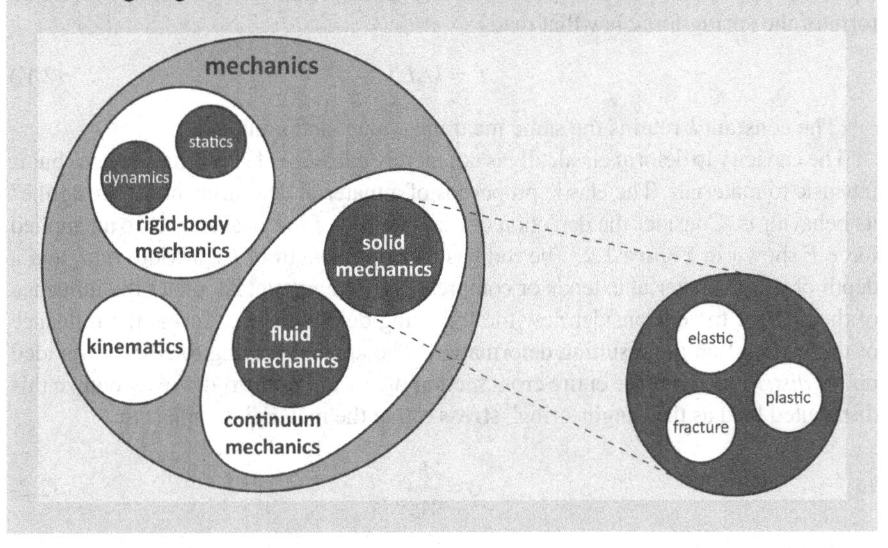

The mechanical properties of materials are important to us for a number of reasons. First, they are "universal" in that *every* component in an engineered system will be subject to some amount of force. Because of this, mechanical properties enter into most engineering designs explicitly, as opposed to other (e.g., thermal, electrical, and optical) properties that may remain unconsidered or considered to be of little importance. The analysis of forces and their effect on materials is the study of the **mechanics of materials**. Second, lifespan and safety considerations related to materials depend strongly on their mechanical behavior. Components can wear out rapidly or even fail catastrophically under undesirable mechanical conditions, and the envelope of conditions deemed safe depends on the specific material. Finally, understanding and applying the mechanical properties of solids relates to other principles introduced throughout the engineering curriculum, such as engineering statics and **stress analysis**.

Consider the role that forces and materials play in the extension of a solid yet deformable object – a spring. The relationship between the extension x and the resulting restoring force F in the spring is given by the spring-force law, also called **Hooke's Law**:

$$F = -kx \tag{2.1}$$

The extension is proportional to the applied force, and the constant of proportionality is k, called the **spring constant.** k has units $[k]$ = [force/length] = N/m. **Newton's Laws** of mechanics tell us the relationships between forces and the motion of objects in space, and Hooke's Law tells us that there is a simple relationship between forces and extensions in springlike objects themselves. The familiar physical scenario is depicted in **Figure 2.1**. A spring of initial length L_0 extends under the application of a force F. The applied force F produces an **extension** ΔL. Or, considered differently, extending the spring by ΔL requires a force F. The restoring force that develops is equal and opposite: $-F$. We describe this mechanical situation by a slightly modified form of the spring-force law that reads

$$F = k\Delta L \tag{2.1$'$}$$

(The constant k retains the same meaning, value, and units.)

The capacity to deform elastically is not merely a feature of springlike devices but is intrinsic to materials. The elastic properties of a material determine how "springlike" its behavior is. Consider the depiction of a square rod of material subject to an applied force F shown in **Figure 2.2**. The rod has an initial length of L_0, a width of a, and a depth of a. The material extends or compresses by an amount ΔL under the influence of the applied force. Consider now the following definitions that reveal the influence of the material on the resulting deformation. The applied loading force F is "divided up" or *distributed* over the entire cross section $A_0 = a \times a$ of the rod. We recognize this distributed load as the "engineering" **stress** σ that the material is subject to:

$$\sigma = \frac{F}{A_0} \tag{2.2}$$

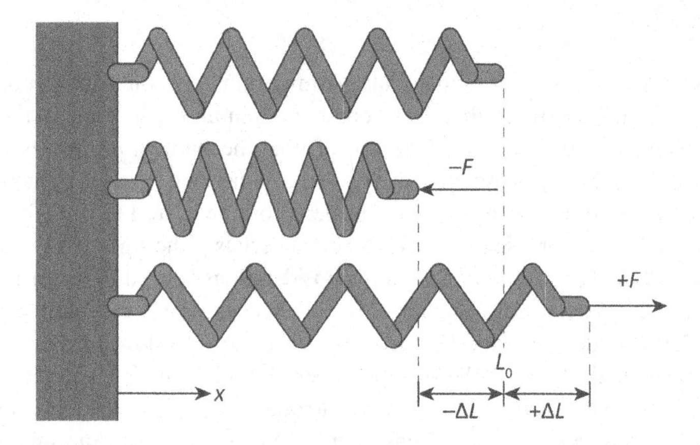

FIGURE 2.1 The geometry of spring forces and extensions. The extension ($+\Delta L$) or compression ($-\Delta L$) of a spring of normal length L_0 depends linearly on the applied force: $\Delta L \propto F$.

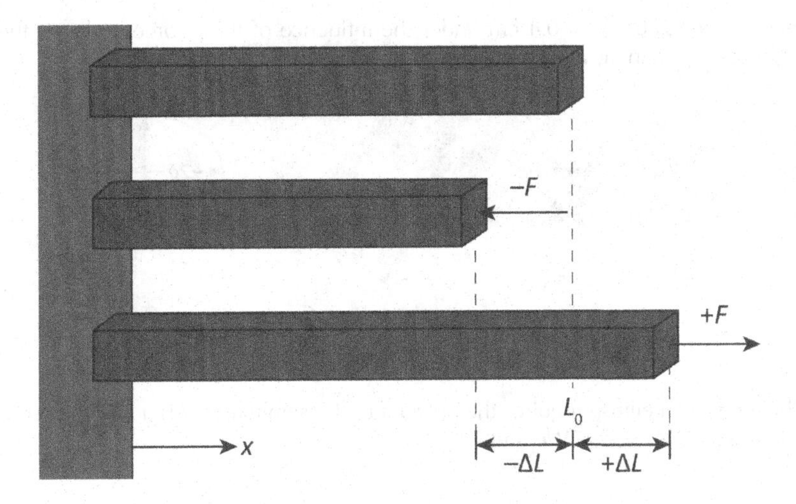

FIGURE 2.2 Extension and compression of a deformable rod of material. Unloaded, the rod has length L_0. When loaded in tension (or compression) by a force $+F$ (or $-F$), the rod length changes by $+\Delta L$ (or $-\Delta L$). The relationship between the force and the extension is determined by Hooke's law.

Let's assume that this stress is constant, or *uniform*, over the cross-section perpendicular to the rod; this is a good approximation for small cross sections. Using the engineering stress σ, we may account for the effect of a force on the material in a way that is essentially independent of the shape/cross section of the material. Similarly, we can represent the deformation in the material in a way that is independent of its overall length/size. Since the extension is ΔL, we may compute the *relative* extension ε as

$$\varepsilon = \frac{\Delta L}{L_0} \tag{2.3}$$

We call this relative extension the "engineering" **strain** in the material. It is frequently expressed as a percentage:

$$\%\varepsilon = \frac{\Delta L}{L_0} \times 100 \tag{2.3'}$$

We call the deformation in **Figure 2.2 normal deformation**, since the forces act along the axis of the rod, normal to the rod's cross section.

Example 2.1:

Suppose that you have an elliptical cylinder, as shown. The major axis of the elliptical cross section $2a = 6.0$ cm, and the minor axis $2b = 4.0$ cm. What is the cross-sectional area A_0 of the cylinder? If an axial tensile force $F = 25$ kN is applied to the cylinder, what is the engineering stress σ (assumed uniform)? If the cylinder's length extends by ΔL = 0.4 cm under the influence of these forces, what is the engineering strain (as a percentage)?

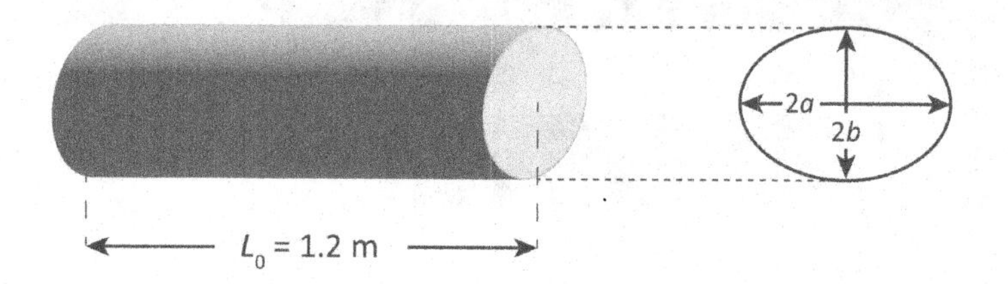

SOLUTION

The area of an ellipse requires the product of the semimajor axis a and the semi-minor axis b: $A_0 = \pi ab$. We find

$$A_0 = \pi ab = \pi(3.0 \text{ cm})(2.0 \text{ cm}) = \underline{\mathbf{19\ cm^2}}$$

For the given force F = +25,000 N (i.e., a tensile force), the stress is now

$$\sigma = F/A_0 = (25{,}000 \text{ N})/(19 \text{ cm}^2) = 1300 \text{ N/cm}^2 = 1.3\times 10^7 \text{Pa} = \underline{\mathbf{13\ MPa}}$$

For an elongation of ΔL = +0.4 cm, we obtain a strain of

$$\%\varepsilon = \Delta L / L_0 \times 100 = (0.004 \text{ m})/(1.2 \text{ m})\times 100 = \underline{\mathbf{0.3\%}}$$

From the definition of the stress σ (**Equation 2.2**), we have

$$\sigma = \frac{F}{A_0}$$

Substituting for F from **Equation 2.1′** gives

$$\sigma = \frac{k\Delta L}{A_0}$$

Multiplying the numerator and denominator by L_0 and rearranging terms:

$$\sigma = \frac{k\Delta L_0}{A_0}\times\frac{\Delta L}{L_0}$$

Finally, using the definition of the strain ε from **Equation 2.3** produces

$$\sigma = \frac{k\Delta L_0}{A_0}\varepsilon$$

What is important to note about this result is that kL_0/A_0 is a *constant*. This means that the stress is proportional to the strain: $\sigma \propto \varepsilon$. We call the constant of proportionality in this relationship the **Young's modulus** E. E is similar to the spring constant k, but it has different units: $[E]$ = [force/length^2] = N/m^2 = Pa. Using the Young's modulus, we can rewrite the relationship between stress and strain as

$$\sigma = E\varepsilon \tag{2.4}$$

This new formula is also Hooke's Law but reorganized for the description of materials specifically. **Equation 2.4** also describes the situation when the sense of the applied load is reversed. The stress that develops is no longer extensional but compressive: $\sigma = -F/A_0$. Since the modulus E is a positive quantity, the associated strain is a negative quantity as well.

Since the parameter σ reflects the amount of distributed force that the material is subject to and the parameter ε describes the resulting change in shape, the constant E must capture the role of the *material* in determining how much force produces how much deformation. Different materials will therefore present different deformations for similar stresses. We therefore recognize the parameter E as a materials property that captures something about the elastic response of the material. By ascribing different values of E to different materials, we can establish quantitatively the basis for their different behaviors when subject to forces that produce extension.

Another consideration regarding the deformation behavior of materials is that not all deformation is strictly extensional. Some other important deformation modes are shown in **Figure 2.3**. If the deformation of an object occurs exactly as depicted in **Figure 2.2**, then its volume will *increase* as a result. As it turns out, the volume increase associated with the extension ΔL is offset by a contraction of the material along directions transverse to ΔL. This effect is shown in **Figure 2.3(a)**. A block of material deformed along the horizontal axis contracts along the vertical axis. The vertical contraction corresponds to a transverse strain $\varepsilon_{\text{trans}}$ according to

$$\varepsilon_{\text{trans}} = \frac{\Delta h}{h_0} = -\nu\frac{\Delta L}{L_0} = -\nu\varepsilon \tag{2.5}$$

where ε is the extensional strain. The parameter ν is called **Poisson's ratio** of the material and is related to how much overall volumetric change is possible. Materials that are *incompressible* have $\nu = ½$, and materials generally have $0.1 < \nu < 0.5$. Since this transverse deformation is recovered when the extensional force is released, this "Poisson effect" is another type of elastic response.

Figure 2.3(b) shows a related mode of deformation that results when a material is compressed by an "isostatic" stress σ_h, meaning a uniform stress $-\sigma_h$ is applied from every direction. This uniform stress is identified as pressure p, where $p = -\sigma_h$. The elastic relationship here is

$$p = -B\frac{V}{V_0} \tag{2.6}$$

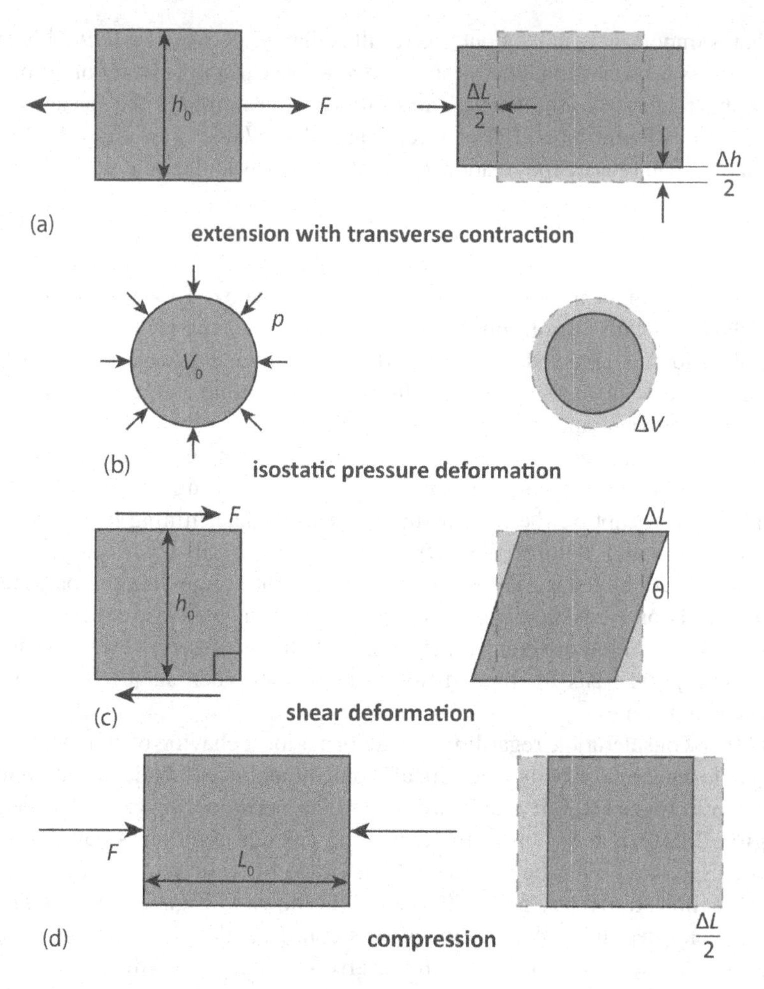

FIGURE 2.3 Other deformation modes of a loaded material. (a) When subjected to normal forces that cause extension along one (the horizontal) dimension, there is an accompanying change Δh in the transverse (the vertical) dimension. (b) A uniform (or hydrostatic) pressure p applied to an object produces a uniform contraction ΔV. (c) Shear forces, forces that are not normal to the cross section, produce deformation that does not make the material longer or shorter but rather induces some slant in the faces. The amount of shear deformation can be assessed with the slant angle θ. (d) Compression is similar to tension, but the senses of the forces and deformations are opposite.

where B is called the **bulk modulus** and $\Delta V/V_0$ (i.e., the change in volume over the undeformed volume) is then a "volumetric strain". Also consider that, since the density $\rho = m/V$ and the mass m is constant during deformation, we must have

$$\Delta V = \Delta\left(\frac{m}{\rho}\right) = -m\frac{\Delta\rho}{\rho_0^2}$$

And therefore

$$p = -B\left(-m\frac{\Delta\rho}{\rho_0^2}\right) \Big/ \left(\frac{m}{\rho_0}\right) = B\frac{\Delta\rho}{\rho_0} \tag{2.6$'$}$$

This type of elastic behavior associated with the bulk-modulus property is familiar; it is the behavior of a compressible fluid whose density increases with the hydrostatic pressure.

As shown in **Figure 2.3(c)**, a block of material is subjected to **shear deformation**. Shear forces – forces that are applied transversely to a material object's cross section – produce a deformed configuration that is "slanted" (similar to sliding a deck of cards on a table from the top). The faces transverse to those the force is applied to are inclined by an angle θ relative to the original upright orientation. The shear force F results in a shear stress τ when normalized to the area A_0 of the cross section that the force is applied to:

$$\tau = \frac{F}{A_0} \tag{2.7}$$

The shear strain γ associated with the deformation is

$$\gamma = \frac{\Delta L}{h_0} = \tan\theta \tag{2.8}$$

This shear strain is related to the applied shear stress according to an expression similar to **Equation 2.4**

$$\tau = G\gamma \tag{2.9}$$

Equation 2.8 is a straightforward modification of Hooke's Law to apply to shear deformation; the relationship between the stress and the strain is linear and determined by the materials property G, called the **shear modulus**.

The set of properties E, B, G, and ν are collectively the **elastic moduli** of a material. Some elastic moduli of typical materials are given in **Table 2.1**. Also, the moduli provided in **Table 2.1** are not necessarily independent quantities. When the material involved has the quality of being isotropic, meaning the values of the properties do not depend on the material's orientation, the moduli are related to one another. For instance, in an isotropic material

$$G = \frac{E}{2(1+\nu)} \quad \text{and} \quad B = \frac{E}{3(1-2\nu)} \tag{2.10}$$

This means that if you have any *two* of the elastic moduli for an isotropic material, you know or can calculate all *four*.

TABLE 2.1
Elastic Moduli of Selected Materials at Room Temperature

Material	E [GPa]	G [GPa]	ν [unitless][a]	B [GPa]
Iron	208	81	0.291	166
Unalloyed ("plain-carbon") steel	200	78	0.28	152
Aluminum	67	25	0.345	72
Aircraft aluminum	72.4	27.2	0.33	71
Pure silica glass	73	31	0.17	37
Soda-lime glass	66	27	0.22	39
Copper	128	46.8	0.308	111
Alumina (Al_2O_3)	380	150	0.26	260
Silicon carbide (SiC)	400	168	0.19	215
Low-density Polyethylene (LDPE)	0.15	0.051	0.48	1.25
High-density Polyethylene (HDPE)	1.1	0.38	0.46	4.6
Polycarbonate	2.3	0.85	0.35	2.6
Bisphenol A Epoxy	2.4	0.91	0.32	2.2
Isoprene ("natural") rubber†	0.0046	0.0015	0.499	0.77

[a] Poisson's ratio is typically represented by the Greek letter "nu". †The elastic behavior of rubber cannot typically be captured by a single property value (see **Problem M2.2**).

Example 2.2:

A circular cylindrical rod of low-carbon steel 10.0 cm long and 4.0 cm in diameter has an axial compressive load of 90.0 kN applied to it. Determine (a) the compressive stress σ; (b) the total longitudinal deformation, δ; (c) the total transverse deformation δ_t; and (d) the change in volume, ΔV.

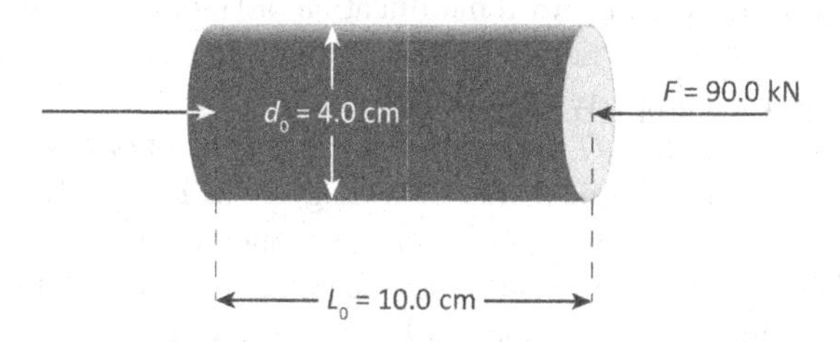

SOLUTION

We recognize the applied force as a force normal to the circular cross section (with area A_0) of the cylinder, directed inward, i.e., in the (–) sense. This means that the axial stress on the cylinder is

$$\sigma = -\frac{F}{A_0} = -\frac{F}{\pi(d_0/2)^2} = -\frac{90{,}000\ \text{N}}{\pi(0.020\ \text{m})^2} = -7.2\times10^7\ \frac{\text{N}}{\text{m}^2} = \underline{\mathbf{-72\ MPa}}$$

How much deformation results from this stress? The axial strain established in the rod is

$$\varepsilon = \frac{\sigma}{E} = \frac{-72 \text{ MPa}}{200{,}000 \text{ MPa}} = -3.6 \times 10^{-4}$$

and so we can calculate the associated axial *contraction* as

$$\delta = \varepsilon L_0 = \left(-3.6 \times 10^{-4}\right)(0.10 \text{ m}) = -3.6 \times 10^{-5}\text{m} = \underline{\mathbf{-0.36\ mm}}$$

The transverse deformation δ_t is then

$$\delta_t = -\nu\varepsilon d_0 = -0.28\left(-3.6 \times 10^{-4}\right)(0.040 \text{ m}) = 4.0 \times 10^{-6} \text{ m} = \underline{\mathbf{0.0040\ mm}}$$

i.e., an *extension*. Finally, to obtain the overall change in volume, consider that the deformed area

$$A = \pi\left(\frac{d_0 + \delta_t}{2}\right)^2 = \pi(1-\nu\varepsilon)^2\left(\frac{d_0}{2}\right)^2 = (1-\nu\varepsilon)^2 A_0$$

and the deformed length $L = L_0 + \delta = (1 + \nu\varepsilon)L_0$. This gives

$$\Delta V = V - V_0 = AL - V_0 = (1-\nu\varepsilon)^2(1+\nu\varepsilon)A_0L_0 - V_0 = \left[(1-\nu\varepsilon)^2(1+\nu\varepsilon) - 1\right]V_0$$

and we compute

$$\Delta V = -2.0 \times 10^{-8} \text{ m}^3 = \underline{\mathbf{-0.020\ cm^3}}$$

Example 2.3:

What is the value of the bulk modulus B for an isotropic Al alloy with $G = 28$ GPA and $\nu = 0.33$?

SOLUTION

Equation 2.10 provides relations for B and G in terms of E and ν. Though both B and E are unknown to us, these two relationships are sufficient to determine them. We solve to obtain

$$E = 2G(1+\nu)$$

and

$$B = \frac{E}{3(1-2\nu)} = \frac{2G(1+\nu)}{3-6\nu}$$

Evaluating this result with the given properties produces

$$B = \frac{(28\text{ GPa})(1+0.33)}{3-6(0.33)} = \underline{\mathbf{73\ GPa}}$$

An important consideration when employing these values is that they are meant to apply for small deformations/loads only; materials are only elastic up to a certain point. For common engineering materials, this limit will be a strain of around $\%\varepsilon < 0.1\%$. An additional consideration is that as an object deforms its cross-sectional area A can depart significantly from the initial value A_0. It means that the engineering stress $\sigma = F/A_0$ will become an increasingly inaccurate representation of the distribution of the load F over the current cross-sectional area A changes. For small deformations ($\%\varepsilon < 0.1\%$), this inaccuracy is negligible, and the engineering stress is suitable for Hooke's law calculations. As deformation becomes larger, a better representation of the stress in a body is given by the "true" stress σ_t:

$$\sigma_t = \frac{F}{A} \tag{2.11}$$

Similar considerations apply to the engineering strain $\varepsilon = \Delta L/L_0$. The quantity $\Delta L = L - L_0$ refers to the "extension" or the overall change in length determined from the "endpoints" of the deformation: L and L_0. A superior representation describes the strain in terms of small increments of deformation that accumulate to produce the overall deformation:

$$\varepsilon \approx \frac{L_1 - L_0}{L_0} + \frac{L_2 - L_1}{L_1} + \frac{L_3 - L_2}{L_2} + \cdots + \frac{L_n - L_{n-1}}{L_{n-1}} + \frac{L - L_n}{L_n} = \sum_i \frac{(\Delta L)_i}{L_i}$$

Then, in the limit as the individual $(\Delta L)_i$s becomes small:

$$\varepsilon_t = \int_{L_0}^{L} \frac{dL}{L} = \ln \frac{L}{L_0} \tag{2.12}$$

The strain value ε_t is called the "true" strain. The true stress and the true strain are important quantities in the description of materials subject to large deformations, particularly those encountered when the material is deformed to its limits. Such conditions frequently develop during a **tensile test**.

2.3 CONSTITUTIVE BEHAVIOR AND STRUCTURE/PROPERTY RELATIONSHIPS

We can organize much of the information pertaining to the mechanical responses of materials using **stress-strain (σ vs. ε) curves**. In these curves, the stress response σ is depicted as the dependent variable, and the normal strain ε is given as the independent variable. The σ vs. ε curves of three types of material are illustrated schematically in **Figure 2.4**. For the materials in **Figures 2.4(a)** and **(b)**, the relationship between σ and ε is *linear* when the deformation is elastic, according to Hooke's Law, and so the elastic portion of a material's deformation is plotted as a straight line (with slope E) at low ε. These elastic deformation regimes are associated with low

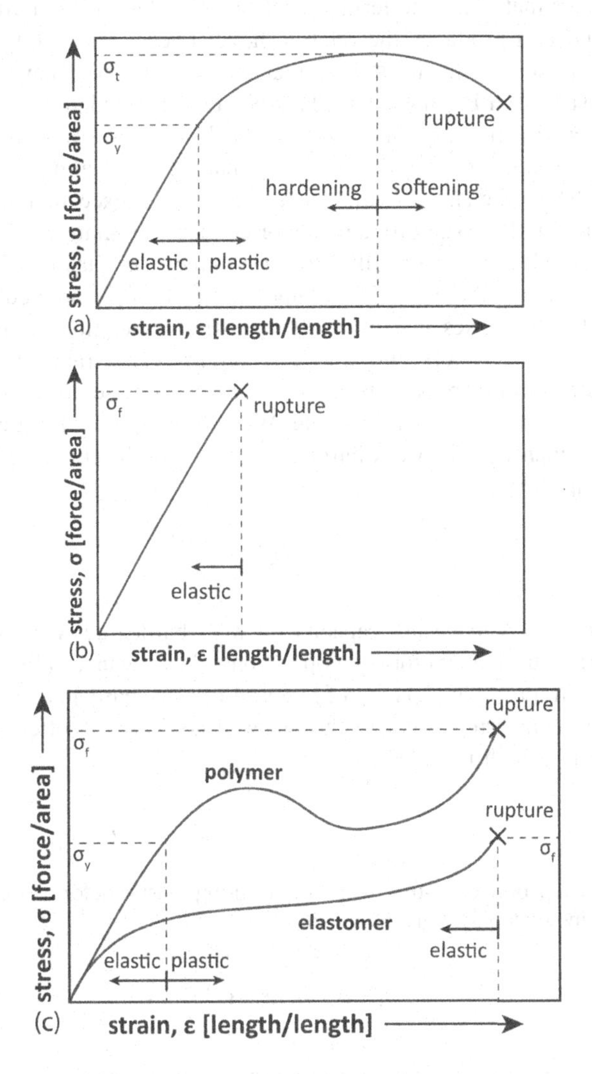

FIGURE 2.4 Stress vs. strain (σ vs. ε) response of idealized materials. Initially, as the material extends, the stress increases linearly, according to Hooke's Law. (a) In a plastically deformable (ductile) material, when the deformation level in the material exceeds the elastic limit, the deformation changes from elastic to plastic, with a corresponding change in the shape of the curve according to the principles of work hardening. Eventually, the stress the material can sustain reaches a limit. (b) In a brittle material, no significant plastic deformation is present. Instead, the material ruptures (fractures) at a limiting stress. (c) Polymers exhibit different deformation characteristics, and elastomers exhibit a nonlinear elastic response.

stresses and strains. A contrasting case is illustrated in **Figure 2.4(c)**. This type of σ–ε response is associated with polymeric elastomers. These materials can sustain elastic deformations to large strains (on the order of 500%), and the stress-strain relationship is more complicated than a straight line.

The graphs in **Figure 2.4** are depictions of the overall **constitutive behavior** of the particular material. The term "constitutive" refers to the built-in response of the material that produces the relationship between the loading parameter σ and the deformation parameter ε. Consider the constitutive behavior of the typical metal illustrated in **Figure 2.4(a)**. This behavior is divided into three separate regions associated with different behaviors. The low-stress/low-strain behavior we recognize as the elastic region, and we know that the material obeys Hooke's Law in this range. When the elastic limits of the material are exceeded (a phenomenon called "yielding"), the constitutive behavior changes qualitatively. This deformation is no longer elastic or springlike but is permanent or "plastic". The test specimen will not return to its original state upon removal of the applied force. In this region of moderate stresses and strains, the constitutive relationship is not that of a straight line with a fixed slope but rather a curved line. This trajectory indicates the stress associated with increased strain continues to increase during **plastic deformation**, but according to a different rule. We call this purely plastic deformation effect **work hardening**. The work-hardening portion of the curve typically follows a parabolic trajectory:

$$\sigma_t \propto \varepsilon_t^n \tag{2.13}$$

where $0.1 < n < ½$. A material's capacity for work hardening is an important consideration during mechanical forming operations like drawing, rolling, and forging. There is also a limit to the capacity of material to work harden, and if this limit is exceeded, the engineering stress in the material decreases ("softening") until the material experiences ultimate failure.

Example 2.4:

The relationship between stress and strain during plastic deformation for many polycrystalline materials is given by

$$\sigma_t = \sigma_y + K\left(\varepsilon_t - \varepsilon_y\right)^n$$

where K, n, and the yield stress σ_y and strain at yield ε_y are constants for a particular material. What is the energy absorbed by the specimen during plastic deformation if the material fails at a strain of $\varepsilon_t = \varepsilon_f$?

SOLUTION

The plastic deformation domain on a stress-strain curve is from the yield strain ε_y to the strain at failure ε_f. The energy absorbed (U), sometimes called the "toughness" or "resilience", in this range is given by the area under the stress-strain curve:

$$U = \int_{\varepsilon_y}^{\varepsilon_f} \sigma_t \, d\varepsilon_t$$

$$= \int_{\varepsilon_y}^{\varepsilon_f} \left[\sigma_y + K\left(\varepsilon_t - \varepsilon_y\right)^n\right] d\varepsilon_t$$

$$= \left(\sigma_y \varepsilon_t\right)\Big|_{\varepsilon_y}^{\varepsilon_f} + \frac{K}{n+1}\left(\varepsilon_t - \varepsilon_y\right)^{n+1}\Big|_{\varepsilon_y}^{\varepsilon_f}$$

$$= \sigma_y\left(\varepsilon_f - \varepsilon_y\right) + \frac{K}{n+1}\left(\varepsilon_f - \varepsilon_y\right)^{n+1}$$

The material's constitutive behavior, as captured by the stress-strain curve, is at the engineering level of description. This behavior should bc explainable from principles associated with the fundamental level of description, but how? Let's take a more specific question as a jumping-off point: what structural features and physics determine the elastic modulus E of a material? We recall the forces that exist between molecules from **Chapter 1** in **Figure 2.5**. The force relationship $F(r)$ between neighboring atoms is shown schematically in **Figure 2.5(a)**. The force between molecules $F(r_0) = 0$ at their equilibrium separation r_0. For small displacements $\Delta r = r - r_0$ about r_0, the change in the force is approximately *linear*: $F \propto \Delta r$. That is, the forces between molecules displaced slightly from their accustomed equilibrium location resemble those present in a spring. This fact lends itself to the construction of a **model**, or simplified physical description, of the elastic response of materials. This model is shown in **Figure 2.5(b)**. The molecules in the material are individually connected by springlike interactions, and so the bulk material can extend and flex like any spring through the extension and flexure of the individual elastic bonds. When the applied force is removed, the intermolecular spring forces restore the material to its original configuration.

Consider now that the number of springlike bonds in a material is large and distributed uniformly throughout the bulk solid. How many bonds are there? Each bond has a nominal length of r_0, so in a typical cross section, the number of bonds N per unit area in the material is

$$N = \frac{\text{\# bonds}}{\text{area}} = \frac{1}{r_0^2} \tag{2.14}$$

If each bond produces a force $F = k\Delta r$ when deformed, then the equivalent stress σ in the material is

$$\sigma = \frac{\text{force}}{\text{bond}} \times \frac{\text{\# bonds}}{\text{area}} = F \times N = \frac{k\Delta r}{r_0^2} \tag{2.15}$$

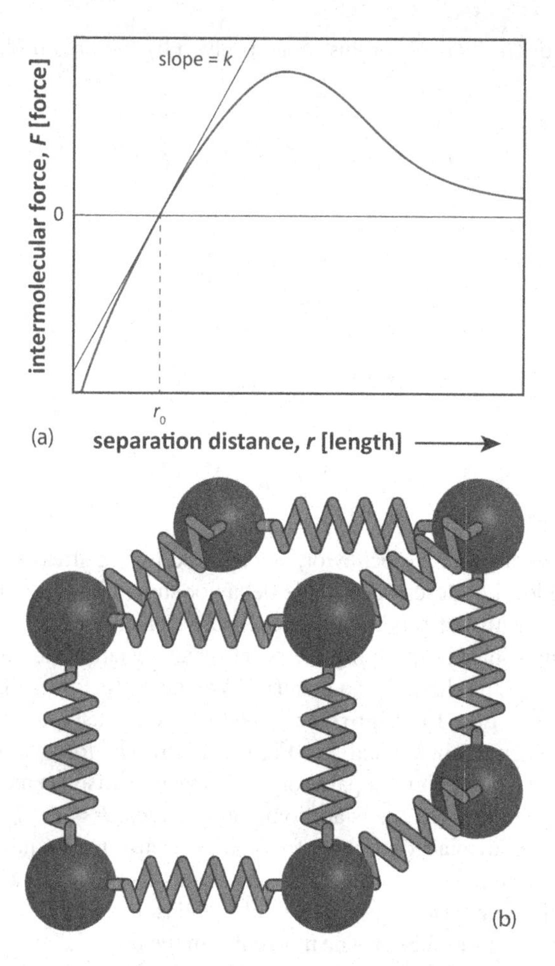

FIGURE 2.5 The origin of elasticity in typical materials. (a) shows the intermolecular force curve $F(r)$; this force is approximately linear (with effective spring constant k) near the equilibrium separation r_0. This principle gives rise to the simple model of a material shown in (b). In this model, the molecules in a material are connected by imaginary springs, and these springs can extend or contract under the influence of forces applied to the material. These forces will also restore a slightly deformed material to its initial shape.

The relative displacement (i.e., a strain) of each atom from its equilibrium position is

$$\varepsilon = \frac{\Delta r}{r_0} \tag{2.16}$$

and so we can now calculate

$$E = \frac{\sigma}{\varepsilon} = \left(\frac{k\Delta r}{r_0^2}\right) \Big/ \left(\frac{\Delta r}{r_0}\right) = \frac{k}{r_0} \tag{2.17}$$

This is another important result. It physically connects the microscopic parameters – intermolecular-bond force response (k) and bond geometry/density (r_0) – and the observed macroscopic property E. The takeaway here is that the stiffer and denser the bonds in a material, the higher the elastic modulus.

2.4 APPLICATION OF MATERIALS PROPERTIES

The elastic moduli collectively describe the elastic mechanical response of a material to (small) applied normal and shear deformations. These mechanical properties are measured using a number of different experimental techniques, and we say that performing these measurements characterizes the material. Ignoring the interesting and important details of these characterization techniques for now (see **Chapter 3**), we may ask: of what use is the information that they produce? How can we employ the revealed properties in engineering analysis and design?

As a grounded example of the influence of a material's properties on its engineering behavior, consider how the properties can be used in the analysis of loaded structures. **Figure 2.6** shows a weightless horizontal **beam** that is embedded in two vertical supports and loaded at its midpoint by a force F. A rigorous mechanics-of-materials analysis applied to the beam can reveal how much deformation we expect. The distance δ that the beam is deflected in the y direction depends on the position x along the beam. In this case, the analysis gives

$$\delta(x) = \frac{F}{48EI}\left(3x^2L - 4x^3\right) \tag{2.18}$$

In this beam-bending equation, I is the 2nd moment of area of the beam (a constant related to its cross-sectional shape), and E is the Young's modulus of the beam material. Some 2nd moments of area for typical beam geometries are given in **Table 2.2**,

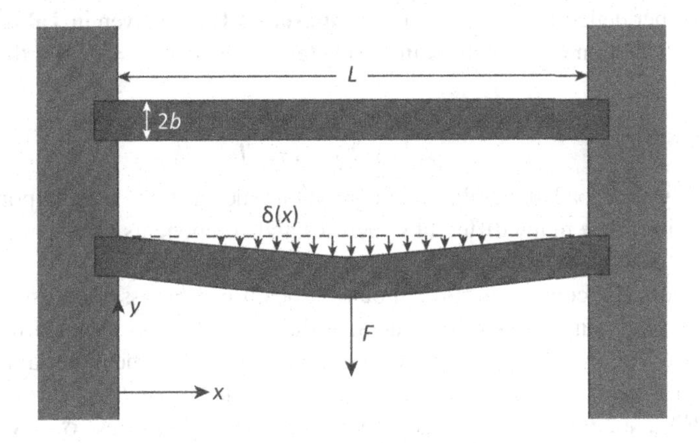

FIGURE 2.6 Deflection in an embedded beam of length L loaded by a weight F at the center. The downward y deflection δ of a beam of a particular geometry depends on the position x and the load F, as well as the properties and geometry of the beam.

TABLE 2.2
2nd Moment of Area for Selected Cross-sectional Shapes

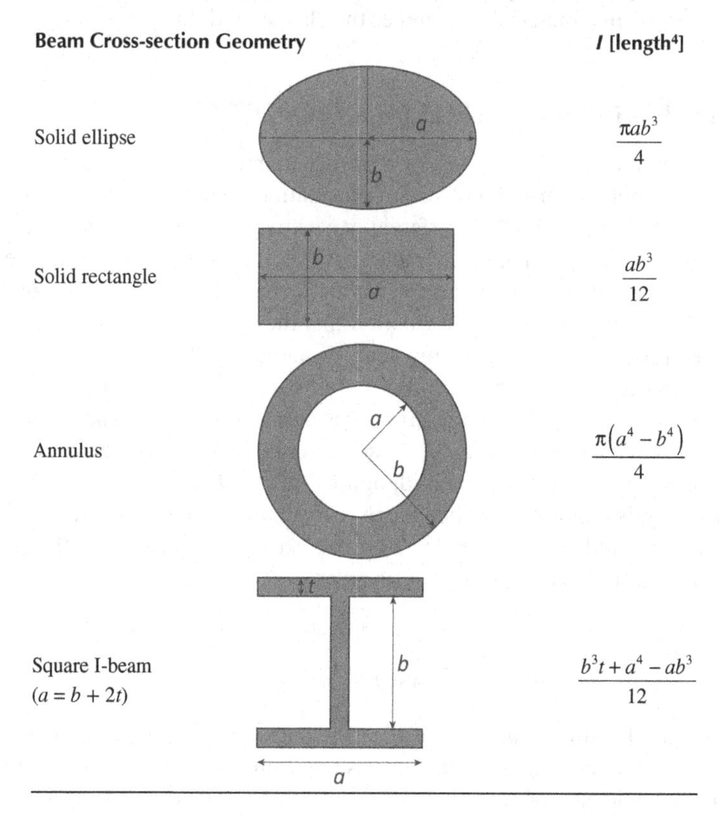

Beam Cross-section Geometry	*I* [length⁴]
Solid ellipse	$\frac{\pi ab^3}{4}$
Solid rectangle	$\frac{ab^3}{12}$
Annulus	$\frac{\pi(a^4 - b^4)}{4}$
Square I-beam ($a = b + 2t$)	$\frac{b^3t + a^4 - ab^3}{12}$

and some other deflection formulae, like **Equation 2.18**, are given in **Table 2.3**. The maximum deflection δ_{max} obtained in the center of the beam ($x = L/2$) is then

$$\delta_{max} = \delta\left(\frac{L}{2}\right) = \frac{FL^3}{192EI} \tag{2.19}$$

The analysis of loaded members using relations such as these is an important area of engineering since many different structures and components of structures can be represented or modeled as beams.

In addition, the conditions throughout the beam can be assessed using a stress analysis. A stress analysis seeks to identify the level of stress attained throughout the structure and, therefore, whether or not the limitations of the material have been exceeded. **Figure 2.7** shows the stress at particular locations (x, y) throughout the beam of **Figure 2.6** (depicted in the undeformed state). The stress σ may be either tensile (+) or compressive (–) and attains a maximum magnitude $\sigma_{max} = FLb/8I$ at the ends and center of the beam. The normalized stress ($= \sigma/\sigma_{max}$) varies between the limits $-1 \leq \sigma/\sigma_{max} \leq 1$. From the stress analysis, we can ascertain whether or not the

TABLE 2.3

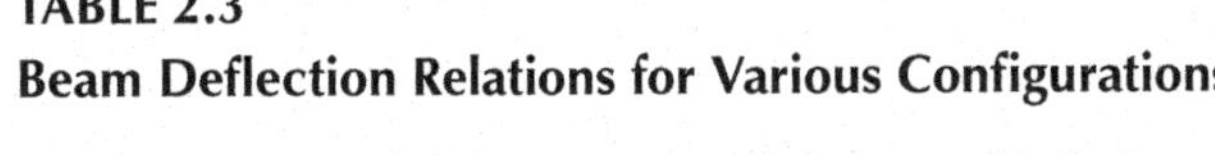

Beam Deflection Relations for Various Configurations

Beam Configuration and Loading	Deflection δ [length]	Stress σ_{max} [force/length²]
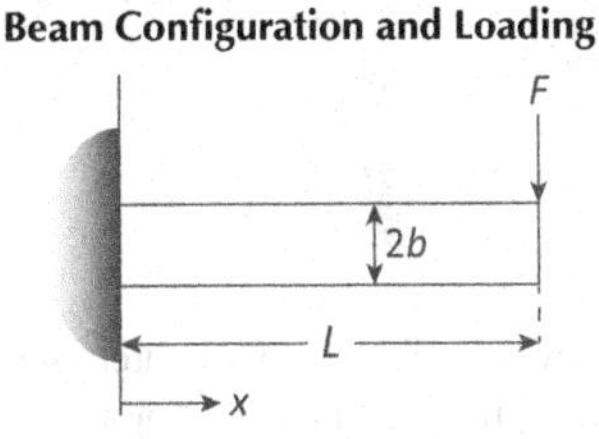 Embedded at one end (cantilever)	$\delta(x)=\frac{F}{6EI}\left(3x^2L-x^3\right)$ $\delta_{max}=\delta(L)=\frac{FL^3}{3EI}$	$\pm\frac{FLb}{I}$ @ $x=0$
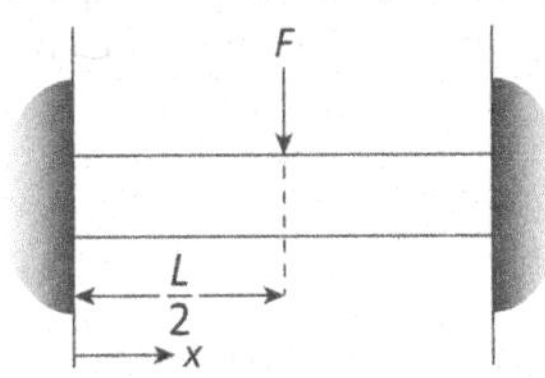 Embedded at both ends	$\delta(x)=\frac{F}{48EI}\left(3x^2L-4x^3\right)$ $\delta_{max}=\delta\left(\frac{L}{2}\right)=\frac{FL^3}{192EI}$	$\pm\frac{FLb}{8I}$ @ $x=0, L/2, L$
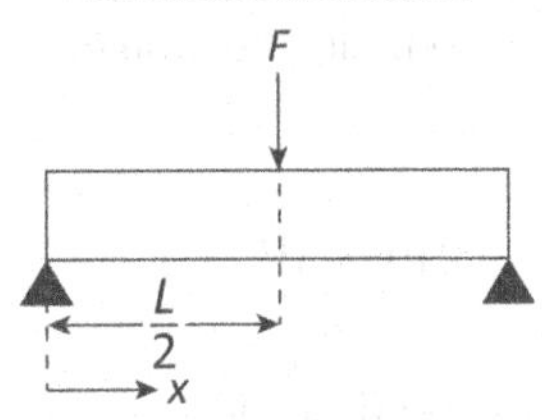 Supported at both ends	$\delta(x)=\frac{F}{48EI}\left(3L^2x-4x^3\right)$ $\delta_{max}=\delta\left(\frac{L}{2}\right)=\frac{FL^3}{48EI}$	$\pm\frac{FLb}{4I}$ @ $x=L/2$
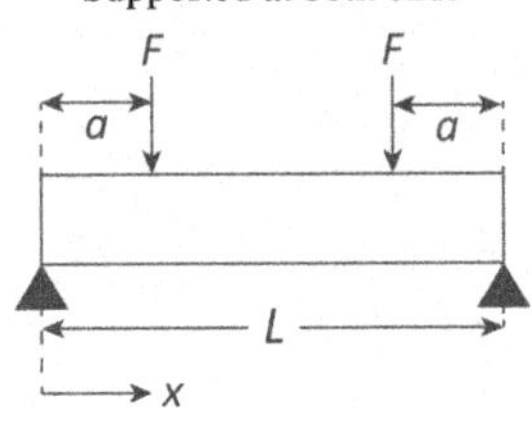 Symmetrical loading	$\delta(x)=\frac{F}{6EI}\left(3aLx-3a^2x-x^3\right)$ $(0\le x\le a)$ $\delta(x)=\frac{F}{6EI}\left(3aLx-3ax^2-a^3\right)$ $(a\le x\le L-a)$ $\delta_{max}=\delta\left(\frac{L}{2}\right)=\frac{F}{24EI}\left(3aL^2-4a^3\right)$	$\pm\frac{Fab}{I}$ @ $a\le x\le L-a$

stress levels in the loaded structure are expected to exceed the material limitations by comparison with the material's strength values.

Example 2.5:

Suppose that the beam from **Example 2.1** is embedded at one end and loaded transversely at the other by a force F = 25 N. What is the deflection at the unembedded end? The beam is made of polycarbonate.

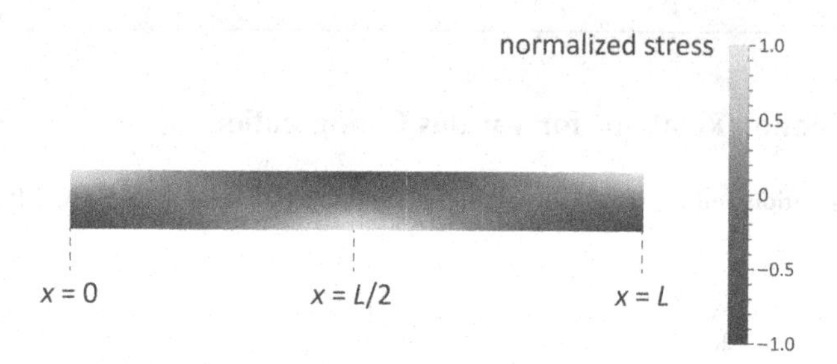

FIGURE 2.7 Stress analysis of a beam of length L loaded by a force F at the center (see **Figure 2.6**). The normalized stress level at a particular location, indicated by the color, can be either tensile or compressive. The stresses provided by the stress analysis can be compared to the material's strength to determine if this combination of loading and geometrical factors is expected to exceed the material limitations.

SOLUTION

We recognize this beam and loading geometry as those of the first case in **Table 2.2** and the first case in **Table 2.3**. From **Table 2.1**, we have $E = 2.3$ GPa for polycarbonate. Since $a = 3.0$ cm and $b = 2.0$ cm, we can compute the moment of area I as

$$I = \frac{\pi ab^3}{4} = \frac{\pi(3.0\text{ cm})(2.0\text{ cm})^3}{4} = 19\text{ cm}^4 = 1.9 \times 10^{-7}\text{ m}^4$$

Since we wish to know the deflection at the unembedded end (i.e., for $x = L = 1.2$ m), we find

$$\delta_{\text{max}} = \delta(L) = \frac{F}{6EI}\left(3L^3 - L^3\right) = \frac{FL^3}{3EI}$$

With the given values and computed value of I, this is

$$\delta_{\text{max}} = \frac{(25\text{ N})(1.2\text{ m})^3}{3\left(2.3 \times 10^9\text{ Pa}\right)\left(1.9 \times 10^{-7}\text{ m}^4\right)} = 0.033\text{ m} = \underline{\mathbf{33\ mm}}$$

Example 2.6:

The stress $\sigma = \sigma(x, y)$ in a loaded beam with deflection $\delta(x)$ is given by

$$\sigma(x,y) = -E\frac{d^2\delta}{dx^2}y$$

Show that σ_{max} for the third beam configuration of **Table 2.3** is

$$\sigma_{\text{max}} = \pm\frac{FLb}{4I}$$

SOLUTION

Starting with $\delta(x)$, we obtain

$$\frac{d^2}{dx^2}\delta(x) = \frac{d^2}{dx^2}\left[\frac{F}{48EI}\left(3L^2x - 4x^3\right)\right] = -\frac{F}{2EI}x$$

We can now find the normal stress at any location in the beam:

$$\sigma(x,y) = -E\frac{d^2\delta}{dx^2}y = -E\left(-\frac{F}{2EI}x\right)y = \frac{F}{2I}xy$$

The stress is highest when $x = L/2$ and $y = \pm b$ so

$$\sigma_{\max} = \sigma\left(\frac{L}{2}, \pm b\right) = \frac{F}{2I}\left(\frac{L}{2}\right)(\pm b) = \pm\frac{FLb}{4I}$$

which is the same as the formula provided.

2.5 CLOSING

The way we understand materials from a fundamental science perspective is different from the way we understand materials from an applied or engineering perspective. These two levels of description appear at first to narrate completely different types of information about a material but are in fact connected by definite physical relationships. We introduced and defined a number of important properties of materials – the elastic mechanical properties – and described their role in describing quantitatively the response of materials to forces. We then discussed how these properties are inherited from the fundamental level of description by introducing a model that captures the essential connections. Finally, we illustrate how these properties are employed in important structural applications.

2.6 CHAPTER SUMMARY

Key Terms

beam
bulk modulus
constitutive behavior
elastic moduli
extension
Hooke's Law
mechanical properties
mechanics of materials
model
Newton's Laws
normal deformation
plastic deformation
Poisson's ratio
shear deformation
shear modulus
spring constant
strain
stress
stress-strain (σ vs. ε) curves
stress analysis
tensile test
work hardening
Young's modulus

Important Relationships

$F = -kx$	(Hooke's law for springs)
$F = k\Delta L$	(modified Hooke's law)
$\sigma = \dfrac{F}{A_0}$	(normal engineering stress)
$\%\varepsilon = \dfrac{\Delta L}{L_0}$	(normal engineering strain)
$\%\varepsilon = \dfrac{\Delta L}{L_0} \times 100$	(normal engineering strain (%))
$\sigma = E\varepsilon$	(Hooke's law for materials (normal))
$\varepsilon_{trans} = -\nu\varepsilon$	(Poisson effect)
$p = -B\dfrac{V}{V_0}$	(isostatic pressure deformation)
$p = B\dfrac{\Delta\rho}{\rho_0}$	(compressibility relation)
$\tau = \dfrac{F}{A_0}$	(engineering shear stress)
$\gamma = \tan\theta$	(engineering shear strain)
$\tau = G\gamma$	(Hooke's law for materials (shear))
$G = \dfrac{E}{2(1+\nu)}$ and $B = \dfrac{E}{3(1-2\nu)}$	(moduli relationships)
$\sigma = \dfrac{F}{A}$	(normal true stress)
$\varepsilon_t = \ln\dfrac{L}{L_0}$	(normal true strain)
$\sigma_t \propto \varepsilon_t^n$	(work-hardening law)
$E = \dfrac{k}{r_0}$	(Young's modulus from bond properties)

2.7 QUESTIONS AND EXERCISES

Concept Review

C2.1 How would you describe the difference between the mechanical response of a *structure* and the mechanical response of a *material*?

C2.2 There is no equivalent to the "true" stress or strain in shear deformation. Why do you think this is the case?

C2.3 In the calculation of the density of a crystalline solid (from unit-cell size and contents) and the calculation of the Young's modulus of a crystalline solid (from intermolecular-force and bond-density information), we have two examples of how the structure and the properties of materials are related. However, not every material is crystalline. What is an example of a structure/property relationship that is different in a non-crystalline material?

Discussion-forum Prompt

D2.1 Consider typical applications that are associated with your field of engineering. Choose one of these applications and consider what kind of mechanical conditions materials in that application are subject to. Write a forum post outlining your application, the material(s) involved, and your reasoning about the origin and magnitude of the forces. Read the posts of the other students in the course, identify one involving the same or similar material, and comment on how the difference in conditions necessitates different design considerations.

PROBLEMS

P2.1 Consider a circular-cylindrical specimen of an Al steel alloy 15 mm in diameter and 75 mm long that is pulled in tension. Determine its elongation when a load of 20,000 N is applied.

P2.2 The maximum shear stress τ_{max} that a circular cylinder of material is subject to during twisting deformation produced by an applied torque T is

$$\tau_{max} = \frac{TR}{J}$$

where R is the radius of the cylinder and $J = \pi R^4/2$ is the "polar moment of inertia". A circular cylindrical bar of a steel alloy (E = 250 GPa, ν = 0.30) with length ℓ = 200 mm and diameter d = 12 mm is subjected to a torque T = 15 N m. What is the maximum shear stress? What is the maximum shear strain?

P2.3 A circular aircraft aluminum cylinder with an initial diameter of 20.0 mm is compressed elastically until its diameter is 20.2 mm. If its length after deformation is 88 mm, what was the initial length?

P2.4 Two polymer specimens are tested, one in compression and one in tension, in a way that produces an engineering strain $\varepsilon = +2.0$ in the first and $\varepsilon = -2.0$ in the second. If the specimens have an initial length ℓ_0 = 8.0 cm, what are the changes in length in each specimen? What are the corresponding true strains ε_t in each specimen? Comment on the difference between the two types of strains.

P2.5 A copper specimen with an original length ℓ_0 = 20. cm is stretched until it has a length ℓ_1 = 20.1 cm. Then it is stretched further until its length is ℓ_2 = 20.2 cm. What are the engineering strains associated with these two deformation steps individually? What is the sum of these strains? What are the two true strains for each of the two steps? What is this sum? What are the engineering strain and true strain associated with the *net* deformation of the two steps combined? Comment on the differences between these values and the sums above.

P2.6 Suppose you have a simple cantilever beam from **Table 2.3**. The beam is 0.50 m long and has a circular cross section of diameter 3.0 cm. If the beam is made of a material with $E = 200$ GPa and must not deflect at the end by more than $\delta_{max} = 4.0$ mm, what is the maximum load F that can be applied?

MATLAB® Exercises

M2.1 The plastic stress-strain data for materials that follow the form given in **Example 2.4**:

$$\sigma_t = \sigma_y + K(\varepsilon_t - \varepsilon_y)^n$$

can be linearized by writing

$$\ln(\sigma_t - \sigma_y) = \ln K + n\ln(\varepsilon_t - \varepsilon_y)$$

Written this way, ln σ vs. ln ε data have the form $y = mx + b$, where the slope m is the work-hardening exponent n and the intercept b is the logarithm of K. For the data given below, compute n and K, using the `log()` and `polyfit()` functions.

$\varepsilon_t - \varepsilon_y$ [m/m]	$\sigma_t - \sigma_y$ [MPa]
0.05	65
0.10	100
0.15	120
0.20	140
0.25	145

M2.2 The nonlinear deformation characteristics of elastomers are sometimes described in terms of the "stretch ratio" λ given by

$$\lambda = \frac{\ell}{\ell_0} = \varepsilon + 1$$

The relationship between stress and strain ("Hooke's law for rubber") is

$$\sigma = \frac{E_0}{3}\left(\lambda - \frac{1}{\lambda^2}\right)$$

Suppose a synthetic rubber has $E_0 = 3.0$ MPa. Plot the stress-strain curve for this material out to a strain of 150%.

M2.3 Consider the symmetrically loaded beam of **Table 2.3**. Plot the deflection of a steel beam ($E = 200$ GPa) with a square cross section 2.0 cm × 2.0 cm. Take $L = 2.0$ m, $a = L/5$, and the load $F = 3500$ N. What is the maximum stress on the beam?

3 Materials Testing and Validation *From Uncertainty to Assurance*

LEARNING OBJECTIVES

After completing this chapter, you should be able to:

1. List several motivations for materials characterization and explain their importance.
2. List and describe the important components of a tensile test platform.
3. Describe the importance of a technical standard and list some of the typical components of a standard.
4. Differentiate between the "engineering" and "true" values of stress and strain and compute one from the other.
5. Evaluate a stress-strain curve to obtain important materials properties: Young's modulus, yield strength, tensile strength, and strain to failure.
6. List and describe the primary defects present in crystalline materials.
7. Describe qualitatively and quantitatively the relationship between defect content and material strength. Compute these strengths provided the relevant materials data.
8. Define the hardness property and differentiate between its various quantitative and semiquantitative measures.
9. Describe the essential principles of microscopy required for metallography.
10. List and describe the steps in metallographic preparation.
11. Use an intercept method to measure grain size from a micrograph.

3.1 THE IMPORTANCE OF MEASUREMENT

Engineers of all varieties are frequently tasked with determining the characteristics of a particular material. A more or less complete characterization of a material requires measurement of the relevant properties and a detailed inspection of its structure. From this characterization, the structure/property principles can be described and/or applied to the situation at hand. These activities are important for engineers to perform for several possible reasons:

1. You might want to determine the characteristics of a material that is brand new (or that you personally have never worked with before) to see how they compare to the other materials you use.

DOI: 10.1201/9781003214403-4

2. You might want to evaluate whether a sample of a batch of freshly processed material meets the minimum desired requirements as part of a **quality-control** process.
3. You might want to test the performance-related aspect of a material to verify that it meets the specifications for a given application.
4. You might want to analyze the material characteristics of a component that has failed in service or during testing, i.e., a **forensic analysis**.

Being able to address these kinds of concerns about the implementation of materials is essential to deploying them in components and structures. You may ask: what kind of experimental strategy and what kind of experimental tools are required for these tasks? Unsurprisingly, the answers are different for different circumstances and different properties. Since we cannot discuss them all in this book, we will survey some of the most important measurement and characterization principles and techniques.

3.2 TENSILE TESTING

The idea that materials exhibit a particular constitutive behavior was introduced in **Section 2.3**. The plots in **Figure 2.4** show how different materials respond when subject to various amounts of deformation; the response is recorded in the features of the stress-strain curve. We expect that different materials will have different responses, of course, but we need some way to capture those differences in a small number of property values that we can tabulate, compare, and utilize in calculations.

The experiment that reveals constitutive behavior of a material in the form of a stress-strain curve is the tensile test. During a tensile test, a specimen is extended along one axis by a load frame. This extension, converted into a strain via **Equation 2.3** or **Equation 2.12**, becomes the independent variable of the experiment. Simultaneously, force applied to the specimen is measured by a load cell. The force reading, converted into a stress via **Equation 2.2** or **Equation 2.11**, is then the dependent variable. The combination of a load frame (that applies the elongation) and a load cell (that measures the force) along a common axis is the most common configuration for a test apparatus. This apparatus is called a **tensile tester** or "universal testing machine" and is depicted schematically in **Figure 3.1**. The mobile section of the load frame, called the crosshead, is driven by a motor (or sometimes hydraulics). The load cell is typically mounted on the crosshead and connected to the specimen via a clamp.

APPLICATION NOTE – ASTM STANDARDS

Many materials test methods appear to be straightforward in terms of the physical parameters that they control and measure, but an analysis of the engineering details reveals complications that can arise when the tests are conducted with slightly different parameters. Specimen size and geometry, calibration standards, testing rate, pre-test preparations, analysis strategies, etc. can vary

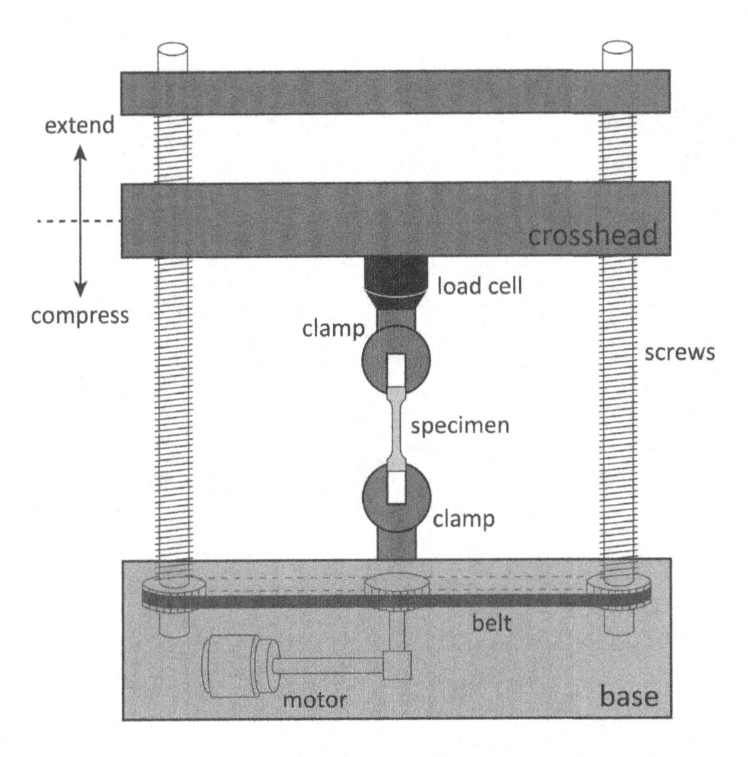

FIGURE 3.1 Schematic of an electromechanical tensile tester. The load frame has a fixed base that the specimen is clamped to. The other end of the specimen is clamped onto the mobile crosshead, and the crosshead is driven upward (to produce tension) or downward (to produce compression). The load cell is entrained with the crosshead, clamps, and specimen and can resolve the force acting along the tensile axis.

between implementations of tests that nominally measure the same properties. Given these complications and the variation in measured properties that result, what can be done to improve the uniformity of the values we employ in our engineering design?

A partial answer to the question is the development and dissemination of test standards. The standard is a document that is produced and accepted by workers in the field as containing the important experimental considerations for the types of tests that it encompasses. ASTM International is a nonprofit organization that collects these community-produced standards and maintains a repository of them. Whenever new materials or experiments come into common usage, the associated experimental considerations are encoded in a new standard. While the use of ASTM standards is in general voluntary, sometimes testing for contract or government work is required to conform to the standards so that the results possess a well-established basis for validation and comparison. An example title page for an ASTM standard is shown below.[1]

This international standard was developed in accordance with internationally recognized principles on standardization established in the Decisi on on Principles for the Development of International Standards, Guides and Recommendations issued by the World Trade Organization Technical Barriers to Trade (TBT) Commi ttee.

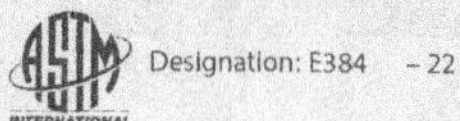

Designation: E384 – 22

Standard Test Method for
Microindentation Hardness of Materials [1]

This standard is issued under the fixed designation E384; the number immediately following the designation indicates the year of original adoption or, in the case of revision, the year oflast revision. A number in parentheses indicates the year oflast reapproval. A superscript epsilon () indicates an editorial change since the last revision or reapproval.

This standard has been approved for use by agencies of the U.S. Department of Defense.

1. Scope*

1.1 This test method covers determination of the microindentation hardness of materials.

1.2 This test method covers microindentation tests made with Knoop and Vickers indenters under test forces in the range from 9.8×10^{-3} to 9.8 N (1 to 1000 gf).

1.3 This test method includes an analysis of the possible sources of errors that can occur during microindentation testing and how these factors affect the precision, bias, repeatability, and reproducibility of test results.

1.4 Information pertaining to the requirements for direct verification and calibration of the testing machine and the requirements for the manufacture and calibration of Vickers and Knoop reference hardness test blocks are in Test Method E92.

Note 1—While Committee E04 is primarily concerned with metals, the test procedures described are applicable to other materials.

1.5 Units—The values stated in SI units are to be regarded as standard. No other units of measurement are included in this standard.

1.6 This standard does not purport to address all of the safety concerns, if any, associated with its use. It is the responsibility of the user of this standard to establish appropriate safety, health, and environmental practices and determine the applicability of regulatory limitations prior to use.

1.7 This international standard was developed in accordance with internationally recognized principles on standardization established in the Decision on Principles for the Development of International Standards, Guides and Recommendations issued by the World Trade Organization Technical Barriers to Trade (TBT) Committee.

2. Referenced Documents

2.1 ASTM Standards:[2]

C1326 Test Method for Knoop Indentation Hardness of Advanced Ceramics
C1327 Test Method for Vickers Indentation Hardness of Advanced Ceramics
E3 Guide for Preparation of Metallographic Specimens
E7 Terminology Relating to Metallography
E92 Test Methods for Vickers Hardness and Knoop Hardness of Metallic Materials
E140 Hardness Conversion Tables for Metals Relationship Among Brinell Hardness, Vickers Hardness, Rockwell Hardness, Superficial Hardness, Knoop Hardness, Scleroscope Hardness, and Leeb Hardness
E175 Terminology of Microscopy (Withdrawn 2019)[3]
E177 Practice for Use of the Terms Precision and Bias in ASTM Test Methods
E691 Practice for Conducting an Interlaboratory Study to Determine the Precision of a Test Method
E766 Practice for Calibrating the Magnification of a Scanning Electron Microscope
E1268 Practice for Assessing the Degree of Banding or Orientation of Microstructures
E2554 Practice for Estimating and Monitoring the Uncertainty of Test Results of a Test Method Using Control Chart Techniques
E2587 Practice for Use of Control Charts in Statistical Process Control

2.2 ISO Standard:[4]

ISO/IEC 17025 General Requirements for the Competence of Testing and Calibration Laboratories

[1] This test method is under the jurisdiction of ASTM Committee E04 on Metallography and is the direct responsibility of Subcommittee E04.05 on Microindentation Hardness Testing. With this revision the test method was expanded to include the requirements previously defined in E28.92, Standard Test Method for Vickers Hardness Testing of Metallic Material that was under the jurisdiction of E28.06

Current edition approved Oct. 1, 2022. Published November 2022. Originally approved in 1969. Last previous edition approved in 2017 as E384 – 17. DOI: 10.1520/E0384-22

[2] For referenced ASTM standards, visit the ASTM website, www.astm.org, or contact ASTM Customer Service at service@astm.org. For Annual Book of ASTM Standards volume information, refer to the standard's Document Summary page on the ASTM website.

[3] The last approved version of this historical standard is referenced on www.astm.org.

[4] Available from International Organization for Standardization (ISO), 1, ch. de la Voie-Creuse, Case postale 56, CH-1211, Geneva 20, Switzerland, http://www.iso.org.

*A Summary of Changes section appears at the end of this standard

1

Typically, the components of an ASTM standard include:

- The scope of materials and properties that the standard applies to.
- The important definitions and mathematical relationships that are required to interpret the test.
- Test-specimen specifications and preparation strategy.
- Essential apparatus specifications, calibration practices, and recommended settings.
- Other sources of experimental error and recommended statistical practices.

ASTM standards can be obtained through library resources.

In many cases, a test specimen with a regular cross section will do. Examples of standardized tensile specimens are shown in **Figure 3.2**. The specimens are obtained by shaping stock materials or they are taken from finished components and cut to a suitable geometry. The specimens have flared ends and a uniform, central "gauge" section. Since deformation is concentrated in the gauge section, it is the material in this portion of the specimen that overwhelmingly influences the stress/strain measurement. The length of this gauge region G is frequently taken as the L_0 value in the calculation of the engineering strain. The gauge region is sometimes fitted with a

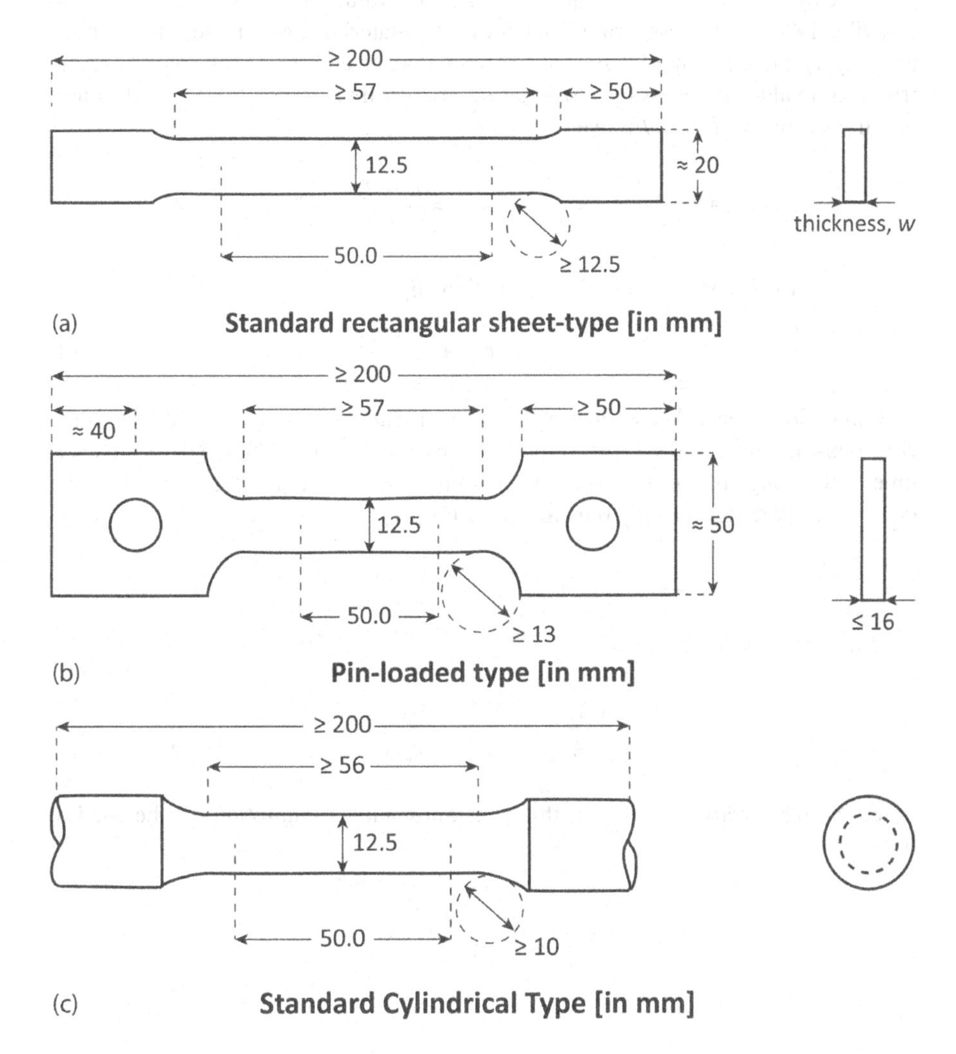

FIGURE 3.2 Tensile-test specimen configurations. The choice of configuration depends on the geometry of the component that the material was sampled from and the capabilities of the tensile tester (thicker specimens require larger forces to extend). (a) A typical flat rectangular specimen (also has a half-size version). (b) A specimen configuration for pin-based clamps. (c) A standard cylindrical specimen.

device called an extensometer to measure the specimen length. In this case, the initial span of the extensometer is L_0 rather than the G. Instrumented monitoring of the crosshead displacement or the extensometer extension provides the instantaneous length L. When the crosshead is driven at a constant displacement rate/velocity V, the specimen elongates at a constant rate $d(\Delta L)/dt = V$. The rate V is typically chosen to be low (~2 cm/min for metallic materials). This low rate of deformation establishes a condition in the material called "quasistatic" loading and helps to control for some dynamic processes that can influence the test results.

The engineering stress and engineering strain values are typically sufficient to describe the overall stress-strain behavior of the material. Though these parameters are only accurate for small deformations, they may be converted into true stress and true strain values if necessary. Consider the true strain $\varepsilon_t = \ln(L/L_0)$. Since the engineering strain $\varepsilon = (L - L_0)/L_0$, we can write

$$\varepsilon + 1 = \frac{\Delta L}{L_0} + 1 = \frac{L - L_0}{L_0} + 1 = \frac{L - L_0}{L_0} + \frac{L_0}{L_0} = \frac{L}{L_0}$$

Since $\varepsilon_t = \ln(L/L_0)$, we now obtain the relationship

$$\varepsilon_t = \ln(\varepsilon + 1) \tag{3.1}$$

Using **Equation 3.1**, engineering strain and true strain may be interconverted. Now consider the initial volume of the gauge region $V_0 = A_0L_0$ and the deformed volume of the gauge region $V = AL$. If we assume that the volume of the gauge region does not change significantly during elongation, i.e.

$$V_0 = V$$

at all times, we can compute

$$\varepsilon + 1 = \frac{L}{L_0} = \left(\frac{V}{A}\right) \Big/ \left(\frac{V_0}{A_0}\right) = \left(\frac{V}{A}\right) \times \left(\frac{A_0}{V_0}\right) = \left(\frac{V_0}{A}\right) \times \left(\frac{A_0}{V_0}\right) = \frac{A_0}{A}$$

Since the true strain is $\sigma_t = F/A$, this constant-volume assumption can be used to find

$$\sigma_t = \frac{F}{A} = \left(\frac{F}{A_0}\right) \times \left(\frac{A_0}{A}\right)$$

or

$$\sigma_t = \sigma(\varepsilon + 1) \tag{3.2}$$

This relationship, good for uniform constant-volume deformation during testing, relates the true stress to the engineering stress.

Example 3.1:

Suppose that you have a standard cylindrical bar specimen like that of **Figure 3.2(c)**. During testing, the clip-on extensometer indicates an overall extension ΔL = 2.00 mm. At this level of deformation, the force readout shows 10.0 kN. What are the engineering stress and strain at this time? What are the true stress and strain?

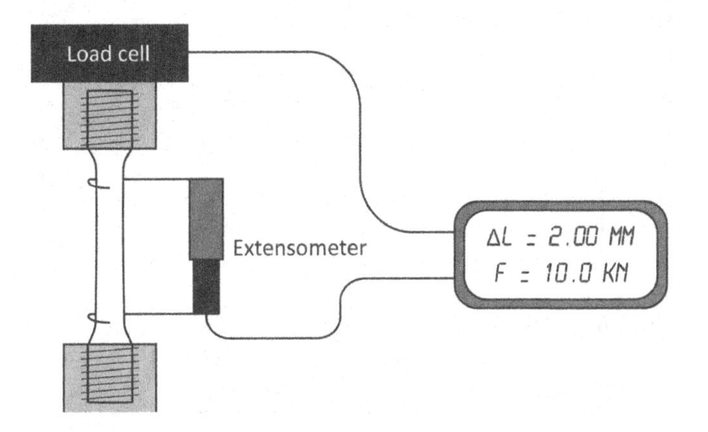

SOLUTION

The engineering stress and strain are determined by the instantaneous load, instantaneous deformation, and the original specimen geometry. Since the initial diameter of the specimen is d_0 = 12.5 mm, we obtain an engineering stress of

$$\sigma = \frac{F}{A_0} = \frac{F}{\pi(d_0/2)^2} = \frac{10 \text{ kN}}{\pi(6.25 \text{ mm})^2} = 0.0815 \frac{\text{kN}}{\text{mm}^2} = \underline{\mathbf{81.5\ MPa}}$$

and since the gauge length L_0 = 50.0 mm, we get an engineering strain of

$$\varepsilon = \frac{\Delta L}{L_0} = \frac{2.00 \text{ mm}}{50.0 \text{ mm}} = 0.0400 \frac{\text{mm}}{\text{mm}} = \underline{\mathbf{4.00\%}}$$

The corresponding true stress and true strain values can be obtained using **Equations 3.1** and **3.2**:

$$\sigma_t = \sigma(\varepsilon + 1) = (81.5 \text{ MPa})(0.0400 + 1) = \underline{\mathbf{84.5\ MPa}}$$

and

$$\varepsilon_t = \ln(\varepsilon + 1) = \ln(0.0400 + 1) = 0.0392 \frac{\text{mm}}{\text{mm}} = \underline{\mathbf{3.92\%}}$$

(Note that the true strain is a bit larger than the engineering strain since the instantaneous area is *smaller* than the initial area. At this level of deformation, the engineering strain and true strain are close to one another.)

Examples of stress-strain data for a common "structural" steel are shown in **Figure 3.3**. The full stress-strain curve, covering a range of strains from ε = 0

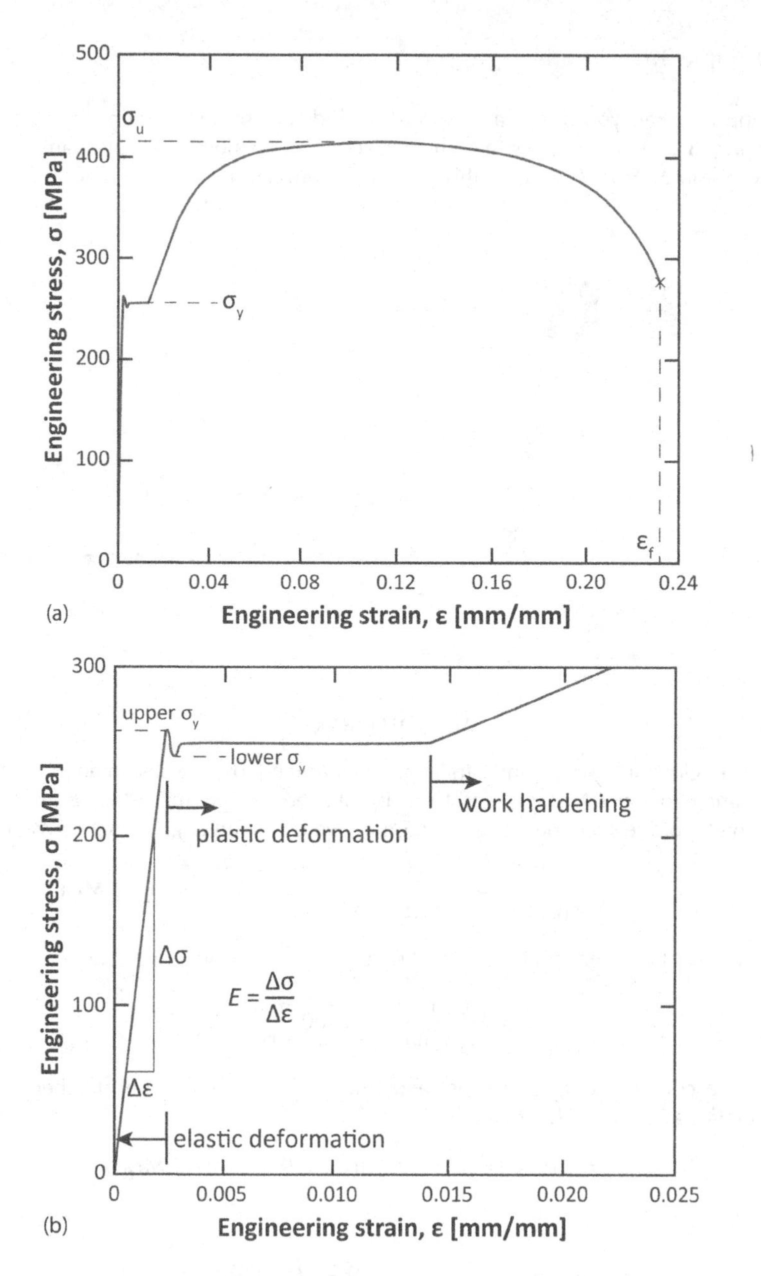

FIGURE 3.3 Stress-strain curve (engineering values) for an "A 36" structural steel. The full behavior is illustrated in (a). The material's strength values – yield strength σ_y, ultimate tensile strength σ_u, and strain-to-failure ε_f – are all indicated as the limits of elastic deformation, maximum stress, and maximum deformation, respectively. A detailed view of the low-strain behavior is in (b). The segregation of the yield point into "upper" and "lower" values is common in some alloys. In this alloy, there is also an extended region of constant-stress plastic deformation preceding the onset of work hardening. Prior to yielding, the stress-strain curve reveals the elastic behavior of the steel in the form of the elastic modulus.

to $\varepsilon = \varepsilon_f$ is shown in **Figure 3.3(a)**. ε_f is the engineering strain at rupture, sometimes called the **ductility**. Some other important characteristic properties of this alloy are indicated in the figure: the **yield strength** σ_y and the ultimate **tensile strength** σ_u. Frequently, as design or quality-control engineers, we are interested in a material's *limitations* with respect to the stresses and strains that it can sustain without a change in behavior. The elastic moduli describe the elastic mechanical behavior of a particular material, and this elastic behavior is associated with low levels of force or deformation. (The constitutive relationships in this regime are typically linear according to **Equations 2.4**, **2.5**, and **2.7**.) At higher stresses and strains, the material's behavior can change qualitatively. These limiting or transitional values are referred to as **strengths**. Some data on material strengths are provided in **Table 3.1**.

The yield strength σ_y reflects the level of stress at or above which the deformation in the material will be *permanent*. We call such permanent deformation plastic deformation to distinguish it from springlike, elastic deformation. Knowledge of σ_y is applied in engineering design (where it is necessary to know the circumstances under which the material is subject to permanent changes), materials processing (where it is necessary to know the force required to reshape an object into a different form), and materials qualification (where the measured property value can be compared to a required minimum or standard value). The elastic/plastic transition is reflected in the overall stress-strain behavior of a material as a departure from the linear elastic portion of the curve. In **Figure 3.3(b)**, we observe this transition as a sharp peak or crest in the trajectory of the elastic deformation portion of the curve, called the "upper yield stress".

TABLE 3.1
Limitations or Strengths of Materials

Material	σ_y or σ_f [MPa]	σ_u [MPa]	ε_f [%]
Iron	130	265	45
Low-carbon, unalloyed steel	315	420	39
High-carbon, unalloyed steel	585	965	12
Aluminum	40	75	60
Aircraft aluminum	395	495	13
Pure silica glass	–[a]	4480[b]	0[c]
Soda-lime glass	–[a]	3930[b]	0[c]
Copper	33.3	209	33.3
Alumina (Al_2O_3)	–[a]	500	0[c]
Silicon carbide (SiC)	–[a]	300	0[c]
Low-density Polyethylene (LDPE)	–[a]	15	600
High-density Polyethylene (HDPE)	–[a]	30	800
Polycarbonate	–[a]	62	110
Bisphenol A Epoxy	–[a]	50	6

[a] There is little-to-no yielding in these materials, so σ_y and σ_t coincide. [b]Measured in compression. [c]Plastic deformation is minimal in brittle materials.

In materials that possess some ductility, the material changes behavior when the stress in the material exceeds the limiting value of the yield strength. When this occurs, plastic deformation begins to accrue alongside elastic deformation, as shown in **Figure 3.3(b)**. The physical principles governing plastic deformation are sophisticated in comparison to the principles governing Hookean elastic deformation, but we can represent the behavior of many materials in this regime using a somewhat universal relationship. Outside of some initial constant-stress deformation immediately following arrival at the yield stress, the relationship between stress and strain during plastic deformation is not typically a straight line with a fixed slope. Rather, a bending line indicates the stress associated with increased strain continues to increase during plastic deformation, according to a different rule. We call this purely plastic deformation mode work hardening. The work-hardening portion of the σ–ε curve typically follows a *parabolic* trajectory:

$$\sigma_t = \sigma_y + K\varepsilon_{pl}^n \tag{3.3}$$

where $\varepsilon_{pl} = \varepsilon_t - \varepsilon_{el} \approx \varepsilon_t$ is the plastic strain, σ_y is the initial, undeformed yield stress, and K and n are constants for a given material. Typically, $0.10 < n < 0.55$. Also note that it is necessary to represent the stress and strain parameters in this relationship using the true values since the relationship applies over a range of deformations where the engineering values do not represent the parameters accurately.

Example 3.2:

You have two pieces of copper that have been deformed to true-strain levels of ε_1 = 1.0% and ε_2 = 2.0%. The stresses associated with these strains were σ_1 = 60. MPa and σ_2 = 72 MPa. If σ_y for this material is 33 MPa, what are the work-hardening parameters K and n?

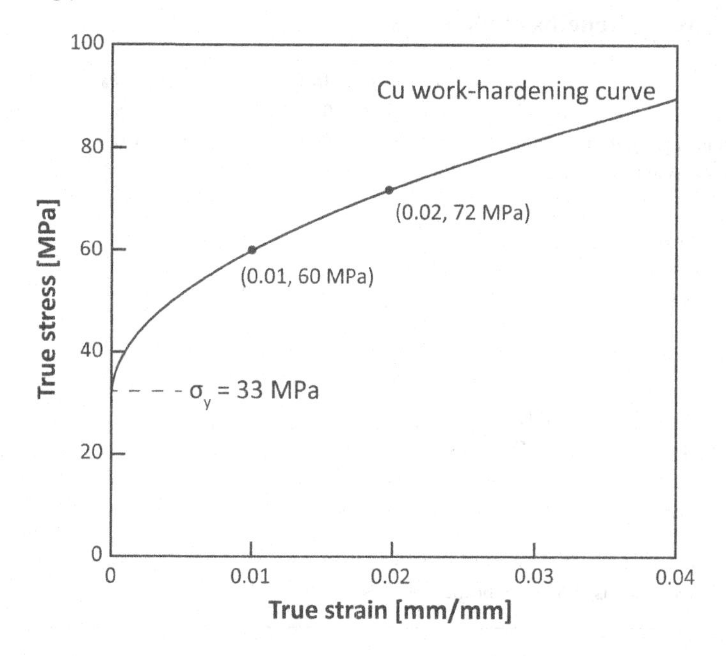

SOLUTION

From the data provided, we have two relationships that both agree with **Equation 3.3**:

$$\sigma_1 = \sigma_y + K\varepsilon_1^n \quad \text{and} \quad \sigma_2 = \sigma_y + K\varepsilon_2^n$$

From these two relationships, we may obtain the values of the unknowns K and n. Consider:

$$\frac{\sigma_2 - \sigma_y}{\sigma_1 - \sigma_y} = \frac{K\varepsilon_2^n}{K\varepsilon_1^n} = \left(\frac{\varepsilon_2}{\varepsilon_1}\right)^n$$

Taking the logarithm of both sides gives

$$\ln\left(\frac{\sigma_2 - \sigma_y}{\sigma_1 - \sigma_y}\right) = \ln\left(\frac{\varepsilon_2}{\varepsilon_1}\right)^n = n\ln\left(\frac{\varepsilon_2}{\varepsilon_1}\right)$$

and we find

$$n = \ln\left(\frac{\sigma_2 - \sigma_y}{\sigma_1 - \sigma_y}\right) \Big/ \ln\left(\frac{\varepsilon_2}{\varepsilon_1}\right) = \ln\left(\frac{72-33}{60-33}\right) \Big/ \ln\left(\frac{0.010}{0.020}\right) = \underline{\mathbf{0.53}}$$

Substituting this value of n back into one of the relationships above gives

$$\sigma_1 = \sigma_y + K\varepsilon_1^n \rightarrow K = (\sigma_1 - \sigma_y)\varepsilon_1^{-n} = (27\ \text{MPa})(0.010)^{-0.53} = \underline{\mathbf{310\ MPa}}$$

Alternately, a numerical solver may be employed. In MATLAB®:

```
>> syms K n
>> s1 = 60; s2 = 72; e1 = 0.010; e2 = 0.020; sy = 33;
>> eqns = [s1 == sy + K*e1^n, s2 == sy + K*e2^n];
>> soln = vpasolve(eqns, [K n], [100 0.50]);
>> double(soln.K)
ans =
310.7372
>> double(soln.n)
ans =
0.5305
```

A material's behavior during work hardening is an important consideration in mechanical forming operations like drawing, rolling, and forging. There is a limit to the amount of work hardening that is possible in a given material; this limit is the ultimate tensile strength σ_u (sometimes just the "tensile strength"). During plastic deformation, not only hardening effects but also "softening" effects are present. The specimen has an innate tendency to deform inhomogeneously and to develop a **neck**, as shown in **Figure 3.4**. The development of a neck concentrates stress, but work hardening suppresses its development. As work hardening diminishes with

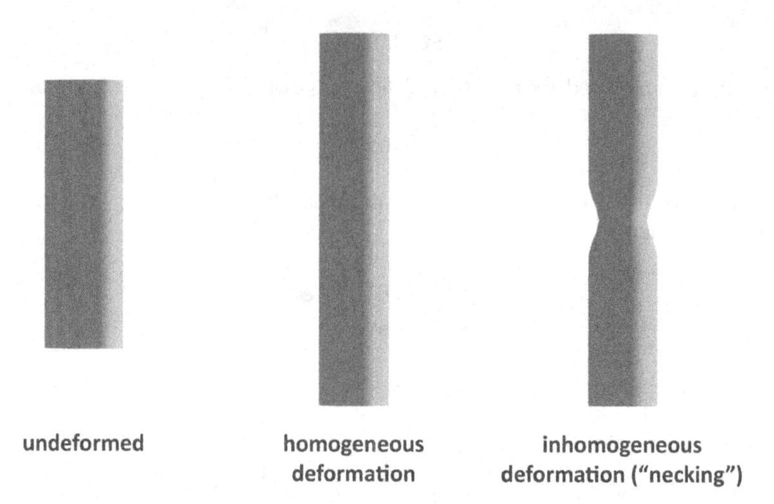

FIGURE 3.4 Different deformation types are encountered during a tensile test. The undeformed specimen gauge region tends to extend with uniform cross section during work hardening. The development of a localized narrowing of the cross section (the neck region) indicates the end of work hardening.

deformation, softening effects predominate, and the slope of the stress-strain curve reverses. That is, at the onset of necking

$$\frac{d\sigma}{d\varepsilon} = 0 \tag{3.4}$$

This reversal occurs at σ_u. The tensile strength thus appears to be the highest stress level attained on the σ–ε curve. Said differently, the tensile strength represents another intrinsic limitation that the material is subject to: the maximum stress that it can sustain before concentrated deformation proceeds rapidly to rupture.

Some materials don't have yield strengths that appear as sharp discontinuities like that of the steel, as shown in **Figure 3.3**. In these cases, a different criterion must be employed to determine σ_y. The most common criterion gives what is called the 0.2% offset yield strength. Consider the stress-strain curves of **Figure 3.5**. **Figure 3.5(a)** provides two curves for two different aluminum alloys. Both of these materials obviously possess drastically different properties, but in both curves, the transition from elastic deformation to plastic deformation is not well defined. The offset strategy for determining the yield strength in such a material is illustrated in **Figure 3.5(b)**. The "offset curve" is constructed with the same slope (= E) as the measured curve but is shifted along the strain axis by 0.2% (= 0.002 m/m). The intersection of the offset curve with the σ–ε curve indicates the yield stress.

Example 3.3:

Using the stress-strain behavior of the "2024" Al alloy shown in **Figure 3.5(a)**, determine the Young's modulus, yield strength, tensile strength, and ductility.

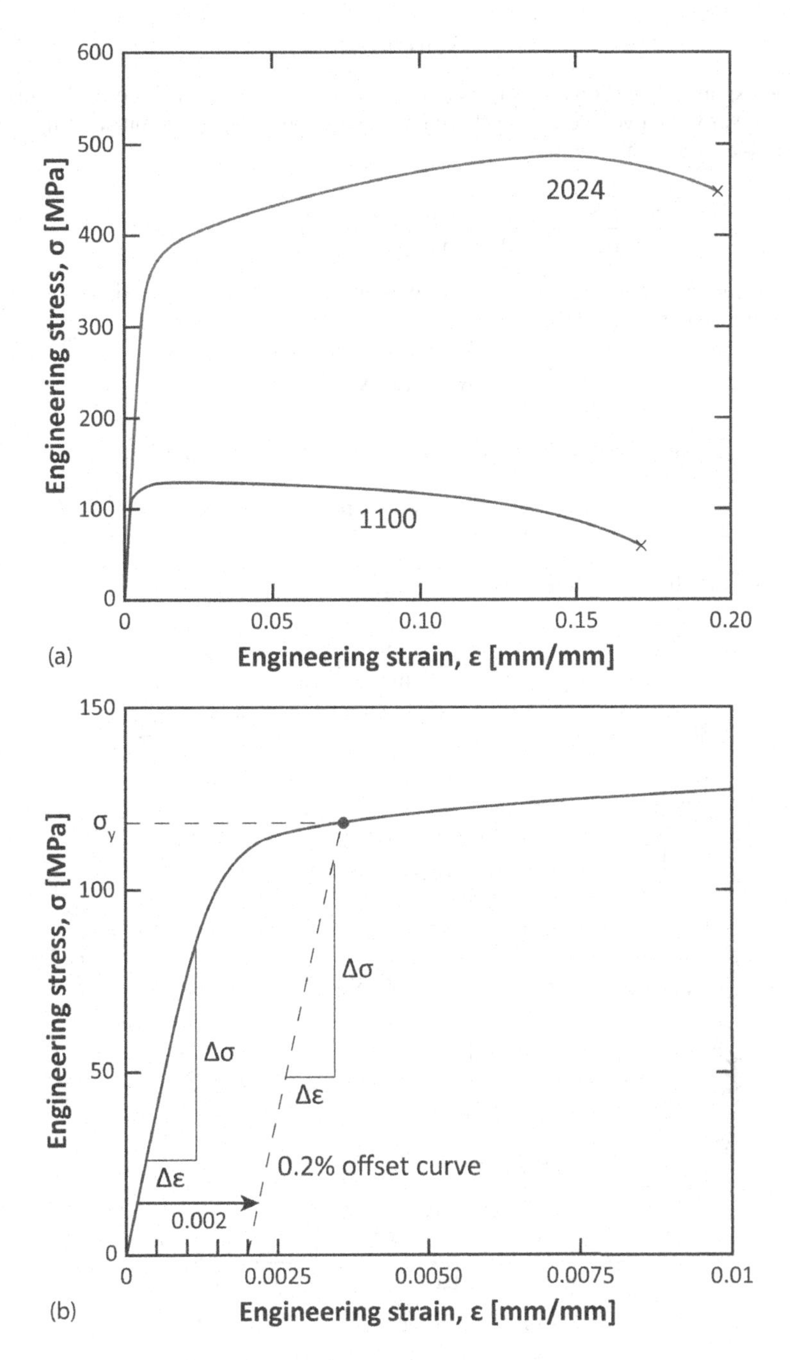

FIGURE 3.5 Stress-strain curves for two Al alloys. In (a), the full curves are shown; their elastic moduli are similar, but their strengths are different. A close-up of the elastic portion of the Al 1100 curve alongside a 0.2% offset curve is in (b). The intersection of offset with the stress-strain curve indicates the location of the yield point and the corresponding yield strength.

SOLUTION

The expanded elastic/yield region of the curve is shown below. The 0.2% offset curve has been overlaid on the chart; it has the same slope as the initial portion of the σ–ε curve. This line has slope

$$E = \frac{\text{rise}}{\text{run}} = \frac{245 \text{ MPa}}{0.0035 \text{ mm/mm}} = 7\underline{0},000 \text{ MPa} = 70. \text{ GPa}$$

The intersection of the offset curve with the σ–ε curve corresponds to an engineering yield stress σ_y of

$$\sigma_y = \underline{\mathbf{360.\ MPa}}$$

We similarly identify the tensile strength σ_u from **Figure 3.5(a)** as

$$\sigma_u = \underline{\mathbf{490\ MPa}}$$

and the strain to failure ε_f as

$$\varepsilon_f = 0.195 \frac{\text{mm}}{\text{mm}} = \underline{\mathbf{19.5\%}}$$

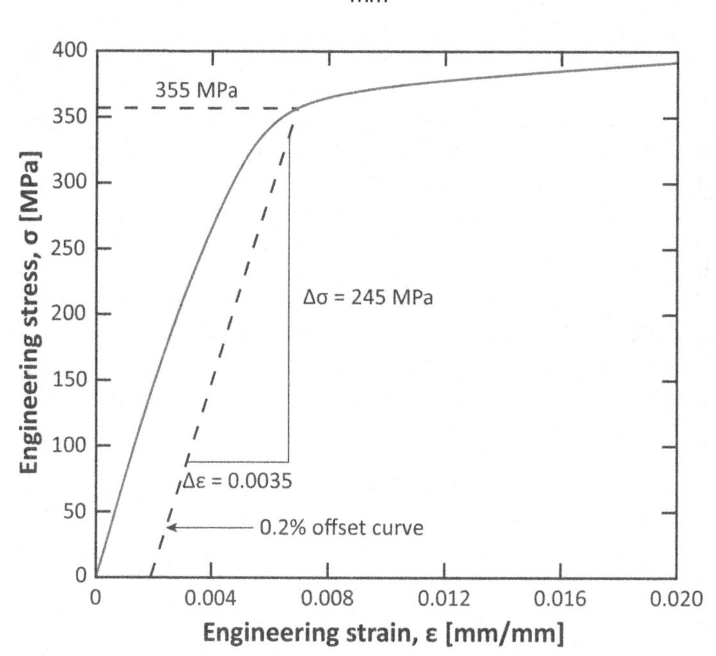

In materials that lack the capacity for plastic deformation – i.e., brittle materials – the work hardening and softening stages are absent. Complete failure or **fracture** occurs soon after the material departs the elastic regime of behavior. Such behavior is shown in **Figure 3.6**. Fracture is a physical process by which a material separates

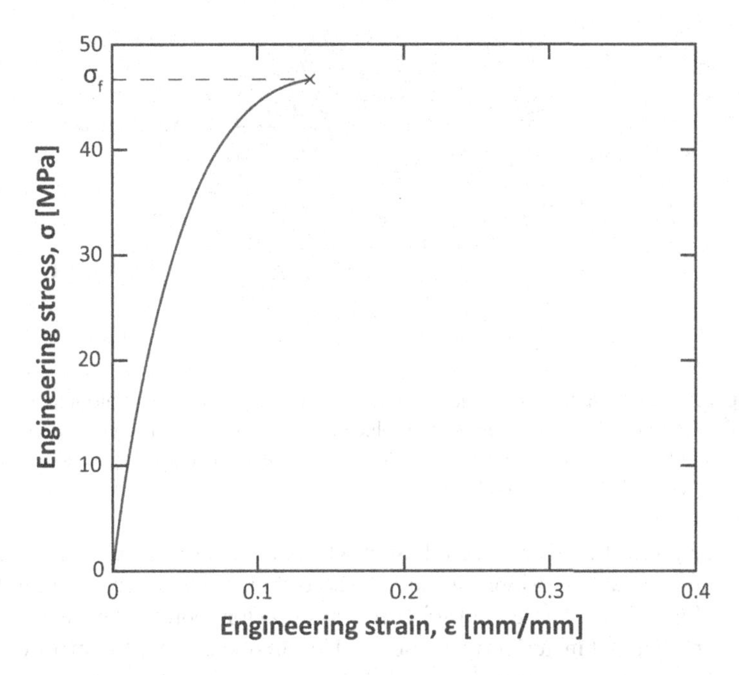

FIGURE 3.6 Stress-strain curve for Al_2O_3, a brittle material. The yield stress σ_y and the fracture stress σ_f are roughly coincident. Plastic deformation is almost entirely absent.

into two or more pieces. The stress at which this fatal qualitative change occurs in susceptible materials is the **fracture strength** σ_f. It should be intuitively obvious to engineering students that this is an important quantity when working with brittle materials.

3.3 DEFECTS AND STRENGTH

In **Section 1.6**, we introduced the concepts that connect the crystal structure of a material with the material's density. In **Section 2.3**, we discussed the connection between structure and elastic behavior. Let's now consider the physical origin of a material's limiting properties, such as the yield strength and tensile strength. Though these properties are determined by relatively complicated physical phenomena, they have a common structural explanation in the **defect** content of the material. A defect is a feature of a material's structure that departs from the "ideal" arrangement. The concept is mostly applied to crystalline materials that have a more or less regular structure, but the concept can be extended to amorphous materials as well.

Defects come in many types. For example, defects situated at a single lattice position, shown in **Figure 3.7**, can take the form of **vacancies**, **substitutions**, or **interstitials**. All materials contain vacancies in the form of "uninhabited" lattice positions. They are a consequence of the tendency toward disorder, or entropy, present in a material. The other defects occur when atoms of a different type are introduced into the crystal, forming a mixture or, when the constituents are primarily metallic, an

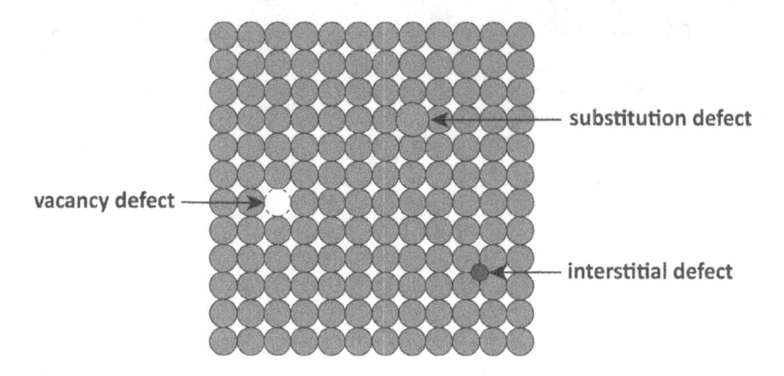

FIGURE 3.7 Single–lattice-site defects. A vacancy is the absence of an atom at a normally occupied lattice site. A substitution is the replacement of a native atom by one of a different species at a lattice site. An interstitial is a different (or potentially native) species inserted into the lattice at a non-lattice location ("interstice").

alloy. They are distinguished according to whether the non-native atom replaces a native atom at a lattice position (substitutional) or lodges in a "gap" between lattice positions (interstitial). These kinds of lattice position (or "point") defects are important in many internal materials processes such as atomic transport ("diffusion") and **solid-solution strengthening**.

The defects that influence the yield stress are a lattice rearrangement called a **dislocation**. Consider the plastic deformation of an ideal/defect-free crystal shown in **Figure 3.8**. In order for the shear deformation induced by the applied stress τ to be permanent, there must be a shift of the top half of the crystal with respect to the bottom half; the atoms across the middle plane **A–A′** relocate to a new stable configuration. This new configuration is "slipped" compared to the original, and this slip process is the **mechanism** by which crystals can adopt new shapes via plastic shearing at the

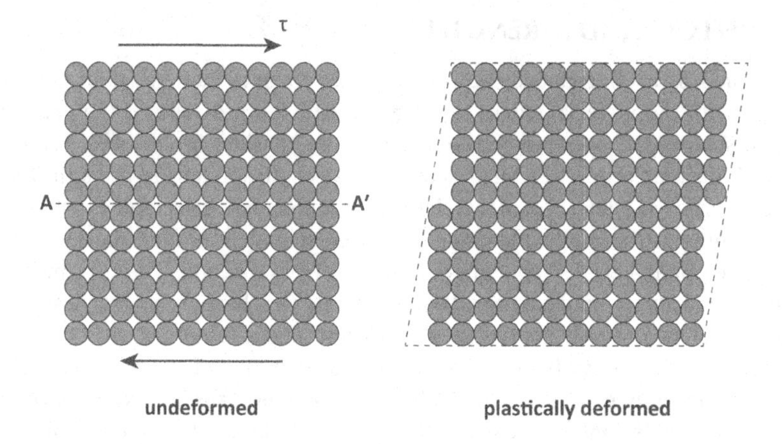

FIGURE 3.8 Idealized mechanism for plastic deformation in individual crystals. Shifting the atoms above plane **A–A′** to the right and the atoms below plane **A–A′** to the left produces a new, stable, sheared configuration. The crystal has been deformed permanently.

microscale. Arbitrary rearrangements/deformations of the crystal are possible if this process occurs on multiple planes simultaneously throughout the crystal.

This simple slip mechanism in a perfect crystal is not the only possible mechanism. In fact, the mechanism of **Figure 3.8** requires rather intense stresses (and therefore large inputs of mechanical energy) to reshape crystals. **Figure 3.9** depicts a mechanism that is similar in that shear stresses produce an increment of slip, but

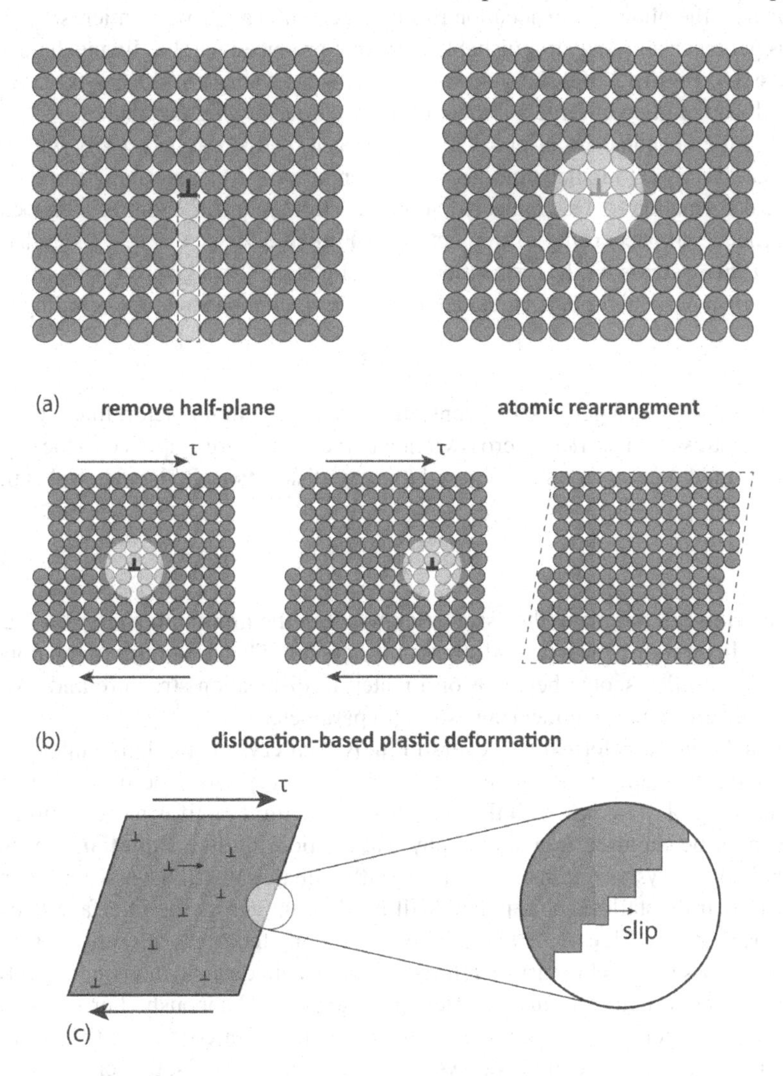

FIGURE 3.9 Low-stress mechanism for plastic deformation involving dislocation defects. (a) shows the structure of the defect; atoms relax around a missing half-plane in the crystal, forming a persistent misalignment. In (b), the dislocation glides in its plane in response to an applied shear stress. The dislocation can exit the crystal, leaving behind a crystal with a slipped structure. As shown in (c), the ensemble glide of many dislocations produces the cumulative plastic deformation. Though macroscopically this deformation appears smooth, the operation of the underlying mechanism requires discrete dislocation slip events.

different in that the mechanism requires the involvement of the dislocation defect. The nature of the defect is illustrated in **Figure 3.9(a)**. It is not a defect located at a single lattice site, like a vacancy, but rather distributed along the edge of the plane of atoms marked with "⊥". Hence, this dislocation defect takes the form of a *line* that wends its way through the crystal.[2] The dislocation defect is not a static feature of the crystal but can move (or glide) in response to an applied shear stress τ. **Figure 3.9(b)** shows how the glide of a dislocation through a crystal can produce microscopic slip that is geometrically indistinguishable from that produced via the slip mechanism of **Figure 3.8**. **Figure 3.9(c)** shows how the macroscopic shape of the crystal is reconfigured via the glide of large numbers of dislocations moving together.

Since the phenomenon of plasticity is governed by dislocation activity within the material, we wish to understand how the essential property describing the onset of plastic deformation (σ_y) is connected. Take the stress required to move a dislocation and produce slip according to the picture in **Figure 3.9b** as $\tau = \tau_c$. τ_c is called the **critical shear stress** for the dislocation in its "slip plane". In a given crystal, the shear stress τ on a particular plane is related to the overall applied stress σ according to

$$\tau = M\sigma$$

where $0 \leq M \leq ½$ is a geometrical constant (called the Schmid factor) that "resolves" the *shear* stress on a particular cross section to the *tensile* stress applied to the crystal. The threshold normal stress σ_c required to move dislocations in the material is then

$$\sigma_c = \frac{\tau_0}{M} = \sigma_y \tag{3.5}$$

In **Equation 3.5**, we make the association between the minimum stress σ_c required to move dislocations and the yield stress of the crystal. This is therefore a relationship between the microscopic behavior of a material (dislocation structure and motion) and the macroscopic "engineering" strength parameter σ_y.

Though plastic deformation occurs in individual crystals predominantly via the dislocation mechanism, not all materials are single crystals. The plastic deformation behavior of materials with the polycrystal structural motif is more complicated and cannot be captured in a simple physical relationship like **Equation 3.5**. Since the individual crystals/grains in a polycrystalline material will have different orientations, their crystallographic planes will not line up across the interface between them, as shown in **Figure 3.10**. This makes resolving the stress on well-defined slip planes impossible. Yield must therefore become a collective phenomenon involving slip on different planes in many different crystals simultaneously. For this reason, we identify in polycrystalline materials sets of planes that contribute the most slip; these sets of planes are called **slip systems** (see **Application Note**, below). Because the nature of plasticity in polycrystalline systems is so complicated, we expect that yield occurs when the applied stress σ exceeds a definite yield stress:

$$\sigma > \sigma_y$$

where the value of the yield stress is inferred from an experiment.

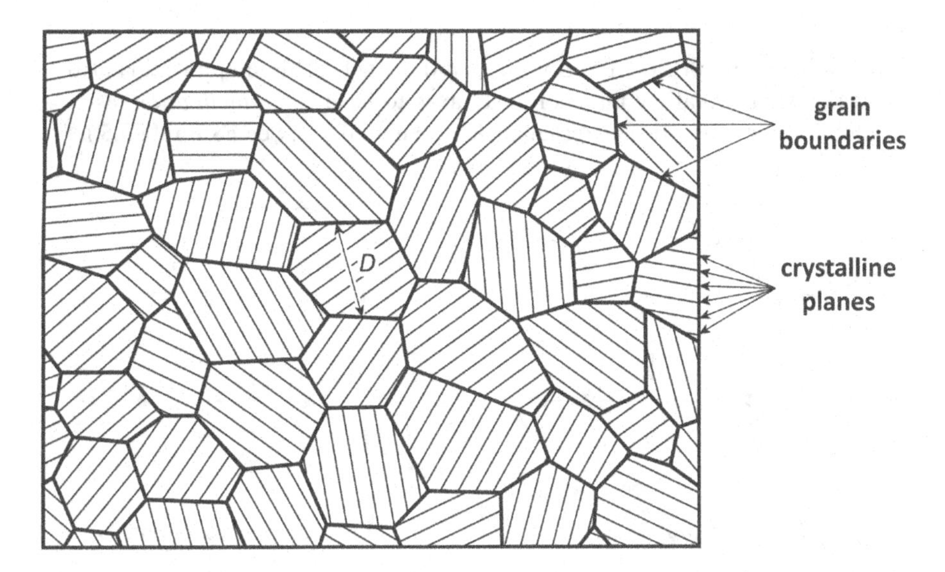

FIGURE 3.10 Grain structure of a polycrystalline material. The individual crystals/grains have different orientations, and the intersection of two or more grains occurs at a grain boundary. The grains have a characteristic size (their diameter, approximately), called the grain size *D*.

DISLOCATIONS AND CRYSTALLINE PLANES

Dislocations are constrained to travel on particular planes within a crystal (e.g., plane **A–A′** in **Figure 3.8**). What is the geometry of these planes, and how do we distinguish and catalog them? Mathematically, a plane incorporates three non-collinear points, and different sets of points define planes with different orientations. When describing a plane that passes through a crystal, it is sufficient to describe how the plane passes through a single unit cell, since each unit cell incorporates all the important features of the crystal. (A lengthier description of lattice geometry is presented in **Appendix B**.)

Consider two different coordinate systems attached to the unit cell shown below. The unit cell has lattice parameters a, b, and c, and the origin of the (right-handed) x, y, and z coordinate systems is placed at a corner of the cell. In "normal" Euclidean coordinates, we can identify locations in the unit cell, such as the corner points, and label them using coordinates based on the system of axes. Consider the following simplification: when we wish to describe locations in a unit cell with known lattice parameters, we may take the lengths a, b, and c as the *unit distances* along the x, y, and z axes, respectively. When we do this, the values of the coordinates change, but the overall geometry of the cell does not. These "crystallographic" coordinates are shown on the right. Within the single unit cell, the possible values of the coordinates are $0 \leq x \leq 1$, $0 \leq y \leq 1$, and $0 \leq z \leq 1$. (This simplification also works when we allow the coordinate axes to have some *inclination*, as is required for cells whose internal angles are not all 90°. See **Example 3.5**.)

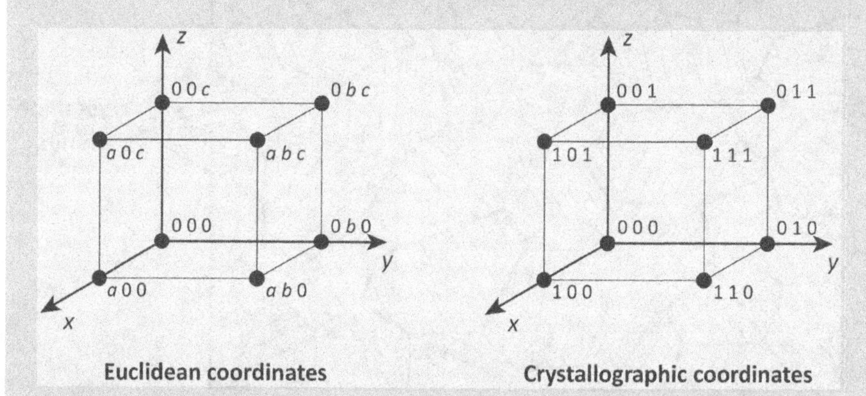

The orientation of a plane in the crystal may be described simply using these crystallographic coordinates and the axes that they are attached to. Any plane can be identified by its **Miller indices** *h*, *k*, and *ℓ*. The Miller indices are unique for a given plane and are related to three points **A**, **B**, and **C** that lie in the plane. To find the values of *h*, *k*, and *ℓ*, follow these rules:

Rules

1. The points **A**, **B**, and **C** are located where the plane intersects the *x*, *y*, and *z* axes in the unit cell: **A** = x_0 0 0, **B** = 0 y_0 0, and **C** = 0 0 z_0. Where x_0, y_0, and z_0 are the respective intercepts.
2. The indices *h*, *k*, *ℓ* of the plane are the reciprocals of the intercepts x_0, y_0, z_0 and are enclosed in parentheses: $(h\ k\ \ell) = (1/x_0\ 1/y_0\ 1/z_0)$.
3. Planes that are parallel to (i.e., do not intersect with) an axis are assigned an intercept of "∞".
4. *h*, *k*, and *ℓ* will always be integers. Multiplication of *h*, *k*, and *ℓ* by the same constant *c* does not change anything about the geometry of the plane.
5. Negative values of *h*, *k*, and *ℓ* may result; those values are identified using an overbar (e.g., "$\bar{2}$" rather than "–2").
6. If the plane passes through the origin 0 0 0 in the standard coordinate system above, shift the coordinate system so that its origin is at different, more convenient corner.

Of all the possible planes present in a crystal, slip tends to occur on those planes where the value of τ_c is the smallest. This typically occurs on planes with the highest density of atoms and the shortest atom-atom separation; these are the slip systems.

Material	Crystal Structure	Slip Plane & Direction
α-Fe Mo W	BCC	(1 1 0) family
Al Cu γ-Fe Ni	FCC	(1 1 1) family
Be Mg α-Ti Zn Co	HCP	(0 0 0 1) family[a]

[a] There are several other possible HCP slip planes, like the $(1\ 0\ \bar{1}\ 0)$ and the $(1\ 0\ \bar{1}\ 1)$ families, that are influential in materials like α-Ti and Zr. See **Appendix B** for a description of how to index planes in hexagonal systems.

Example 3.4:

Find the Miller indices $(h\ k\ \ell)$ for the three planes illustrated below.

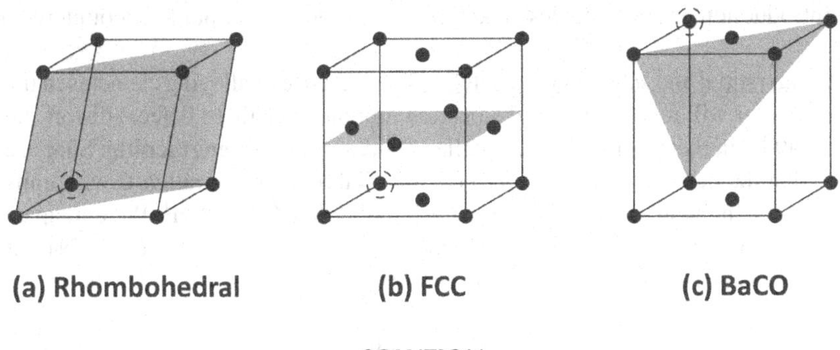

SOLUTION

In the rhombohedral unit cell (a), we take the coordinate origin 0 0 0 as the back, left, and bottom corners. We identify the intersection points **A** = 1 0 0, **B** = 0 1 0, and **C** = "nowhere", and the intersection coordinates $x_0 = 1$, $y_0 = 1$, and $z_0 = \infty$. We now obtain the Miller indices by taking the reciprocals of these values:

$$(h\ k\ \ell) = (1/x_0\ 1/y_0\ 1/z_0) = (1/1\ 1/1\ 1/\infty) = \underline{(\mathbf{1\ 1\ 0})}$$

The unit cell in (b) is face-centered cubic, and we use the origin indicated. We find **A** = "nowhere", **B** = "nowhere", and **C** = 0 0 ½. These give

$$(h\ k\ \ell) = (1/\infty\ 1/\infty\ 1/\tfrac{1}{2}) = (0\ 0\ 2)$$

For the base-centered orthorhombic unit cell in (c), the customary origin in the back, left, and bottom corners is unsuitable since the intersection point **C** would be nonsensical. Instead, we choose the origin shown. All intercept points remain within the unit cell and do not include the origin. We then identify **A** = 1 0 0, **B** = 0 1 0, and **C** = 0 0 –1. The planar indices are then

$$(h\ k\ \ell) = \left(\frac{1}{1}\ \frac{1}{1}\ \frac{1}{-1}\right) = (1\ 1\ \bar{1})$$

Note that the result in (c) depends on the choice of origin. If, for example, we chose the front, right, and top corners, we would compute the indices $(\bar{1}\ \bar{1}\ 1)$. This difference in h, k, ℓ values seems problematic at first, but we also recognize that the $(1\ 1\ \bar{1})$ plane and the $(\bar{1}\ \bar{1}\ 1)$ plane have the same geometric characteristics; they are just being viewed from different *perspectives*. We say that these planes belong to the same "family".

Equation 3.5 tells us how much stress is required to move dislocations and thus produce plastic deformation when the dislocations are *unimpeded* in their motion save for a minimum intrinsic stress σ_0. The presence of obstacles will give rise to an additional required stress σ_s whose value depends on the nature and/or quantity of the obstacles:

$$\sigma_y = \sigma_0 + \sigma_s(\text{obstacles}) \tag{3.6}$$

The additional stress term σ_s tells us how much strengthening is associated with the introduction of the obstacles. The types of obstacles commonly encountered are:

1. <u>Interstitial and substitutional defects.</u> The presence of alloying elements in the crystal will tend to inhibit dislocation motion since these defects distort the lattice in their vicinity. This is called **solid-solution strengthening**. Suppose that the defects are infused into the material to a concentration C with units $[C]$ = [number of defects/volume] = [number/length^3] = #/m^3. We anticipate that the strengthening effect will obey $\sigma_s = \sigma_s(C)$. Though the exact form of $\sigma_s(C)$ can be complicated, at low C we typically observe $\sigma_s \propto \sqrt{C}$, or

$$\sigma_s = k_{ss}\sqrt{C} \tag{3.7}$$

where k_{ss} is a constant that depends on the physics of the dislocation/defect interaction.

2. Other dislocations. Dislocations will impede one another's motion via the distortion they introduce into the lattice. The dislocation content of a material can be summarized using a **dislocation density** $\rho\perp$. This parameter has units $[\rho_\perp]$ = [length of dislocation line/volume] = [length/length^3] = m^2 or cm^2. Additional dislocation density produces strengthening, according to

$$\sigma_s = k_\perp \sqrt{\rho_\perp} \tag{3.8}$$

where $k_\perp$ is a constant. An important and familiar relationship follows from **Equation 3.8** when we recognize that plastic deformation has the effect of *increasing* dislocation line length and hence dislocation density: the more deformation, the more dislocation line. If ε_{pl} is the plastic deformation and we take $\rho_\perp \approx \varepsilon_{pl}$, then

$$\sigma_s = \sigma_0 + k_\perp \sqrt{\rho_\perp} = \sigma_0 + k_\perp \sqrt{\varepsilon_{pl}}$$

which is the work-hardening law of **Equation 3.3** with $n = ½$.

3. Grain boundaries. Grain boundaries are effective obstacles to dislocation motion since dislocations cannot cross them except in instances of heightened stress. Since we expect that the amount of grain boundary will increase as the grain size D decreases, strengthening will *increase* with decreasing D. This grain-boundary strengthening obeys a rule

$$\sigma_s = \frac{k_{GB}}{\sqrt{D}} \tag{3.9}$$

where k_{GB} is a constant. **Equation 3.9** is called the Hall-Petch relationship.

Given the above strengthening mechanisms, we can update **Equation 3.5** to obtain the yield stress of a material as

$$\sigma_s = \sigma_0 + \sigma(C, \rho_\perp, D) = \sigma_0 + k_{ss}\sqrt{C} + k_\perp \sqrt{\rho_\perp} + \frac{k_{GB}}{\sqrt{D}} \tag{3.10}$$

The strengthening principles outlined in (1)–(3) above are more sophisticated than we have presented here and must typically be assessed for individual materials via direct experiments. However, the result in **Equation 3.10** gives us some perspective on how the structural parameters contribute to the overall yield strength of the material via a number of different defect types. We furthermore expect that the tensile strength also has a dependence on such structural features:

$$\sigma_u = \sigma_u(C, \rho_\perp, D)$$

3.4 HARDNESS

The yield and work-hardening behavior of a material are related to another important mechanical property: its **hardness**. Hardness is loosely defined as a material's intrinsic resistance to plastic deformation. The ability of a component to withstand scratching and penetration depends on how hard it is.

TABLE 3.2
The Mohs Scale of Hardness, with Standard Materials and Example Materials

Mohs Number	Mineral Standard	Mineral Name	Example Materials
10	C (Cubic carbon)	Diamond	
9	Al_2O_3 (Alumina)	Corundum	Nitrided steels
8	$Al_2SiO_4(OH^-, F^-)_2$	Topaz	Most cutting tool materials
7	SiO_2 (Fused silica)	Quartz	File steels
6	$KAlSi_3O_8$	Orthoclase	
5	$Ca_5(PO_4)_3(OH^-, Cl^-, F^-)$	Apatite	Machinable steels
4	CaF_2	Fluorite	
3	$CaCO_3$	Calcite	Brasses and Al alloys
2	$CaSO_4 \cdot 2H_2O$	Gypsum	
1	$Mg_3Si_4O_{10}(OH)_2$	Talc	Most plastics

The hardness property is easy to describe in this loose, qualitative way, but we wish to apply a quantitative description so that hardness is a material's property in the same sense as an elastic modulus or density. One framework that attempts to quantify material hardness is called the **Mohs scale**. The Mohs scale places a given material in a ranked list according to which materials it can scratch (i.e., a higher ↑ ranking than) or be scratched by (a lower ↓ ranking than). The ranking or "Mohs number" M is on a scale of 1–10. The Mohs scale was originally developed for mineral classification purposes, but any material with some capacity for plastic deformation can be situated on the scale. Some example materials are ranked alongside the reference materials/minerals in **Table 3.2**.

Though the Mohs scale uses numerical values in its rankings, it is not entirely quantitative in the sense that we wish, where we can tie the property value to measurements of several physical parameters. The numerous materials-related phenomena underlying the hardness property interact in a complicated way, making such measurements difficult, but consider how a material behaves when subjected to a simple **indentation test**. **Figure 3.11** shows the schematic geometry of the test. In **Figure 3.11(a)**, a rigid indenter, in this case a sphere with radius R, is pressed into the surface of a workpiece with a force F. Immediately underneath the indenter, the stresses will cause the material to deform plastically, forming an impression whose shape conforms to that of the indenter. If the material is *soft*, then this residual impression will be *large*; if the material is *hard*, the impression will be *small*.

We recognize that the applied load and the size of the indent are important quantities in the outcome of the indentation experiment. For this reason, we define the **Meyer hardness** property HM as the force applied relative to the *projected* indent area A':

$$\mathrm{HM} = \frac{F}{A'} = \frac{F}{\pi (d/2)^2} \tag{3.11}$$

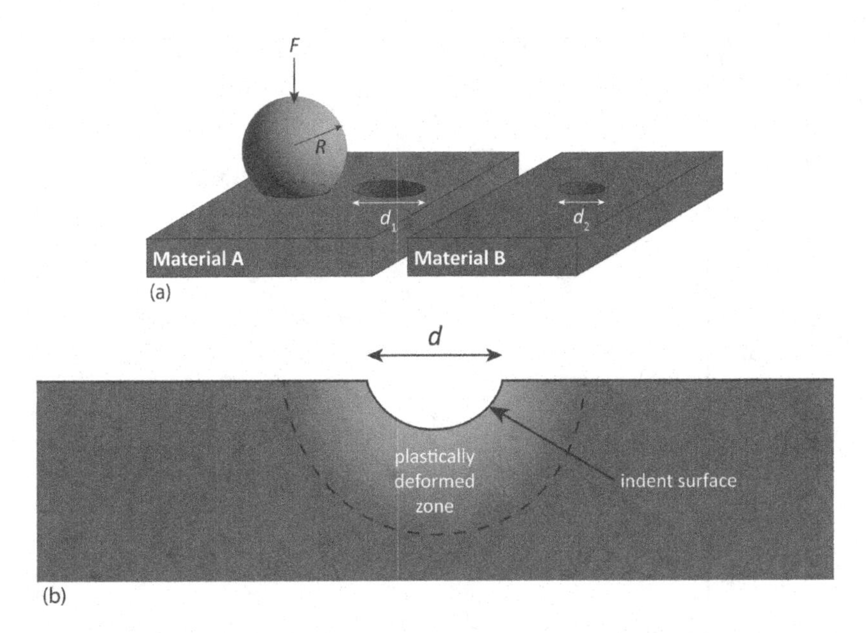

FIGURE 3.11 The essential geometry of an indentation hardness test. (a) An indenter (in this case, a sphere) is impressed into the surface of an object made of Material A. The indenter is pushed into the surface with a loading force *F*. After the withdrawal of the indenter, the impression left behind (formed by plastic deformation) has a certain size d_1. When the same indenter is applied to a different material (Material B) with the same loading force, an impression of size d_2 remains. If $d_2 < d_1$, we say that Material B is harder than Material A. (b) The indent cross section reveals a region of highly deformed material surrounding the indent cavity. During indentation, the plastic deformation behavior of the material in this region determines the hardness.

Alternately, we may compute the **Brinell hardness** HB, a parameter similar to the Meyer hardness but with the force normalized to the total indent surface area *A*:

$$\mathrm{HB} = \frac{F}{A} = \frac{F}{2\pi R^2\left[1 - \sqrt{1 - (d/2R)^2}\right]} \tag{3.12}$$

Though the Meyer hardness and the Brinell hardness produce different numerical values for what is ostensibly the same property of the material, they can still be used to compare the behavior of different materials, so long as it is strictly their HM or HB values being compared. The HM and HB values may be sensibly interconverted in many cases using a geometrical conversion factor based on the indenter geometry, but such conversions are typically only approximate. There is also correspondence between indentation hardness values and Mohs scale values, as shown in **Figure 3.12**. This correspondence is not very uniform given the conceptual differences between the two measures, but an empirical basis for comparison can be obtained from the data.

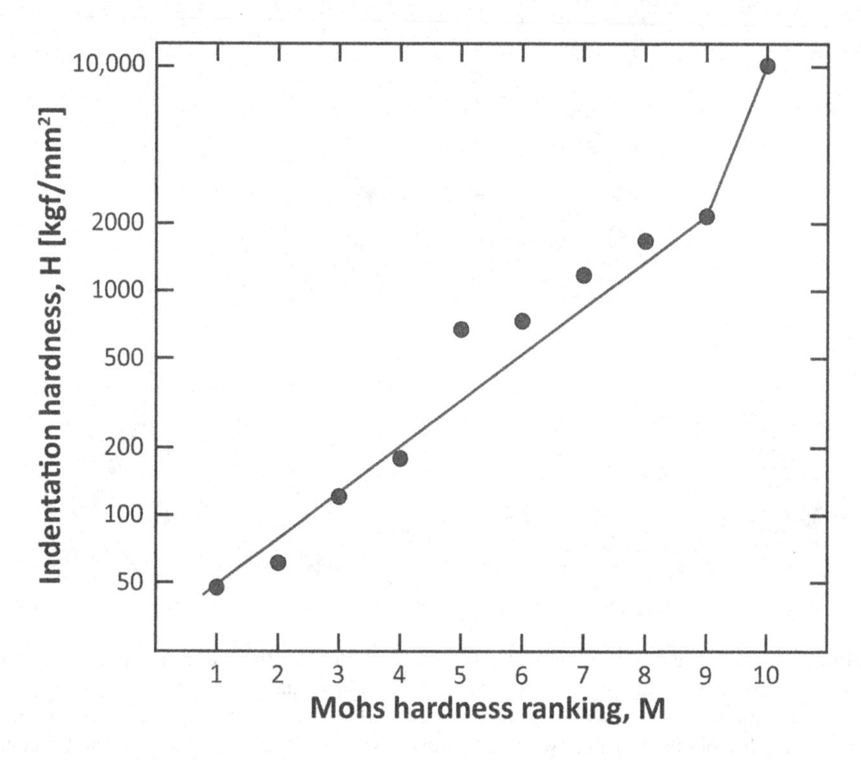

FIGURE 3.12 Measured indentation-hardness values of the minerals in **Table 3.2** compared to their Mohs-scale rankings. The left-hand scale is logarithmic, indicating that measured material hardness varies over several orders of magnitude. An approximate relationship between the measured hardness H and the Mohs number M is $\log H \approx M \log k + C$, where k and C are constants that depend on how H is measured.

The physical quantities that must be measured to obtain HM or HB values are the force F (typically fixed by the test platform) and the indent area A or A' (determined using an optical scope in the case of the Brinell hardness). These measurements can sometimes be difficult or time-consuming, but other types of tests are available. The **Rockwell hardness** HR comes from a simplified test that doesn't require measurement of the residual indent. It has also been adapted to use different scales ("A", "B", "C", ...) to accommodate the large differences in hardness that occur in different types of materials. The different HR values are given as "HRA", "HRB", "HRC", etc., according to the scale. The different scales require different parameters and indenter types and are adapted to different materials and specimen geometries.

Under load, the Rockwell indenter tip penetrates the test piece to a depth of h. For stability and repeatability reasons, the indenter tip is first "preloaded" into the material to a shallow starting depth h_0. Since the area A of an impression left by a regularly shaped indenter is correlated with the depth of penetration, the value of $\Delta h = h - h_0$ can be used as a basis for the determination of a hardness value. Since both the h and

TABLE 3.3
Rockwell Test Parameters

Rockwell Scale	Indenter Type	F [kgf]	H_{max} [unitless]
A	Diamond cone	60	100
B	0.16-cm diameter steel ball	100	130
C	Diamond cone	150	100
D	Diamond cone	100	100
E	0.32-cm diameter steel ball	100	130
M	0.64-cm diameter steel ball	100	130
R	1.27-cm diameter steel ball	60	130

h_0 values are measured in terms of a standard depth increment δ, where $\delta = 0.001$ or 0.002 mm, the units of HR are [HR] = [length/length] = mm/mm = unitless. The value of HR is calculated as

$$\mathrm{HR} = H_{max} - \frac{\Delta h}{\delta} \tag{3.13}$$

In **Equation 3.13**, the H_{max} parameter is the full range (in-depth increments) of the particular Rockwell scale in use. Rockwell parameters associated with several scales are given in **Table 3.3**. Scales A, B, C, and D are typically used for metals, while scales E, M, and R are useful for polymers or thin specimens. Though the Rockwell hardness appears to come from a different kind of measurement than the Meyer or Brinell hardness, they reflect the same property of the material and can be interconverted using empirical relationships. For example, we can estimate

$$\mathrm{HB} = \frac{7300}{130 - \mathrm{HRB}} \frac{\mathrm{kgf}}{\mathrm{mm}^2}$$

for typical steels. There is no general relationship for conversion of hardness values between different systems, so some care must be taken when employing such a conversion.[3]

Intense, localized deformation occurs underneath the indenter tip during an indentation hardness test. This means that this material is subject to conditions that exceed its limitations in the form of the yield and the tensile strengths. This has the implication that hardness experiments are predictive of these other properties. Like conversions between different hardness measures, predicting tensile strengths from hardness values is not theoretically possible, but experimental evidence reveals approximate relationships. **Figure 3.13** shows the empirical relationship between Brinell hardness and tensile strength for a variety of steel and wrought aluminum alloys. The relationship is approximately linear over a wide range of hardness values. The regularity of this relationship permits the use of hardness measurements as a proxy for tensile strength measurements in some cases.

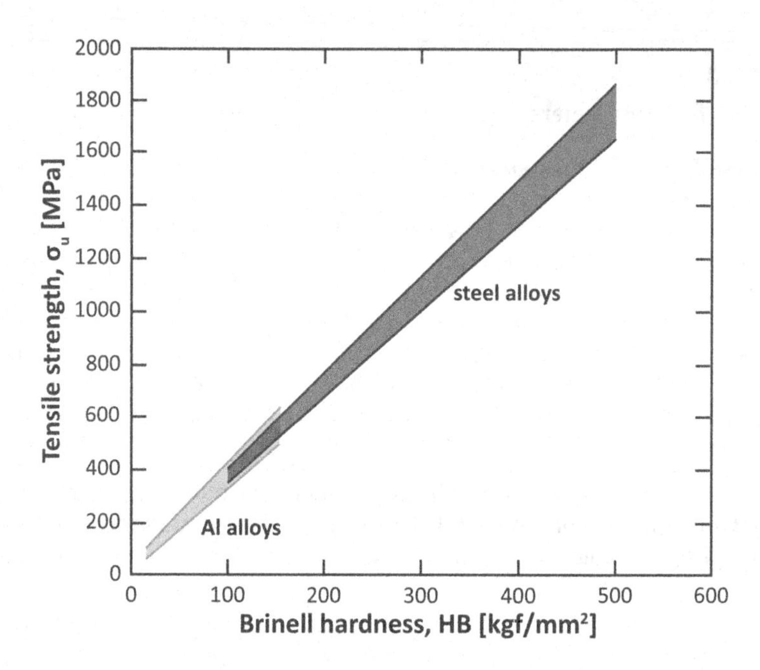

FIGURE 3.13 Trends in strength and hardness for some important alloy systems. Tensile strength is correlated with Brinell hardness in an approximately linear fashion: $\sigma_u \approx 3HB$.

3.5 MICROSCOPY AND METALLOGRAPHY

Much has been said of the mechanical properties of materials and the techniques for determining them, but we also require methods to characterize the structural aspects of materials. The assessment and measurement of the microscopic structural features of metal alloys is called **metallography**, and similar practices similar to those applied to metals can be applied to many types of nonmetallic materials. Since the features of materials that influence their properties are microscopic, the primary tool for structural characterization is the microscope. There are, of course, many different varieties of microscopes, but the devices most frequently encountered in materials characterization are **reflected-light microscopes**, scanning electron microscopes, and transmission electron microscopes. The electron varieties are sophisticated and more common in research settings than in materials-validation and quality-control settings, so we will focus on the operation of light microscopes and what they can reveal about the structure of materials.

Both light and electron microscopes utilize **lenses** in some form. A light microscope uses the transparent convex lenses (typically silica glass) to produce a magnified image of a real object. The essential geometry of magnification is shown in **Figure 3.14**. **Figure 3.14(a)** depicts the optical arrangement. The "thin-lens" equation predicts

$$\frac{1}{f} = \frac{1}{u} + \frac{1}{v} \tag{3.14}$$

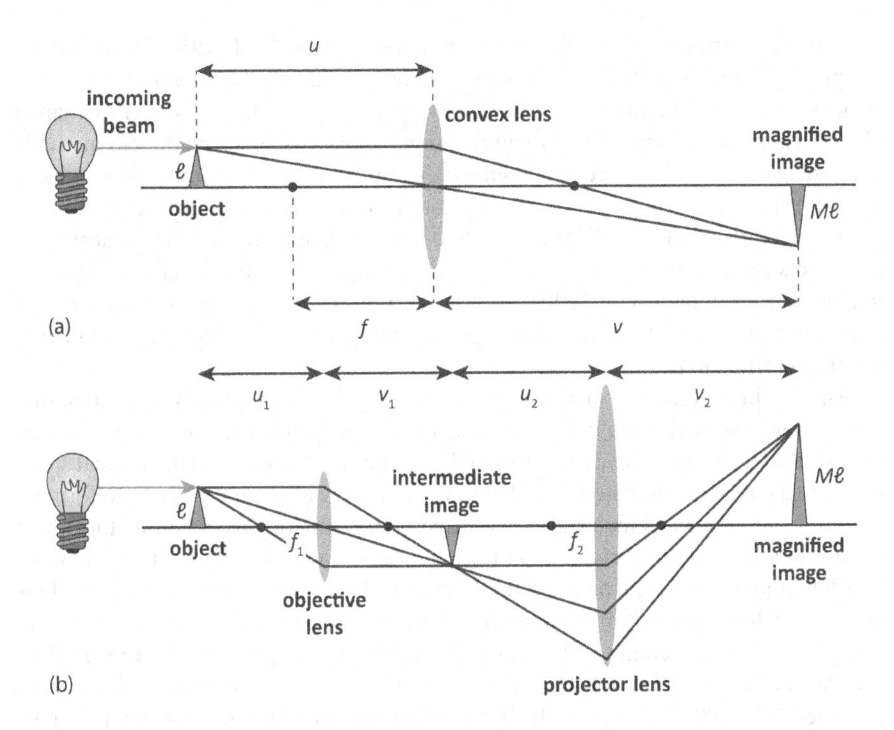

FIGURE 3.14 Magnification of an object via optics. (a) A convex lens with focal length f produces an image of an object with height ℓ placed at a distance $f < u < 2f$ from the lens. This image is at a distance v from the lens and is magnified by a factor M according to **Equation 3.15**. (b) Schematic layout of a two-stage magnifier. Higher magnifications with less distortion can be achieved using this design. Typically, the objective lens is fixed in place, and the other lens can be replaced with one of different f to produce variable magnification.

f is the focal length and is an intrinsic property of the lens. A geometrical analysis indicates that the factor $M = v/u$. Since $v = fu/(u - f)$ from **Equation 3.14**, we obtain

$$M = \frac{v}{u} = \left(\frac{1}{u}\right) \times \left(\frac{fu}{u - f}\right) = \frac{f}{u - f} \tag{3.15}$$

This means that when the object is placed just outside the focal point of the lens (i.e., when $u - f$ is small), the magnification is large.

It is important to point out that all of the points on the image in **Figure 3.14(a)** should be equidistant from the *center* of the lens, i.e., the image is *curved*. Therefore, in this single-lens arrangement, when the image is projected onto a planar surface (like a photographic plate or other photosensitive device), it will be distorted. This unwanted effect makes analysis difficult but can be mitigated using a two-stage arrangement like that shown in **Figure 3.14(b)**. This geometry produces a total magnification of

$$M = \left(\frac{f_1}{u_1 - f_1}\right) \times \left(\frac{f_2}{u_2 - f_2}\right) \tag{3.16}$$

In most instruments, the value of f_2 is fixed, and multiple objective lenses can be swapped to provide different values of f_1. (The various distances u and v are also typically fixed, so the microscope doesn't require any change in its overall shape.) Generally speaking, any level of magnification that is possible using optical methods is possible using a two-stage arrangement. The limit to the resolution of an optical instrument is approximately $\lambda/2$ (the "Abbe limit"), where λ is the wavelength of light. For short-wavelength light of, say, 500 nm wavelength, the smallest feature that can be distinguished (or "resolved") in a microscope is ≈ 250 nm. This means that magnifications greater than 1000× are mostly unnecessary. While they can be used to produce *larger* images, they cannot provide any more *detail* than images taken at lower magnifications.

Another important consideration is that the diagrams of **Figure 3.14** assume that the light necessary for magnification is available in quantity at the surface of and throughout the object. This is an acceptable assumption for very thin biological specimens that can be illuminated from behind and still transmit enough light to form an intelligible image. Most materials are opaque, so illumination from behind the object cannot be utilized in the production of a magnified image. Instead, the material must be illuminated from the front, and the light must be reflected back through the lens stages. Furthermore, this light must be as intense as possible so that the resulting image has sufficient contrast. The reflected-light microscope shown in **Figure 3.15** has the correct configuration for these requirements. This type of microscope uses a condenser lens to concentrate the diffuse light produced by the source and a half-silvered mirror to direct the light at the front of the specimen.

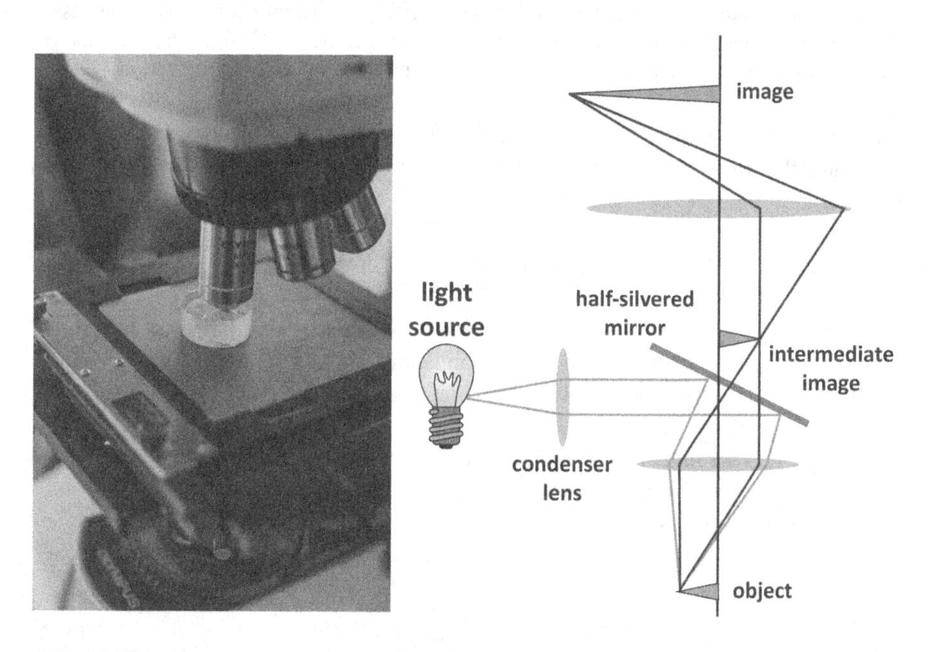

FIGURE 3.15 Schematic layout of a reflected-light microscope.

For the microscopic imaging process to work and the resulting image to provide all the structural information available at that scale, the material must be put into a condition that it

1. is convenient to handle and fits on the microscope stage,
2. is flat so that as much as possible of the image is in focus simultaneously,
3. reflects the maximum possible amount of incoming light, and
4. has distinguishable structural features.

Therefore, the preparation of metallographic specimens requires sectioning and mounting, leveling, polishing, and etching steps.

These steps are shown in **Figure 3.16**. During sectioning and mounting, as shown in **Figure 3.16(a)**, the material to be examined is cut from the component or ingot it originated from. Because only a cross section of the material can be inspected using light microscopy, the sectioning is typically performed so that one of the cut faces is the surface that requires inspection. More often than not, the specimen is mounted on a rigid polymer block to make handling easier. The freshly cut surface is quite rough, as shown in **Figure 3.15(b)**. This surface is unsuitable for microscopic inspection since these nonuniformities will disperse the light from the illumination source and little or no image will appear. **Figure 3.15(c)** shows the surface after polishing. This surface will reflect a significant amount of the light applied to it and produce a bright image. This image will likely not include the features we are interested in. Because it reflects light *uniformly*, the image will appear featureless. The solution is to treat the surface in some way so that the different features of interest are contrasted. Such a treatment (frequently a chemical or electrochemical process) is called etching. For example, the etching treatment depicted in **Figure 3.15(d)** has eroded the surface in the vicinity of the grain boundaries, altering the contrast of the boundaries relative to the rest of the surface.

After treating the surface in a way that reveals the grain boundaries, it becomes possible to measure the grain diameter/grain size D. A comparison of the scale bar on the image to the grain diameters observed (i.e., "eyeballing" the image) gives a rough estimate of D, but this estimate is probably insufficient for precision investigations. A more precise method would be to take a sample of grain diameters from the image and describe D using statistics. An example of such a procedure[5] looks like

1. Obtain an image that includes as many grains as possible but that still has good resolution of the grain boundaries. (Typically, a magnification of ~100× works for most metals.)
2. Draw a line (the longer, the better) across the image.
3. Referring to the image's scale bar (or other suitable calibration mark), determine the real length L of the line. (This determination is assisted by the use of image-analysis software.)
4. Count the number of intercepts or crossings N that the line makes with the grain boundaries in the image. (The endpoints are sometimes counted as 1/2 each.)

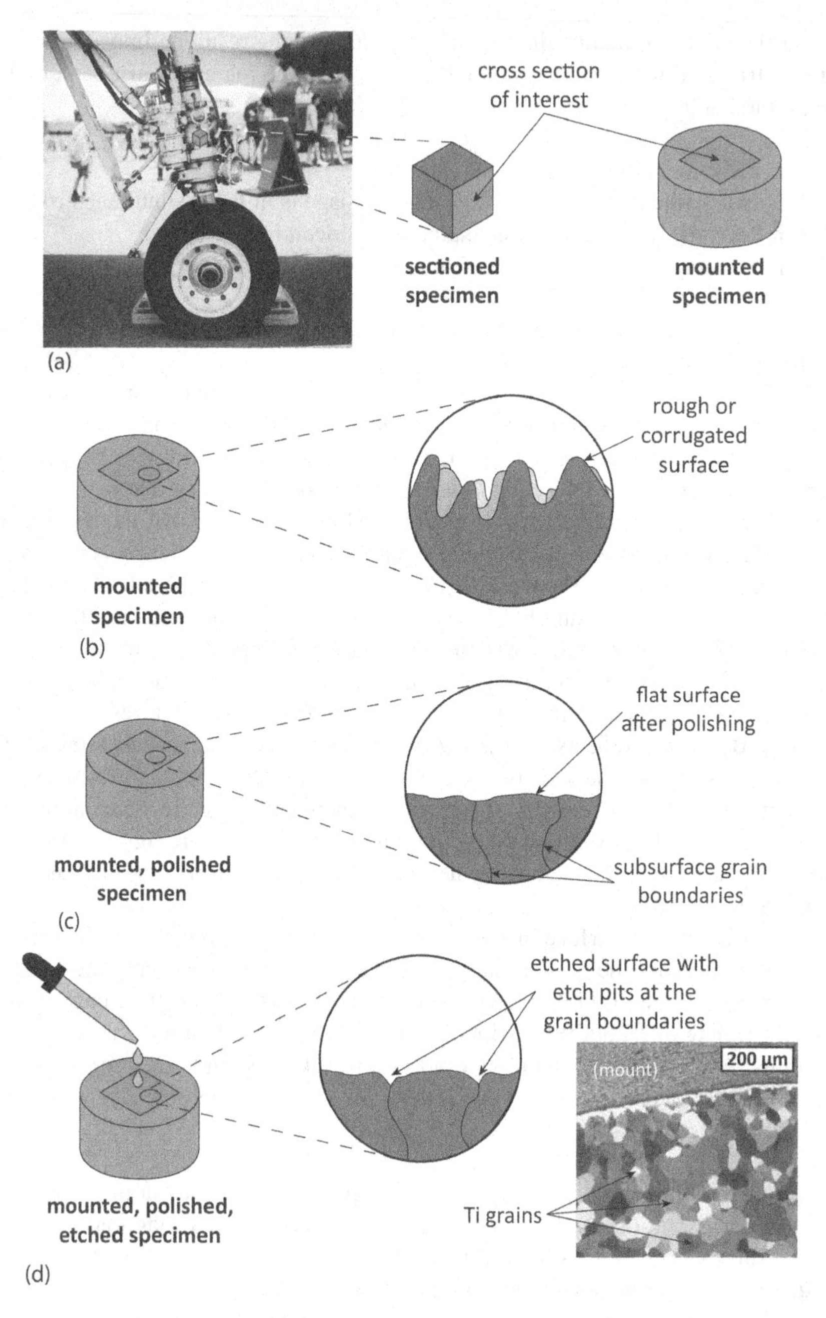

FIGURE 3.16 Step-by-step metallographic specimen preparation. (a) shows the sectioning of the specimen from the original component, followed by mounting in a thermosetting polymer. (b) The rough surface of the exposed specimen is likely very rough, so further preparation is required. (c) Polishing produces a specimen surface that is flat and reflective, and (d) etching reveals contrast between the various microstructural features present on the surface. (Adapted with permission from Vander Voort.[4])

5. Compute the mean lineal intercept length $\overline{\ell}$ for the line:

$$\overline{\ell} = \frac{L}{N} \tag{3.17}$$

6. The mean lineal intercept value $\overline{\ell}$ is a decent estimate of the grain size D:

$$D \approx \overline{\ell} \tag{3.18}$$

They are in general not the same since the grains in the image are never perfectly uniform. In the (ideal) limit of spherical grains, the relationship would be

$$D = \frac{3}{2}\overline{\ell}$$

(For highly irregular grains, different numerical corrections in place of 3/2 can be employed.)

To increase the precision of the overall measurement and establish reasonable bounds on the result, multiple line-based measurements should be made. An example of the method is illustrated in **Example 3.5**.

Example 3.5:

What would you estimate the grain size of the following image?

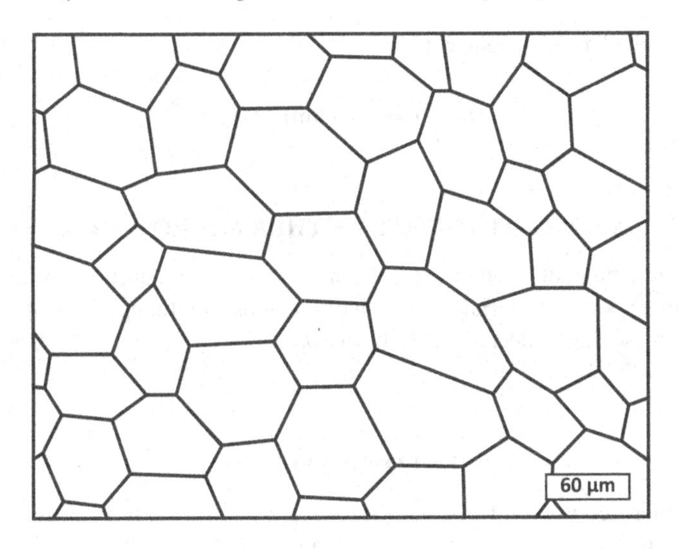

SOLUTION

A line is placed across the image that is long enough to intersect a significant number of grains, as shown below. Based on the scale, we find that $L \approx 492$ µm and intersects 10-grain boundaries, giving an intercept count of

$$N = 10 + \tfrac{1}{2} + \tfrac{1}{2} = 11$$

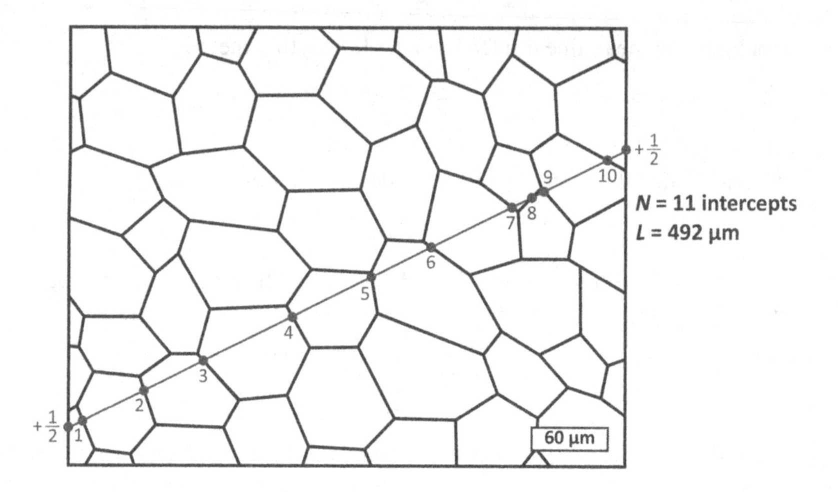

From these values of L and N, we obtain

$$\bar{\ell} = \frac{L}{N} = \frac{492\ \mu\text{m}}{11} = 45\ \mu\text{m}$$

so the grain size is roughly

$$\underline{\boldsymbol{D \approx 45\ \mu\text{m}}}$$

Or, in the limit of spherical grains

$$D = \frac{3}{2}\bar{\ell} = \frac{3}{2}(45\ \mu\text{m}) = \underline{\mathbf{68\ \mu m}}$$

APPLICATION NOTE – OTHER MICROSCOPES

There are a number of other experimental methods for determining the structural features of a material that are worth discussing, though they are in less common use than reflected-light microscopes.

STEREOMICROSCOPES

Light microscopy performed with a reflected-light/metallurgical microscope uses a single eyepiece and provides a magnified image of a flat surface. It is necessary in some cases (e.g., fractography and inspection of finished components) to observe the surfaces of materials with topographic features up close and in a way that provides the viewer 3D perspective. Stereomicroscopes are light-based devices that, because they only provide low magnification (< 50×), can image *unpolished* surfaces. The two eyepieces each provide images of the specimen from slightly different angles, giving the stereomicroscope operator a 3D visual impression. Stereomicroscopes have greater "depth of field" than normal microscopes so that features with different heights can be in focus simultaneously. Longer working distances mean that larger specimens can be accommodated on the stage.

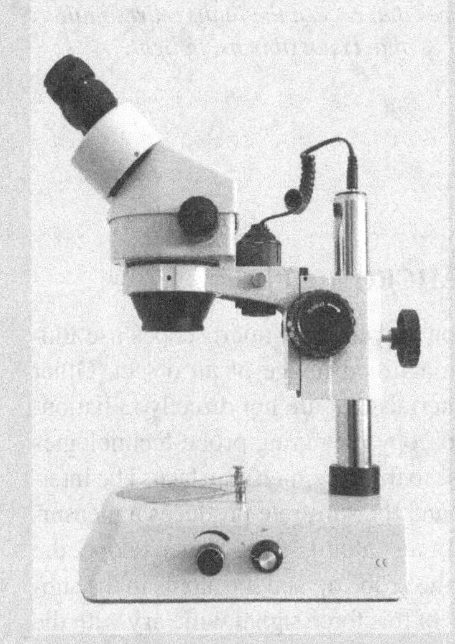

Stereoscopes provide low-magnification, 3D images of a material's surface.

ELECTRON MICROSCOPES

As mentioned above, the limit on the size of features that can be resolved using light is *physically* limited to around 250 nm. If detailed, high-magnification (> 1000×) images are required, and then advanced instruments like electron microscopes are necessary. Using a series of electromagnetic lenses, an electron microscope focuses a beam of electrons, rather than light, on the surface of a specimen. In a "scanning" electron microscope, this beam is scanned across a small region to be magnified. An electron detector catches the electrons that are returned from the region's surface (via inelastic and elastic collisions with the atoms in the sample) and uses those returned electrons (and their energies) to reconstruct information about the surface topography and composition. Scanning electron microscopes can attain magnifications of ~1,000,000× with a resolution limit on the order of 1 nm. Transmission electron microscopes are similar to transmission light microscopes in that the intensity of focused electrons is transmitted through a thin (< 200 nm) specimen. These microscopes can produce magnified images of materials up to 50,000,000×.

Electron microscopes provide magnified images that exceed the limits on resolution that is possible using light.

SCANNING PROBE MICROSCOPES

Light microscopes employ light illumination and electron microscopes use illumination by electron beam to produce a magnified image of an object. Other strategies for the close-up inspection of materials that are not directly radiation-based exist, such as scanning probe microscopy. Scanning-probe technologies use a microscopic probe tip placed very close to the specimen's surface. The interaction between the matter in this probe tip and the substrate produces a measurable effect that is converted into a signal. In an "atomic force" microscope, the intermolecular forces between the atoms in the probe tip and the atoms in the substrate produce the signal, and the magnitude of this force signal will vary with the z distance of the tip from the surface. As the probe tip is scanned across the specimen surface (in, say, the y direction), the surface topography can be "mapped" to high precision. Sensitive scanning-probe techniques can be utilized in the imaging of nanoscopic features and structures and even individual molecules.

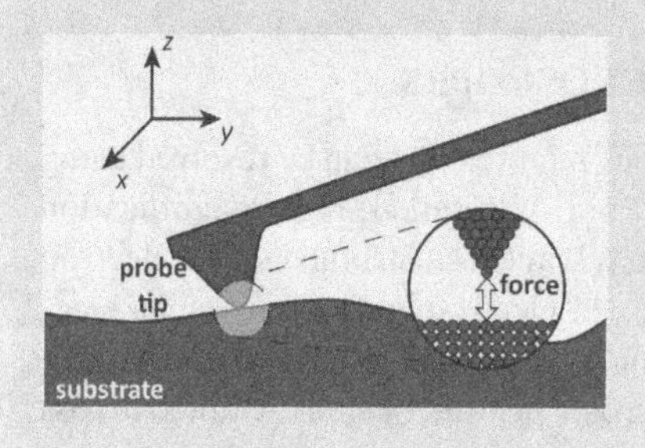

Atomic force microscopes can determine the height of a surface to high precision by measuring substrate/tip interaction forces.

3.6 CLOSING

Measuring the properties of materials – physical, mechanical, or structural – requires a clear understanding of the nature of the property to be measured. Information about the variables involved (σ, ε, magnification, etc.) and the length scales involved indicate the capabilities of the instrument required. The resulting data/property values

(E, H, D, etc.) are important pieces of information that are vital for design decisions and reflect material quality. It is crucial that anyone who wishes to apply these methods perform their measurements in a manner that agrees with common practice; this helps to ensure that results can be compared across experiments and with expected values. (We will revisit the interpretation of measured data in Chapter 6.) Hence, the use of test standards like those published by ASTM becomes an important professional practice.

3.7 CHAPTER SUMMARY

Key Terms

Brinell hardness
critical shear stress
defect
dislocation
dislocation density
ductility
forensic analysis
fracture
hardness
indentation test
interstitial
lens
mechanism
metallography
Meyer hardness
Miller indices
Mohs scale
necking
quality-control
reflected-light microscopes
Rockwell hardness
solid-solution strengthening
substitution
tensile strength
tensile tester
vacancy
yield strength

Important Relationships

$$\varepsilon_t = \ln(\varepsilon + 1)$$ (true strain from eng. strain)

$$\sigma_t = \sigma(\varepsilon + 1)$$ (true stress from eng. values)

$$\sigma_t = \sigma_y + K\varepsilon_{pl}^n$$ (work-hardening behavior)

$$\frac{d\sigma}{d\varepsilon} = 0$$ (tensile-strength criterion)

$$\sigma_y = \frac{\tau_0}{M}$$ (yield strength vs. shear strength)

$$\sigma_y = \sigma_0 + \sigma_s(\text{obstacles})$$ (overall strengthening)

$$\sigma_s = k_{ss}\sqrt{C}$$ (solid-solution strengthening)

$$\sigma_s = k_\perp\sqrt{\rho_\perp}$$ (dislocation strengthening)

$$\sigma_s = \frac{k_{GB}}{\sqrt{D}}$$ (grain-boundary strengthening)

$$\sigma_s = \sigma_0 + k_{ss}\sqrt{C} + k_{\perp}\sqrt{\rho_{\perp}} \quad \text{(overall strengthening)}$$

$$\text{HM} = \frac{F}{\pi(d/2)^2} \quad \text{(Meyer hardness)}$$

$$\text{HB} = \frac{F}{2\pi R^2\left[1 - \sqrt{1-(d/2R)^2}\right]} \quad \text{(Brinell hardness)}$$

$$\text{HR} = H_{\max} - \frac{\Delta h}{\delta} \quad \text{(Rockwell hardness)}$$

$$\frac{1}{f} = \frac{1}{u} + \frac{1}{v} \quad \text{(thin-lens equation)}$$

$$M = \frac{f}{u-f} \quad \text{(single-stage magnification)}$$

$$M = \left(\frac{f_1}{u_1 - f_1}\right) \times \left(\frac{f_2}{u_2 - f_2}\right) \quad \text{(two-stage magnification)}$$

$$\overline{\ell} = \frac{L}{N} \quad \text{(mean lineal intercept)}$$

$$D \approx \overline{\ell} \quad \text{(grain size from } \overline{\ell}\text{)}$$

3.8 QUESTIONS AND EXERCISES

Concept Review

C3.1 The load cell on a tensile test platform has upper and lower limits to the force that it can measure. Dropping below the lower limit produces inaccurate results, and exceeding the upper limit risks damage to the cell. Explain how you would choose the tensile-specimen geometry so that the force range falls within the range of the load cell.

C3.2 **Figure 3.11** shows the plastically deformed regions surrounding an indent placed for hardness testing purposes. Typically, indent samples are performed so that the indents are as far apart from one another as possible on the specimen. Why is this a good practice?

C3.3 Explain the importance of each step in preparing a material specimen for light microscopy.

Discussion-forum Prompt

D3.1 Using campus library resources, search for an ASTM standard related to your field or interests. Post the title of the standard and the details of the standard's scope in the forum. Furthermore, describe the importance of the standard method on your area. Look at the posts of the other students in the course and identify one that you have a question about (e.g., "What kind of specimen is required?") and leave a reply containing your question. Reply to any questions asked under your post with information taken from the standard.

PROBLEMS

P3.1 A tensile specimen like that in **Figure 3.2(b)** is extended by $\Delta\ell = 2.3$ cm. What is the engineering strain? What is the true strain?

P3.2 Frequently, the amount of plastic deformation imparted at low temperature is represented as the "cold work" CW done to the component. For an object with initial cross-sectional area A_0 whose cross section is reduced to A during deformation, the cold work is (as a percentage)

$$\%CW = \frac{A_0 - A}{A_0} \times 100$$

Suppose that you have the rolling operation shown below. If the plate initially has a thickness t_0 and width its does not change during rolling, show that the cold work is given by

$$\%CW = \frac{t_0 - t}{t_0} \times 100$$

If you wish to impart cold work of %CW = 15% to a plate with initial thickness $t_0 = 2.0$ mm, what must you reduce the thickness to?

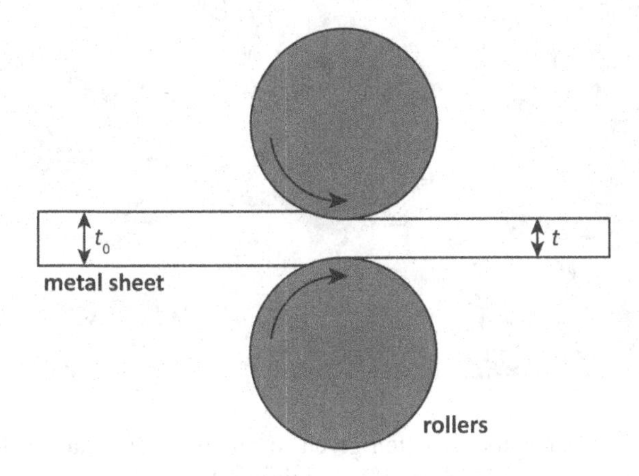

P3.3 What are the indices $(h\ k\ \ell)$ of the plane shown below? ½

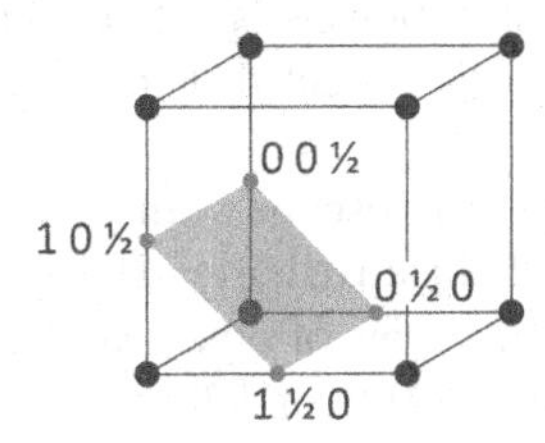

P3.4 Suppose you have the circular indent shown in cross section below. What is the Meyer hardness (HM)? What is the Brinell hardness (HB)?

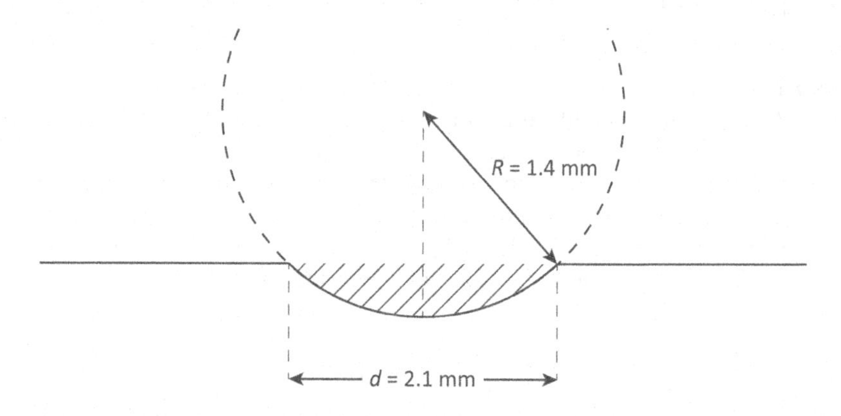

P3.5 Estimate the grain size $D \approx \bar{\ell}$ of the material in the following image. (The scale bar is 4.3 cm on the page.)

P3.6 Dislocation contents are often given in m/cm^3. If you have 1 kg of iron (mass density ρ = 7.874 g/cm^3) containing dislocations with a density of 10^5 m/cm^3, what is the total length of dislocations in the sample?

P3.7 Suppose a material's dislocation density $\rho_\perp$ increased by 25% during deformation. How much do you expect that the material's yield strength will increase?

P3.8 If you have a two-stage microscope with projector/eyepiece focal length f_2 = 50.0 mm and several objective lenses. If the instrument has u_1 = 6.2 mm and u_2 = 44.0 mm, what values of objective-lens focal length f_1 will give you total magnifications of 100× and 200×?

MATLAB Exercises

M3.1 The data file "Exercise M3-1.txt" contains force vs. extension data for a plain-carbon "1045" steel. The diameter of the (circular cylindrical) tensile specimen is d = 12.8 mm and the gauge length is L_0 = 5.08 cm. Plot the engineering stress vs. engineering strain curve for this data set.

M3.2 For the data of **Exercise M3.1**, obtain the modulus, yield strength, ultimate tensile strength, and strain to failure.

M3.3 For the data of **Exercise M3.1**, plot the true stress vs. true strain curve.

NOTES

1. ASTM Standard E384 – 22 "Standard Test Method for Microindentation Hardness of Materials" ASTM International, West Conshohocken, PA, 2003. (www.astm.org.)
2. In addition to the dislocation geometry shown in **Figure 3.8(a)**, called the "edge dislocation," there is another configuration called a "screw dislocation." Materials may contain both edge and screw types, as well as dislocations that are of a "mixed" type with both edge and screw character along a single dislocation line.
3. See ASTM Standard E140 – 12b "Standard Hardness Conversion Tables for Metals Relationship Among Brinell Hardness, Vickers Hardness, Rockwell Hardness, Superficial Hardness, Knoop Hardness, Scleroscope Hardness, and Leeb Hardness."
4. G. F. Vander Voort (Ed.). "Contrast Enhancement and Etching". In: *ASM Handbook, Volume 9: Metallography and Microstructures*. Materials Park, OH, USA: ASM International, Dec. 2004, pp. 294–312.
5. This procedure is described in detail in ASTM Standard E112–13 "Standard Test Methods for Determining Average Grain Size."

4 Making and Modifying Materials

From Raw Material to Finished Product

LEARNING OBJECTIVES

After completing this chapter, you should be able to:

1. List the primary constituents of the various classes of materials.
2. Compute and describe the compositions of various metallic, ceramic, polymeric, semiconductor, and composite systems.
3. Utilize a rule of mixtures to estimate the properties of a material with a single phase or multiple phases.
4. Differentiate between heterogeneous and homogeneous solutions and define the number of phases in each type.
5. Utilize a phase diagram to identify the phases present in a material system under a given set of conditions.
6. Employ phase-diagram tools, like tie lines, the lever rule, and the phase rule, to obtain quantitative information on the nature of a material system.
7. Describe the different diffusion mechanisms and give the microstructural features that influence them.
8. Use Fick's Laws to solve problems in mass transport/diffusion in materials.
9. Describe the nature of phase transformations and estimate their rates/degree of completion using tools like TTT diagrams.
10. Describe the primary materials-processing strategies.

4.1 THE FORMATION AND TRANSFORMATION OF MATERIALS

In **Chapter 1**, we reviewed the principles underlying atomic structure and categorized materials according to their constituents and the forms (or motifs) that those constituents are assembled in. Recall:

- Most metallic alloys are composed predominantly of transition-metal elements and are typically arranged in crystals and polycrystals.
- Ceramics are most frequently oxides or carbides of transition-metal elements. Other ceramics, such as silicates, may or may not contain transition metals like Al or Mg.
- Polymers are based on carbon- and silicon-rich macromolecules that are chemically modified by nonmetallic elements.

DOI: 10.1201/9781003214403-5

- Composites are materials that consist of two or more metallic, ceramic, and polymeric materials consolidated into a common structure.
- Semiconductors are made of elements taken from the region of the periodic table near the "metalloid" region. Frequently, the "plain" or intrinsic semiconductor is treated to have an exceedingly tiny amount of a dopant element to obtain an extrinsic semiconductor.

These definitions include almost all known materials, and every student of engineering likely has some experience working with each of these types.

Consider now: how are different materials to be distinguished within these types? Metal alloys (for example) can be made from many combinations of different elements and subjected to various treatments. Understanding the role of composition and treatment conditions on the structure of metals is the goal of the field of **physical metallurgy**. Similar studies apply in the fields of "physical ceramics" and polymer chemistry and processing.

4.2 SPECIFYING MATERIAL COMPOSITION

The overall *chemical* composition of a system can be described by listing the constituent elements and how much there is of each. The different elements or compounds in the material are the material's **components**, and the amount of each component is specified as a fraction of the total. Recall from **Section 1.1** that we can describe the amount of a component in a multicomponent system in terms of its **mole fraction**:

$$X_1 = \frac{\text{Moles of component 1}}{\text{Total moles of all components}} \tag{4.1}$$

Or, as a percentage

$$X_1 = \frac{\text{Moles of component 1}}{\text{Total moles of all components}} \times 100 \tag{4.1$'$}$$

If you know the mole fraction of each component in the material system, then you can write the overall composition of the material. Note that the sum of all the mole fractions of the various components must be 1:

$$\sum_i X_i = X_1 + X_2 + \cdots + X_n = 1 \quad \text{or} \quad \sum_i X_1 = 100\%$$

The other important way of representing the composition is in terms of **weight fraction** (or equivalently, "mass fraction"). The weight fraction W is the total weight (or mass) of a component divided by the total weight (or mass) of the material. For example,

$$W_1 = \frac{\text{Weight of component 1}}{\text{Total weight of all components}} \tag{4.2}$$

or

$$W_1 = \frac{\text{Weight of component 1}}{\text{Total weight of all components}} \times 100 \tag{4.2$'$}$$

The mole fraction X and the weight fraction W are two ways of representing the amount of a particular component in a material system. Given that the values of both the mole fraction ($0 < X < 1$) and the weight fraction ($0 < W < 1$) appear similar, it is necessary to distinguish them. When specified as percentages, the use of "at%" and "wt%" gives a clear indication of which type of fraction/percentage is applicable. The weight fraction is a useful representation of the composition of a system because it expresses practically how to compose the material, i.e., you must weigh out a certain amount of each component on a scale in the given proportion.

Example 4.1:

Suppose you have a ladle of $1\underline{0}00$ kg of liquid Al and wish to make an alloy with 8.0 wt% Cu. How much Cu do you need to add to the ladle?

SOLUTION

The weight fraction of Cu in the alloy W_{Cu} will be determined by

$$W_{Cu} = \frac{w_{Cu}}{w_{tot}} = \frac{w_{Cu}}{w_{Cu} + w_{Al}}$$

where w_{Al} is the weight of Al and w_{Cu} is the weight of Cu. These weights are related to the masses m_{Al} and m_{Cu} of the respective components: $w_{Al} = m_{Al}g$, etc. This means

$$W_{Cu} = \frac{m_{Cu}g}{m_{Cu}g + m_{Al}g} = \frac{m_{Cu}}{m_{Cu} + m_{Al}}$$

Since the mass of Al is fixed at $m_{Al} = 1\underline{0}00$ kg and the weight fraction of Cu at $W_{Cu} = 0.080$, we solve the above to find

$$m_{Cu} = \frac{m_{Al}W_{Cu}}{1 - W_{Cu}} = \frac{(1\underline{0}00\ \text{kg})(0.080)}{1 - 0.080} = \underline{\mathbf{87\ kg}}$$

If given the composition in terms of mole fractions when the weight fractions are desired, a conversion is required. We may interconvert between these two representations using some data on the component: the atomic weight A. Consider the mole fractions of the components in a two-component system:

$$X_1 = \frac{n_1}{n_1 + n_2} \quad \text{and} \quad X_2 = \frac{n_2}{n_1 + n_2}$$

where the n values are the number of moles of each component. Consider

$$X_1 A_1 g = \frac{n_1 A_1 g}{n_1 + n_2} \quad \text{and} \quad X_2 A_2 g = \frac{n_2 A_2 g}{n_1 + n_2}$$

Then

$$\frac{X_1 A_1 g}{X_1 A_1 g + X_2 A_2 g} = \frac{n_1 A_1 g}{n_1 A_1 g + n_2 A_2 g} = \frac{w_1}{w_1 + w_2} = W_1$$

Thus, we have the general conversion formulas

$$W_1 = \frac{X_1 A_1}{X_1 A_1 + X_2 A_2} \quad \text{and} \quad W_2 = \frac{X_2 A_2}{X_1 A_1 + X_2 A_2} \tag{4.3}$$

with the option to represent as percentages (× 100%) when necessary. Similar to **Equation 4.3**, we can find the corresponding conversions from mole fraction to weight fraction:

$$X_1 = \frac{W_1 A_2}{W_1 A_2 + W_2 A_1} \quad \text{and} \quad X_2 = \frac{W_2 A_1}{W_1 A_2 + W_2 A_1} \tag{4.4}$$

These conversions may be suitably generalized to materials with three or more components. For example, in a three-component system

$$W_1 = \frac{X_1 A_1}{X'} \quad \text{and} \quad W_2 = \frac{X_2 A_2}{X'} \quad \text{and} \quad W_3 = \frac{X_3 A_3}{X'}$$

where $X' = X_1A_1 + X_2A_2 + X_3A_3$ and

$$X_1 = \frac{W_1 A_2 A_3}{W'} \quad \text{and} \quad X_2 = \frac{W_2 A_1 A_3}{W'} \quad \text{and} \quad X_3 = \frac{W_3 A_1 A_2}{W'}$$

where $W' = W_1A_2A_3 + W_2A_1A_3 + W_3A_1A_2$.

Example 4.2:

"Sterling silver" is an alloy mixture of 92.5 wt% Ag and 7.5 wt% Cu. What are the mole fractions of each?

SOLUTION

To find the at% of component 1 (X_{Ag}) from the wt% values of the two components (W_{Ag}, W_{Cu}), we use

$$X_{Ag} = \frac{W_{Ag} A_{Cu}}{W_{Ag} A_{Cu} + W_{Cu} A_{Ag}}$$

The values A_{Ag} = 107.8682 g/mol and A_{Cu} = 63.546 g/mol come from the periodic table. We now compute

$$X_{Ag} = \frac{92.5(63.546\text{ u})}{92.5(63.546\text{ u}) + 7.5(107.8682\text{ u})} = 0.879 = \underline{\mathbf{87.9\ at\%}}$$

To find X_{Cu}, we can employ the same formula as above, transposing "Ag" and "Cu" in the various terms. However, since this is a binary alloy, we can instead apply the fact that $X_{Cu} = 100\% - X_{Ag}$ to obtain

$$X_{Cu} = 100.0\% - 87.9\% = \underline{\mathbf{12.1\ at\%}}$$

Consider now how we might represent the composition of a metal alloy in a compact way. Let's say the alloy has elemental components 1, 2, …, n with mole fractions X_1, X_2, …, X_n. When we write out the full description of an alloy, we might use a format like

$$[\text{component } 1]_{X_1}[\text{component } 2]_{X_2} \ldots [\text{component } n]_{X_n} \tag{4.5}$$

This format gives us a simple and complete way of writing the alloy composition. An expression like "$Ag_{87.9}Cu_{12.1}$" looks something like a chemical formula. In an alloy, however, there aren't typically fixed ratios between components as there are in chemical compounds. An exception to this rule occurs in the case of *intermetallic compounds* or **intermetallics**. Intermetallics are materials that are highly ordered so that the ratios between the amounts of each constituent are more or less fixed. For example, the intermetallic $CuAl_2$ in the Al-Cu alloy system is the stable association of two Al atoms and one Cu atom. This alloy material therefore has an overall composition $Al_{66.7}Cu_{33.3}$ (though some variation is possible). Intermetallics are frequently treated as components themselves (see **Figure 4.4**).

Example 4.3:

Consider the common stainless-steel alloy called "SAE* 304". What is the full alloy description in terms of atomic percentages? What is the alloy description in weight percentages?

Component	Mole Fraction [%]	Weight Fraction [%]
Fe	68.7	69.7
Cr	18.5	17.4
Ni	10.0	10.7
Mn	2.0	2.0
C	0.8	0.2

* SAE = Society of Automotive Engineers.

SOLUTION

A suitable representation of this alloy in at% would be

$$Fe_{68.7}Cr_{18.5}Ni_{10}Mn_2C_{0.8}$$

Sometimes when there is no chance for confusion, the component with the largest mole fraction is left unspecified:

$$Cr_{18.5}Ni_{10}Mn_2C_{0.8}Fe$$

In this case, the balance of the alloy $100\% - X_{Cr} - X_{Ni} - X_{Mn} - X_C = 68.7\% = X_{Fe}$. For the composition in terms of weight, we write

$$69.7Fe17.4Cr10.7Ni2.0Mn0.2C\ (wt\%)$$

In the case of ceramics, the bonding requirements fix the composition in a way similar to intermetallics. For example, in a pure oxide ceramic-like alumina (Al_2O_3), the ratio of moles of Al to moles of O is fixed at 2:3, and nothing more needs to be said about the composition. The carbide ceramic Fe_3C has Fe and C in the ratio 3:1. However, in a mixed ceramic, the two ceramic components can be mixed in different fractions. (See **Example 4.4**.) Such multicomponent ceramics are important as engineering materials, not to mention in geology. It is also possible that different cations can be substituted into a ceramic structure, replacing the original cations in a defect like a substitution. For instance, replacing $2x$ K^+ cations with x Ca^{++} cations in the ceramic KCl gives a formula $K_{1-2x}Ca_xCl$, where x is in some convenient unit like ppm.

Example 4.4:

Suppose you have an alumina-silica "glass ceramic" that is 81% Al_2O_3 chemical units and 19% SiO_2 units. How might you write the composition?

SOLUTION

The different chemical units can be separated using a simple scheme involving parentheses, like

$$(Al_2O_3)_{79}(SiO_2)_{19}$$

The composition of polymer systems is determined by two considerations:

1. the chemical makeup of the individual polymer chemical subunits, called **monomers** (see **Table 4.1**), and
2. the number of monomer units in the macromolecule.

For example, from **Table 4.1**, the chemical formula of the polymer poly(ethylene) (PE) is "$-(C_2H_4)_n-$". The monomer unit is bonded via covalent ("–") bonds to *two* other neighboring units. Because the monomer can bond with two other monomers,

TABLE 4.1
Sample Monomers and Their Chemistry

Material	Other Names	Monomer Unit	Chemical Formula
Poly(ethylene)	PE, Poly(methylene)	H H —C—C— H H	$-(C_2H_4)_n-$
Poly(vinyl chloride)	PVC, vinyl	H Cl —C—C— H H	$-(C_2H_3Cl)_n-$
Poly(propylene)	PP	H CH_3 —C—C— H H	$-(C_3H_6)_n-$
Poly-(tetrafluoroethylene)	PTFE, Teflon	F F —C—C— F F	$-(C_2F_4)_n-$
Polycarbonate[a]	PC	O ‖ CH_3 C— —O—⬡—C—⬡—O CH_3	$-(C_{16}H_{14}O_3)_n-$
Polystyrene[b]	PS	⬡ H —C—C— H H	$-(C_8H_8)_n-$
Poly(methyl methacrylate)	PMMA, acrylic, Plexiglas	OH H C=O —C—C— H CH_3	$-(C_4H_6O_2)_n-$

[a] The "–⬡–" symbol is related to the benzene chemical unit (a hybridized carbon ring); it has the formula "$-C_6H_4-$". [b]The "–⬡" symbol is a "phenyl" group based on the benzene chemical unit; it has the formula "$-C_6H_5$".

the monomers can form the chainlike structure of the macromolecule motif. The number n represents the number of monomer units that make up the polymer macromolecule and is called the **degree of polymerization**. The molecular weight M of the macromolecule is equal to the molecular weight of the monomer A multiplied by the degree of polymerization:

$$M = A \times n \tag{4.6}$$

The units of M are typically given in Daltons: $[M] = [\text{mass}] = \text{u} = \text{Da}$. It is important to recognize that polymer materials are not composed of macromolecules of a single value of M or n, but rather a distribution of values. This being the case, the values typically reported for the polymer molecular weights are the computed means of the distribution.

Example 4.5:

What is the degree of polymerization of a polycarbonate macromolecule with $M = 550{,}000$ Da?

SOLUTION

From **Table 4.1**, the monomer formula is $-(C_{16}H_{14}O_3)_n-$, so we compute the monomer weight (using atomic-weight data from the periodic table) as

$$A = 16A_C + 14A_H + 3A_O = 16(12.011) + 14(1.008) + 3(15.999) = 254.29 \text{ Da}$$

For a total molecule weight of 550,000 Da, we now estimate

$$n = \frac{M}{A} = \frac{550.000}{254.29} = \underline{\mathbf{2200}}$$

In the case of **copolymers**, polymers whose macromolecules incorporate more than one type of monomer, we approach the calculation of M in a similar way. For large macromolecules, the different monomers 1, 2, …, n will be present in mole fractions X_1, X_2, …, X_n. Within the polymer, we can compute an effective monomer weight $\bar{A}$ of

$$\bar{A} = A_1X_1 + A_2X_2 + \cdots + A_nX_n$$

That is, the effective monomer weight is the sum of the individual monomer masses "weighted" according to their frequency. The molecular weight of a typical copolymer macromolecule is then

$$M = \bar{A} \times n \tag{4.6'}$$

(Copolymers are important in many industrial and domestic applications.) Furthermore, bulk polymer materials are composed of many macromolecules, and these molecules do not have uniform sizes. To this end, we define the number-average molecular weight:

$$\bar{M}_n = \frac{\sum_i N_i M_i}{\sum_i N_i} \tag{4.7}$$

where N_i is the number of molecules of mass M_i. Since $\sum_i N_i M_i$ is the total mass m of the specimen and $N = \sum_i N_i$ is the total number of molecules, we obtain

$$\bar{M}_n = \frac{m}{N} \tag{4.7'}$$

as equivalent.

Example 4.6:

Ethylene-propylene rubber (EPR) is a type of synthetic elastomer that is a copolymer of poly(ethylene) and poly(propylene) monomer units. What is the molecular weight M of an EPR sample with $n = 1600$ if the PE and PP monomers are present in a 3:1 ratio?

SOLUTION

From **Table 4.1**, the monomer formula for PE is $-(C_2H_4)_n-$ and the monomer formula for PP is $-(C_3H_6)_n-$. These monomers have weights $A_{PE} = 28.054$ Da and $A_{PP} = 42.081$ Da. We can compute the effective monomer weight A using the fact that $X_{PE} = 75\%$ of the monomers and $X_{PP} = 25\%$ (i.e., 3 PE:1 PP). This gives

$$\bar{A} = X_{PE}A_{PE} + X_{PP}A_{PP} = 0.75(28.054) + 0.25(42.081) = 31.561 \text{ Da}$$

Using the degree of polymerization, we then obtain the "typical" molecular weight

$$M = \bar{A} \times n = (31.561 \text{ Da}) \times 1600 = \underline{\mathbf{5.05 \times 10^4 \text{ Da}}}$$

Semiconductor materials are often doped in order to modify their electrical properties. The doping process involves the addition of impurity elements that "bring along" additional charge carriers into the semiconductor material. Dopants come in two types: ***n*-type dopants** introduce additional negative charge carriers into a material with otherwise balanced "base" number of charges. For example, in a semiconducting Si material (an "intrinsic" semiconductor that is neither conducting nor insulating), the Si atoms form four covalent bonds with their neighbors. Since the Si atoms have four outer-shell "valence" electrons, these valence electrons are sufficient to balance the four-bond requirement. If a different species like P (with *five* valence electrons) is introduced in place of a Si atom, then the excess electron becomes a negative charge carrier in the material. In ***p*-type doping**, positive charge carriers come in with the replacement of Si atoms by atoms of valence 3. (See **Topic 3** for more information on doping processes and influence.) Typically, dopants are introduced at a level of parts-per-billion or parts-per-million. We could represent a material of this composition by writing something like $Si_{100-x}P_x$, where $x \sim 10$ ppm. In highly doped materials, x might be as high as 1 in 10,000.

Example 4.7:

A sample of Si has been doped with Al to the point that the concentration of Al is $X_{Al} = 1.0 \times 10^{-7}$ at%. How many kilograms of Al are there in 1.0 m³ of this doped Si?

SOLUTION

Since we are asked to find the weight/mass of Al in a piece of material, the wt% of P will be useful. Also note that 1.0×10^{-7} at% is the same thing as 1 part-per-billion. So

$$W_{Al} = \frac{n_{Al}A_{Al}}{n_{Al}A_{Al} + n_{Si}A_{Si}} = \frac{1(26.982\ \text{u})}{1(26.982\ \text{u}) + 999{,}999{,}999(28.085\ \text{u})}$$

$$= 9.6 \times 10^{-10} = 9.6 \times 10^{-8}\ \text{wt\%}$$

To within a small error, the density ρ of the doped Si is $\rho \approx \rho_{Si} = 2.3296$ g/cm³. The total mass m of 1.0 m³ = 1.0×10^6 cm³ of the doped material is around

$$m = \rho V \approx (1.0 \times 10^6\ \text{cm}^3)(2.3296\ \text{g/cm}^3) = 2.3 \times 10^6\ \text{g}$$

Hence, the portion of the mass m that is Al (= m_{Al}) is

$$m_{Al} = m \times W_{Al} = (2.3296 \times 10^6)(9.6 \times 10^{-10}) = 0.0022\ \text{g} = \underline{\mathbf{2.2\ mg}}$$

4.3 TYPES OF SOLUTIONS AND PHASE DIAGRAMS

The overall composition of a system tells us what the components are and how much there is of each in the system. This information is useful for formulating materials from their essential ingredients, but by itself it reveals little about how the components are arranged in the resulting material's structure. The components of the material, when blended at the microscale, together constitute a **mixture**. You are already familiar with *liquid* mixtures from chemistry class; two or more chemical components are stirred together in a beaker, producing a substance that can have properties distinct from any of the components individually. For example, adding some salt (NaCl) into water (H_2O) produces a two-component mixture ("brine") that has its own boiling point, density, etc.

There are essentially two ways that the components in the mixture can combine, as illustrated in **Figure 4.1**. The mixture can be homogeneous or heterogeneous.

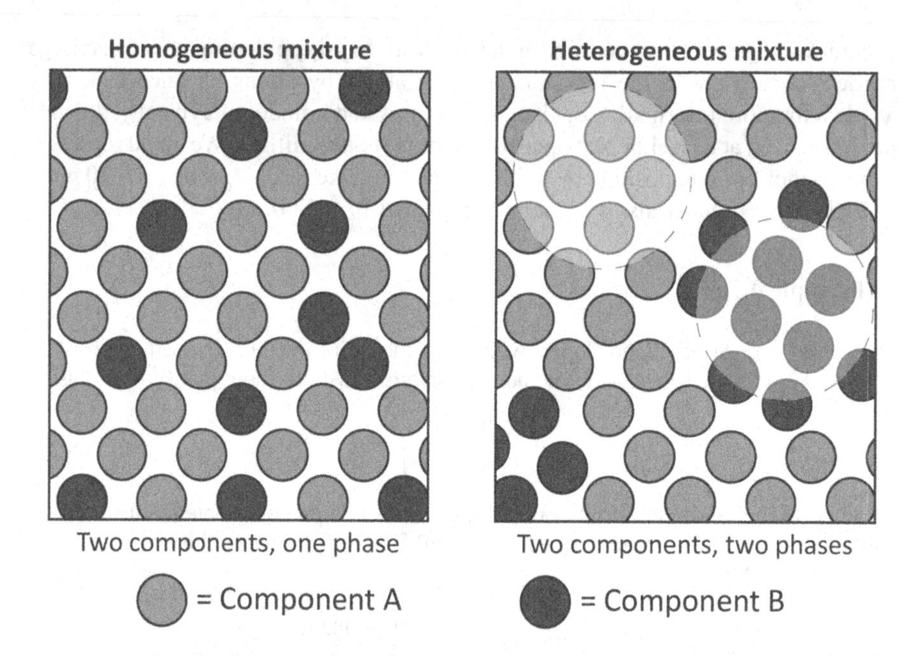

FIGURE 4.1 Comparison of one-phase (homogeneous) and two-phase (heterogeneous) mixtures. Both mixtures incorporate two components and have similar overall compositions, but the components have different distributions. In the homogeneous case, the distribution is uniform. In the heterogeneous case, the different components are concentrated in different regions.

In the homogeneous case, the components are so evenly distributed throughout the mixture that the mixture's properties are *uniform*. In the brine example, small amounts of salt dissolve completely in the water, and it is impossible to distinguish where the salt component ends and the water component begins. Any sample of the mixture will have the same composition and properties as any other sample. In a heterogeneous mixture, the components tend to segregate into distinct regions with different compositions that possess distinct and uniform properties. These regions are frequently themselves homogeneous mixtures, and this means that we must now distinguish between the composition of the mixture overall and the compositions of the individual regions.

APPLICATION NOTE – THE RULE(S) OF MIXTURES

When materials are composed of multiple components, the properties of the material will be a "blend" of the properties of the two components. The rule has the form of a mean or average of the properties of the components, but the mean is weighted by the amount of each component in the mixture. For example, a mixture of A and B that is *mostly* A will have properties that are closer to those of pure A than pure B. The weighting factors in the rule of mixtures are the *volume fractions* ϕ_i of the components.

Suppose that there is a mixture of n different materials with properties $\{P_1, P_2, \ldots, P_n\}$. The overall property E of the mixture will be

$$P = \sum_i P_i \phi_i = P_1\phi_1 + P_2\phi_2 + \cdots + P_n\phi_n$$

Note that, as usual, $\phi_1 + \phi_2 + \ldots \phi_n = 1$. For a two-component mixture, we have

$$P = P_1\phi_1 + P_2\phi_2$$

Also consider that the volume fraction ϕ_1 is related to the masses of the constituents by

$$\phi_1 = \frac{V_1}{V_1 + V_2} = \frac{m_1/\rho_1}{m_1/\rho_1 + m_2/\rho_2}$$

Since the weight fraction $W_1 = m_1/(m_1 + m_2)$, we obtain

$$\phi_1 = \frac{W_1/\rho_1}{W_1/\rho_1 + W_2/\rho_2}$$

and

$$\phi_2 = \frac{W_2/\rho_2}{W_1/\rho_1 + W_2/\rho_2}$$

For example, the density ρ of a mixture of two components "A" and "B" is well represented by

$$\rho = \rho_A\phi_A + \rho_B\phi_B$$

or, equivalently

$$\rho = \frac{1}{W_A/\rho_A + W_B/\rho_B}$$

These relationships emphasize the necessity of having materials data like density, atomic mass, etc. handy when doing practical calculations.

The rule of mixtures given above makes a great deal of sense, but upon further consideration, it requires that the properties E_n combine in a particular way. This is not the only possible combination rule for mixtures. The "inverse rule of mixtures" is

$$\frac{1}{P} = \sum_i \frac{\phi_i}{P_i} = \frac{\phi_1}{P_1} + \frac{\phi_2}{P_2} + \cdots + \frac{\phi_n}{P_n}$$

and represents a different way of combining properties. (Think about the two different rules for combining resistances: one combination rule for series resistances and another for parallel resistances.) For two components, the inverse rule reduces to:

$$P = \frac{P_1 P_2}{\phi_1 P_2 + \phi_1 P_2}$$

Individually, these rules of mixtures are useful in estimating the properties of alloys and composites. These two rules together also establish the *bounds* on the possible values of the combined property P. It may be the case that the combination rule is more complicated than either the rule of mixtures or the inverse rule of mixtures, and the value of P will lie somewhere in between. The possible values for P therefore lie in a region with the two rules as its bounds. This situation is illustrated in the diagram below.

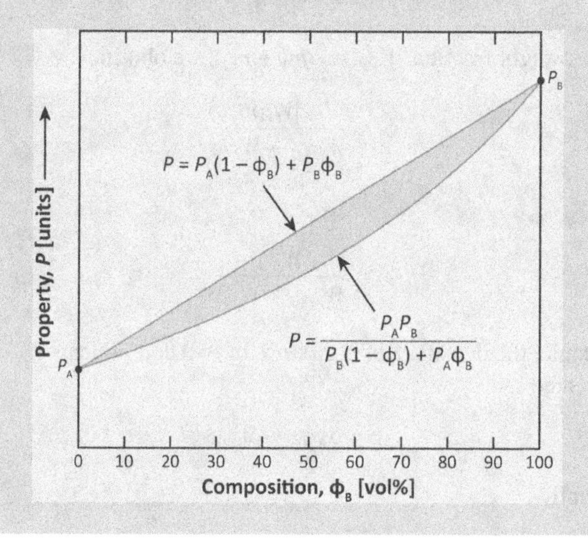

The idea of a chemical mixture applies to the description of materials in their solid state. Recall from **Chapter 1** that an inspection of a material's microstructure reveals the presence of one or more phases. These phases are the "distinct homogeneous regions" of the mixtures described above. When discussing the phases present in a material, we say that a phase is a portion of the material that

1. has a defined boundary (the "phase boundary"),
2. is composed of one or more components, and
3. possesses uniform properties throughout.

The various phase regions occupy a certain amount of volume, and all the phases collectively make up the bulk material. The arrangement of the phases in the bulk material is called its **morphology**.

The phases that are present in a material are a reflection of its thermodynamic **state**. Following thermodynamic principles, the state of a system is established by the values of particular variables:

- the system's volume V,
- the system temperature T,
- the pressure p that the system is subject to, and
- the composition of the system in terms of the mole fractions $X_1, \ldots, X_n$.

This connection between system states and system variables follows from the principles of thermodynamics. For example, a single-component gas exists in a state that is captured by the temperature, pressure, volume, and composition (i.e., number of moles of gas n). As it turns out, only three of these variables need to be known to account for the state completely since they are related by a simple formula: $pV = nRT$. In systems that are condensed (i.e., solid or liquid systems), the volume V is more or less constant, and the overall state of the system is established by a combination of T, p, and X_i.

The system in the above example is in the gaseous phase only for certain values of the variables. At different combinations of temperature and pressure, say, the system may depart from its gaseous identity and adopt a different configuration. The possible configurations (phase makeup) of the system are associated with different states and can be summarized using a **phase diagram**. There are many different types of phase diagrams; some examples of single-component-system diagrams are provided in **Figure 4.2**. You are familiar with the single-component diagram shown schematically in **Figure 4.2(a)**. This phase diagram has axes for the variables p and T, and the coordinate locations (T, p) correspond to a particular state of the system. Depending on the state (T, p), different phases or combinations of phases will be observed. These phase fields are labeled "solid", "liquid", and "gas" according to the phase that the state corresponds to. However, for certain states – the states lying along the boundaries between the phase fields – two phases are observed. Three phases can be observed at the special "triple point", where all three phase fields touch. Changing a system's state may produce a change in the phase present. This change in phase is called a **phase transformation**. The different types of phase transformations that can occur are indicated. For example, lowering the temperature (or raising the pressure) of a liquid below the melting temperature T_m freezes it. The transformation can be represented using an expression similar to that used to represent transformations of chemical units:

$$\mathrm{L} \xleftrightarrow[T=T_\mathrm{m}]{} \mathrm{S} \tag{4.8}$$

This transformation process is more properly called **solidification** and is important in many kinds of materials processing. The transformation is reversible; by raising the temperature above T_m, the material melts.

An example of a phase diagram of a metallic system is in **Figure 4.2(b)**. The metal is iron, and the various phases and boundaries of **Figure 4.2(a)** are present.

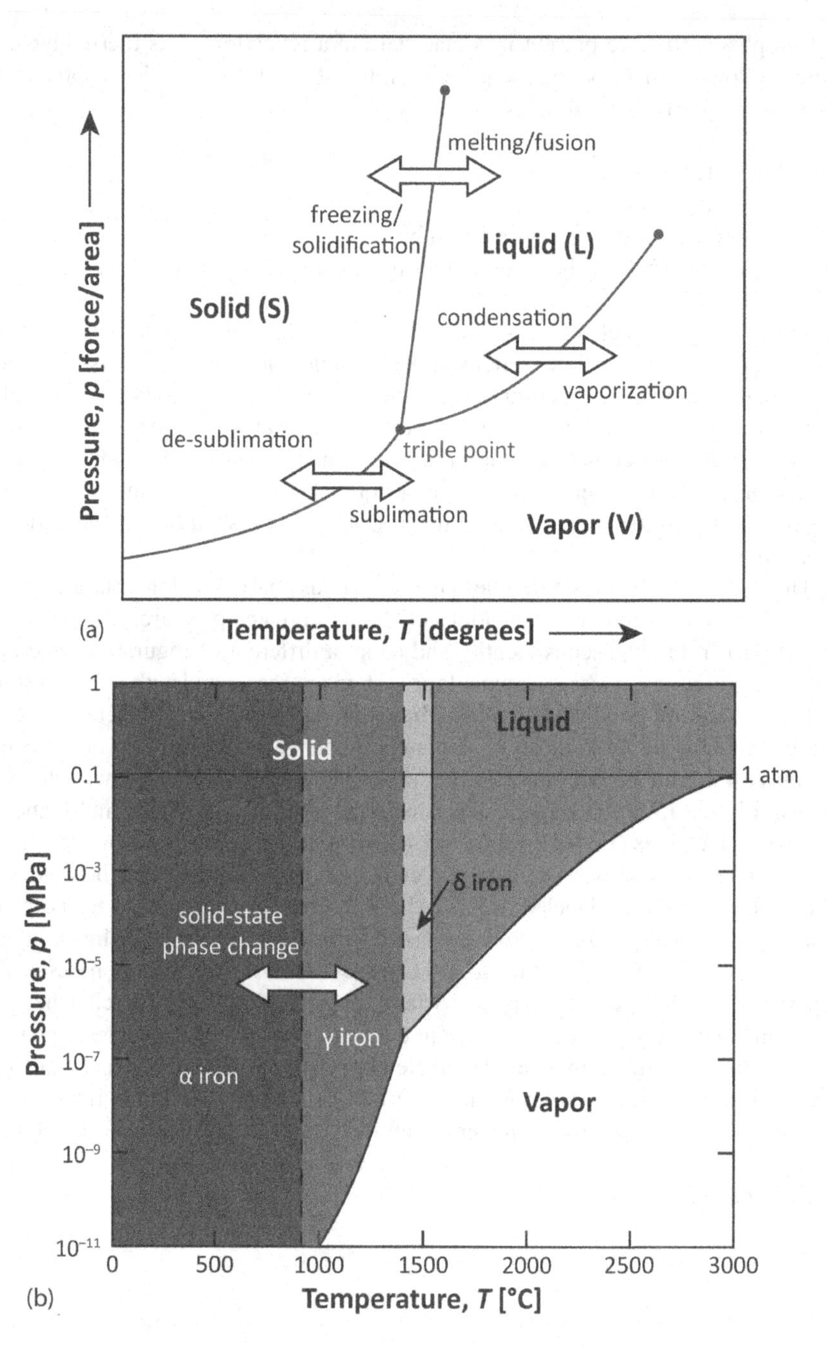

FIGURE 4.2 Sample phase diagrams. (a) is a schematic, single-component diagram, showing three different phases: solid, liquid, and gas. The boundaries between the phases correspond to phase transformations (except at certain *critical points*). (b) is the phase diagram for Fe. Solid, liquid, and gas phases are all present, but Fe can also undergo some solid-state transformations between solids of various structures: α, γ, and δ.

The ranges of temperatures and pressures required depict the useful extent of system behavior much greater than that used in the phase diagram of, say, water (10^{-2} atm $< p < 200$ atm, $-50°C < T < 400°C$). For the most part, metallurgists are only concerned about pressures $p \approx 1$ atm = 101,000 Pa, since most metals are processed or treated under such conditions. Therefore, the most important transformations under consideration for iron and steel metallurgy are solidification/melting and a number of **solid-state transformations** (indicated by the dashed lines in the "solid" phase field). These transformations are associated with changes in the crystal structure of the metal. The phase called "α iron", also known as "ferrite", has a BCC crystal structure. As the temperature rises and exceeds a temperature $T > 912°C$, a change in phase occurs throughout the solid iron. The emergent phase has the FCC structure and is called "γ iron" or "austenite". We write

$$\alpha \underset{912°C}{\longleftrightarrow} \gamma$$

At higher temperatures ($T > 1394°C$), γ iron transforms into the BCC phase "δ iron":

$$\gamma \underset{1394°C}{\longleftrightarrow} \delta$$

This capability of the elemental solid to adopt different structures under different conditions is called **allotropy**. Fe, along with other elements like C, Sn, and Ti, are *allotropes*. Also note that the solid phases present in a phase diagram, because there can potentially be many of them, are labeled using characters from the Greek alphabet: α, β, γ, δ, etc.

The examples in **Figure 4.2** are for single-component systems, i.e., the mole or weight fraction is fixed at 100%. This means that the state of the system is determined only by T and p, and these variables alone (when placed on the axes of a diagram) give us a complete picture of the possible states of the system. More useful to materials scientists and engineers are systems that involve multiple components. In these systems, the mole fractions can all vary over the range of 0–100%. In our study of materials, we will restrict ourselves to systems with two components. Consider a binary system of components A and B. When the mole fraction of B ($= X_B$) is fixed, then the mole fraction of A is determined automatically, as $X_A = 1 - X_B$. In addition, if we fix the pressure of the systems at $p \approx 1$ atm, then the system state can be captured by two variables: X_B (or W_B) and T.

Figure 4.3 is the phase diagram of a two-component system in terms of the variables X_B and T. This **binary phase diagram** is a "map" of the various system states in the same way as the one-component diagrams in **Figure 4.2**. The two components are Au and Ag, and they combine to make an alloy called *electrum*. The value of the composition variable W_{Au} is plotted along the horizontal axis. From **Figure 4.3(a)**, on the left, a composition $W_{Au} = 0\%$ is equivalent to pure Ag ($W_{Ag} = 100\% - W_{Au} = 100\%$). Below the **melting temperature** of Ag (= 962°C), the system is in the solid state; above 962°C, the system is liquid Ag. The same applies to the 100% Au composition on the right, but with a melting temperature of 1064°C. For intermediate

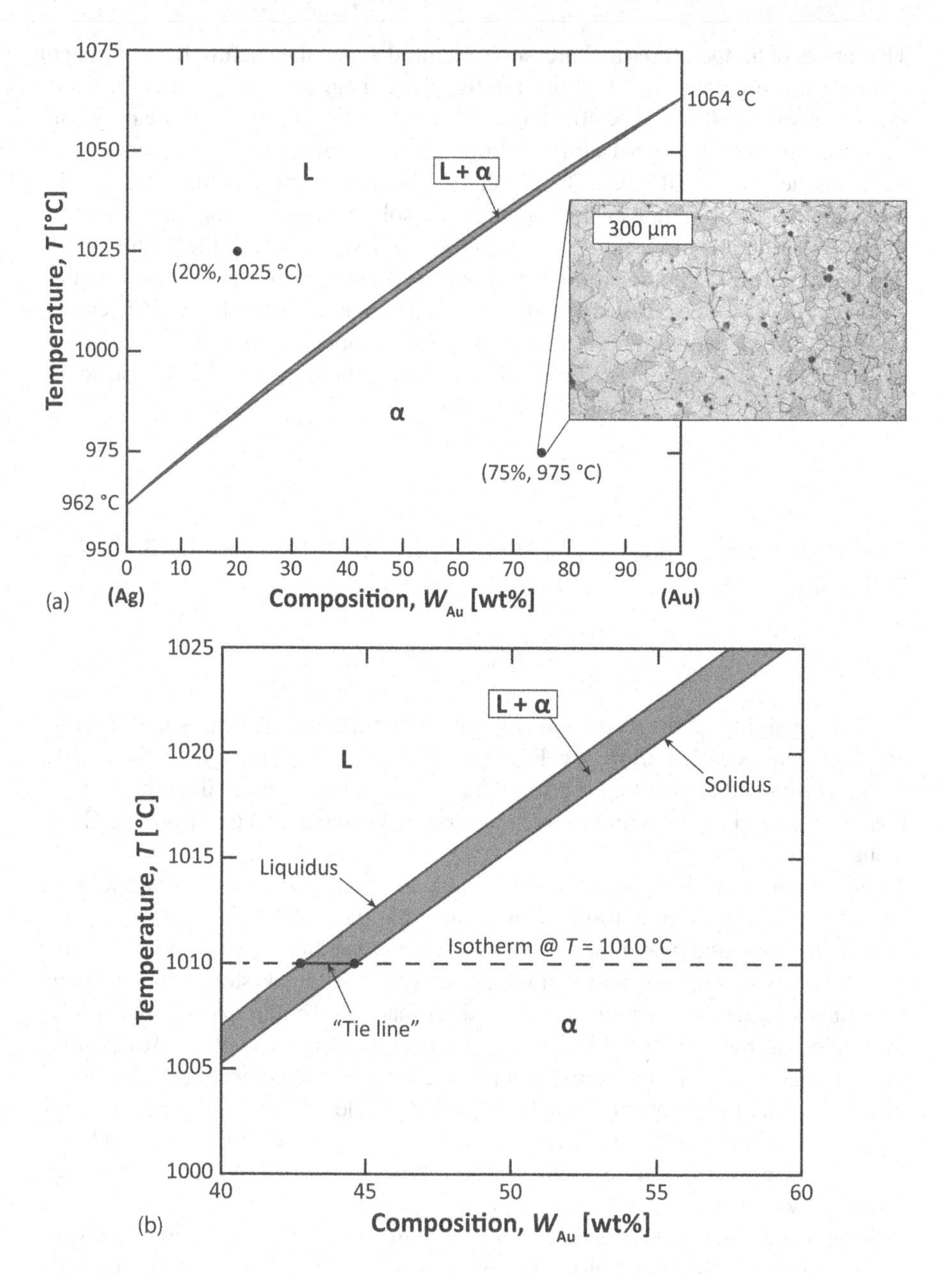

FIGURE 4.3 The phase diagram for electrum, an alloy with components Au and Ag. The phase diagram in (a) indicates the ranges of system composition (X_{Au} = 0 to 1). A liquid mixture "L" is stable at high temperatures, and a solid mixture "α" is stable at low temperatures. The solid α, shown inset, is a single-phase material consisting of numerous grains. (Adapted with permission from van der Lingen et al.[1]) For a narrow range of compositions and temperatures (i.e., the freezing range or "mushy zone"), the L and α phases are equally stable. (b) shows a close view of the diagram around X_{Au} = 50%. The upper limit of stability for the α phase is the liquidus, and the lower limit of stability for the L phase is the solidus.

compositions, the system may be a homogeneous solid solution α, a homogeneous liquid solution L, or *both*. In the separate L and α phase fields, the composition of the system is the same as the overall composition. For example, at a temperature of 1025°C and overall alloy composition $W_{Au} = 20\%$, the state of the system is a single phase (a liquid) with composition W_{Au}. At a temperature of 975°C and overall alloy composition $W_{Au} = 75\%$, the state of the system is a single phase (a solid called α) with composition W_{Au} (depicted inset).

Because this system has two components and those components have different melting temperatures, it is possible for solid α and liquid L to coexist across a *range* of values of T and W_{Au}. In a single-component alloy, it is only possible to have two phases coexisting at a boundary; in a binary system, two phases can coexist in a field. This two-phase field is sometimes called the **freezing range** of the alloy. **Figure 4.3(b)** shows a close-up of the freezing range for compositions near $W_{Au} =$ 50%. The upper boundary of the freezing range is called the **liquidus**. Above the liquidus, the system is entirely liquid. Below the **solidus**, the system is entirely solid. In between the solidus and the liquidus, there are two phases that coexist side-by-side. For this to be the case, the solid phase contains more of the higher melting temperature component (Au), and the liquid phase is more enriched in Ag. That is, the α phase and the L phase have different compositions than each other, and each has different compositions than the overall system. The position and shape of the solidus and liquidus lines reflect this difference, as illustrated in **Example 4.8**. A constant-T trace across the diagram is called an *isotherm*. Where this trace crosses the two-phase field, you have a geometrical construction called a **tie line** or "isotherm". The two endpoints of the tie line are located where the isotherm crosses the liquidus and solidus.

Example 4.8:

Suppose you have an Ag-Au alloy system with an overall composition of $Ag_{60}Au_{40}$. What is the liquidus temperature? What is the solidus temperature? Suppose further that the alloy is at a temperature $T = 1022°C$. What is the composition of the liquid phase? What is the composition of the solid phase?

SOLUTION

To make use of the phase diagram of **Figure 4.3(b)**, we require the composition in wt% (this is most typical for phase diagrams). So we compute

$$W_{Ag} = \frac{X_{Ag}A_{Ag}}{X_{Ag}A_{Ag} + X_{Au}A_{Au}} = 0.45 = 45 \text{ wt\%}$$

It follows that $W_{Au} = 1 - W_{Ag} = 55$ wt%. This is the overall system composition and is marked W_0, below. Using this overall composition, we can easily identify the liquidus temperature $\underline{\boldsymbol{T_L = 1022.5°C}}$ and the solidus temperature $\underline{\boldsymbol{T_\alpha = 1020.5°C}}$ from the respective phase boundaries.

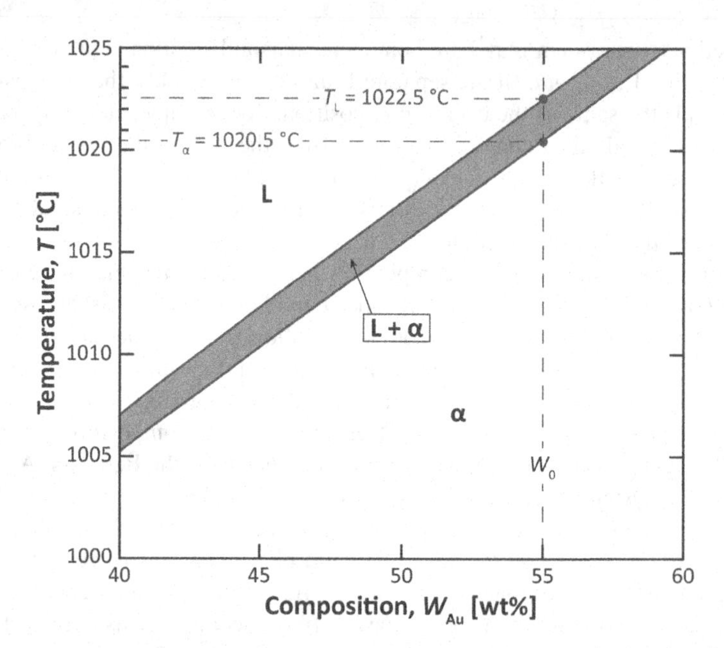

The composition of the liquid phase W_L and the composition of the solid phase W_α can also be determined using the diagram in cases where the system temperature is known. At a temperature of T = 1022°C, the horizontal isotherm is as indicated below. The corresponding tie line intersects the liquidus at one location and the solidus at another location. These intersections indicate the compositions of the phases; the liquidus intersection is at a composition of $\mathbf{W_L = 54.5\%}$ and the solidus intersection is at a composition of $\mathbf{W_\alpha = 56.5\%}$.

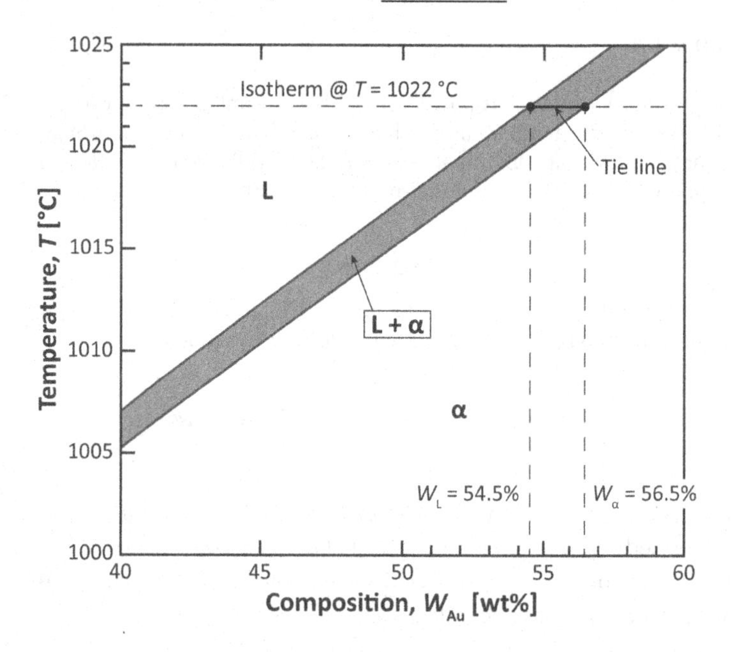

The diagram of the Ag-Au system depicted in **Figure 4.3** is just one of many possible forms a diagram can take. The key feature of this system that gives it its form is the complete solid-state solubility of the two components across all compositions and temperatures up to the liquidus. This is because the silver atoms and the gold atoms share a common crystal structure (FCC) and have similar sizes. When the ability to form a solid solution is limited, then multiple solid phases will coexist in the same material. **Figure 4.4** shows the Al-$CuAl_2$ binary system. The phase diagram is in **Figure 4.4(a)**. Pure Al (with melting temperature $T_m = 660°C$) is along the axis at left, and the κ phase is a homogeneous solid solution of Cu atoms in Al. As the Cu content increases, the solubility limit of Cu in Al is exceeded. This limit depends on the temperature and is called the **solvus**. In passing from the homogeneous phase field κ to the heterogeneous κ + θ phase field, the θ phase **precipitates** from the supersaturated κ:

$$\kappa_{sup} \leftrightarrow \kappa + \theta$$

Here, the θ phase is an intermetallic with nominal composition $CuAl_2$ and a limited homogeneity range. The θ precipitates emerge as the dark flecks embedded in the brighter κ "parent" phase, as shown in **Figure 4.4(a)**. This alloy system is important because the presence of $CuAl_2$ precipitates in a predominantly Al matrix has a strengthening effect that is required in aircraft-component design.

The phase diagram of **Figure 4.4** also has another feature worth discussing: the possibility of **eutectic solidification** in this alloy system. During eutectic solidification, cooling below the **eutectic temperature** T_e produces a transformation

$$L \xleftrightarrow[550\ °C]{} \kappa + \theta$$

That is, the homogeneous liquid phase transforms into a heterogeneous solid. The phases θ and κ solidify *simultaneously* to produce the two-phase mixture indicated below T_e. This co-solidification behavior produces a structure where the two solid phases are arranged in closely spaced stacks. This kind of intermingled *lamellar* structure is shown in **Figure 4.4(c)**. This behavior is likely unfamiliar to you but is a possibility in a two-component system. Also, note the presence of freezing ranges in the form of κ + L and θ + L phase fields. The solid phases that develop at a temperature greater than T_e are called "proeutectic". For example, at system compositions $W_{Cu} < 33$ wt%, we observe

$$\kappa_{pro} + L \xleftrightarrow[550\ °C]{} \kappa_{pro} + \kappa + \theta$$

upon cooling below T_e. Meaning: the proeutectic κ is "carried through" and is not involved in the eutectic solidification process. (This means that the solid κ in the κ + θ field originates from two distinct transformations.) For system compositions of $W_{Cu} = 33$ wt%, called the *eutectic composition*, no proeutectic solids are formed. Finally, in the two-phase field, a tie line can be used to determine the compositions of the θ and κ phases.

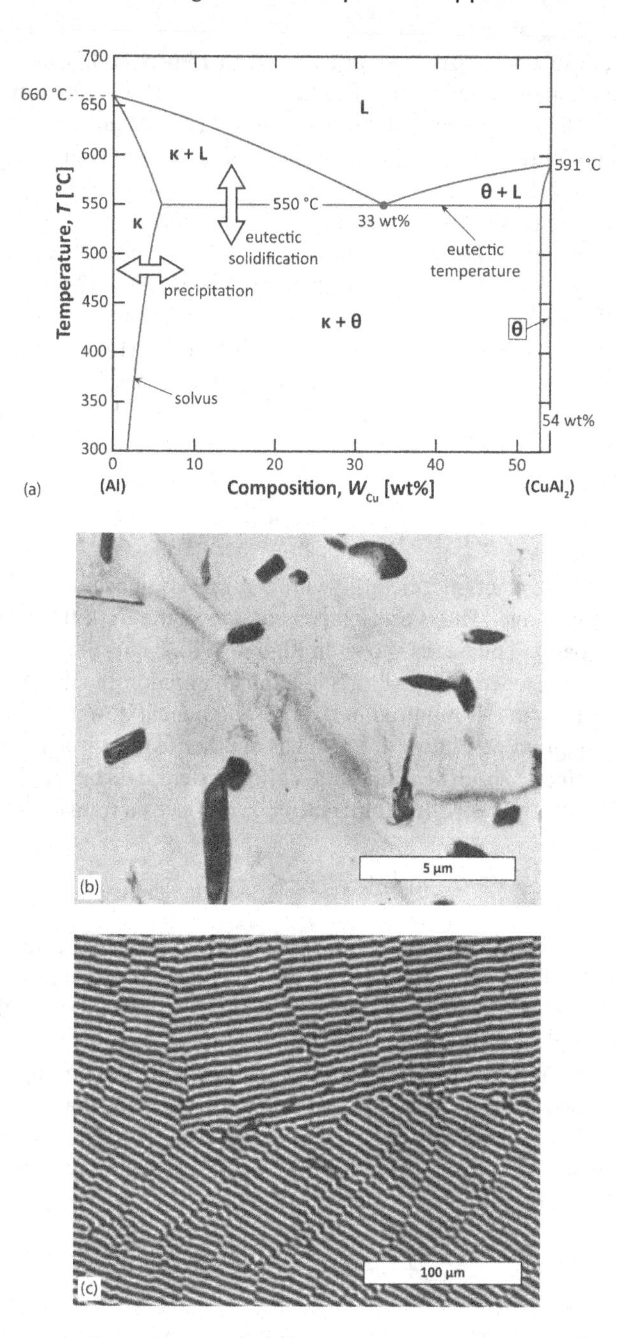

FIGURE 4.4 Features of the Al-Cu binary system. (a) is the Al-Cu phase diagram from 0 to 54 wt% Cu. The right axis corresponds to a system of the pure $CuAl_2$ intermetallic. This binary phase diagram has a precipitation transformation at low Cu content and a eutectic solidification transformation at $T_e = 550°C$. $CuAl_2$ precipitates are shown in (b), (Adapted with permission from Porter, Easterling, and Sherif.[2]) and the lamellar eutectic phase mixture is shown in (c). (Adapted with permission from Stefanescu and Ruxanda.[3])

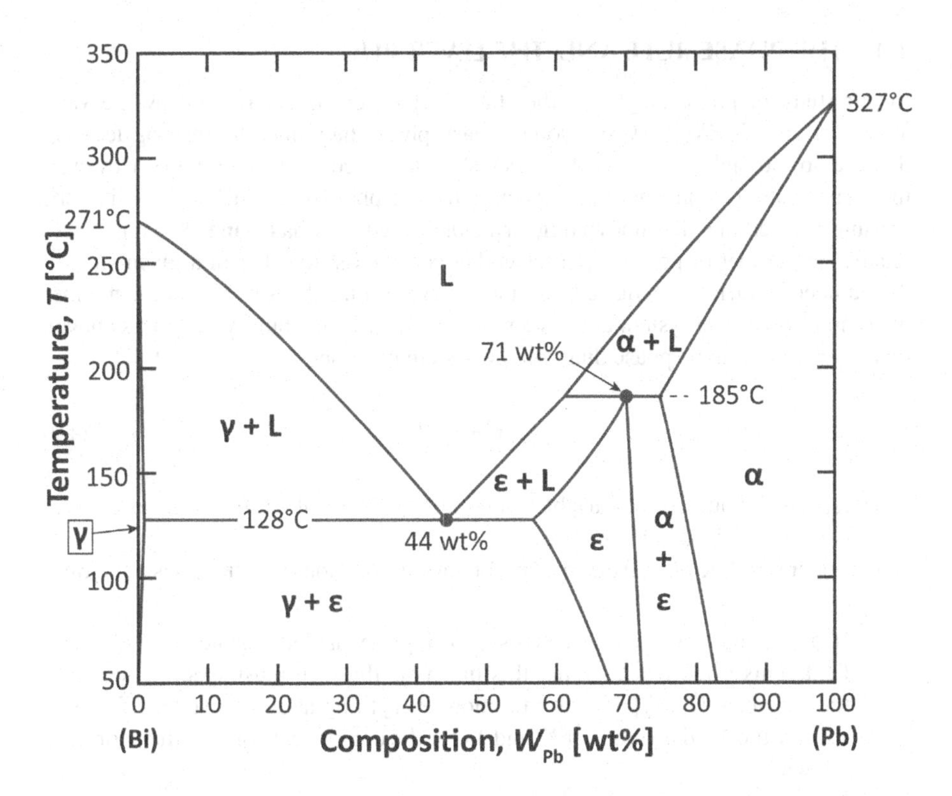

FIGURE 4.5 The Pb-Bi binary system. A peritectic transformation with T_p = 185°C is present, alongside a eutectic transformation with T_e = 128°C.

Yet more sophisticated binary phase diagram arrangements are possible. **Figure 4.5** shows the metallic Pb-Bi system. There are three distinct solid phases present in this system:

1. The α phase (homogeneous mixture of Bi in Pb)
2. The ε phase (HCP intermetallic with homogeneity range)
3. The γ phase (Bi with slight Pb solubility)

The system has an eutectic transformation at 128°C with an eutectic point at W_{Pb} = 44 wt%. There is also a **peritectic transformation** at T_p = 185°C:

$$\alpha + L \xleftrightarrow[185\ ^\circ C]{} \varepsilon$$

This is another possibly unfamiliar type of transformation: a coexisting solid (α) and liquid (L) mutually transform into different solid (ε). The "peritectic point" is located at W_{Pb} = 71 wt%.

4.4 THE PHASE RULE AND THE LEVER RULE

In our study of phase diagrams, the state of a system is determined by the variables p, T, and X_i. We surveyed some binary phase diagrams, discovering features that are not possible (e.g., two-phase fields) in single-component systems. However, there are limitations on what states are possible in a phase diagram. These limitations are summarized in a thermodynamic principle called the **phase rule**. The phase rule relates the number of possible phases φ that can coexist together in a given state to the number of variables required to establish everything about it. If C is the number of components in the system at constant pressure p (like the binary diagrams above), then *everywhere* in the phase diagram the system must obey

$$\varphi = C + 1 - F \quad (4.9)$$

Here, F is the number of variables you must specify (called the "degrees of freedom") in order to establish the state.

Consider what the phase rule implies for various locations within a phase diagram:

- If $\varphi = 1$, your system state lies inside a single-phase/homogeneous/α phase field. This gives $F = 2$. What this means is that you must specify *two* of the variables $\{T, W_\alpha\}$ in order to know everything about the state. (If you specify the X_α, the values of X_1 and X_2 for the components of α both become known.)
- If $\varphi = 2$, then you are either inside a two-phase/heterogeneous/($\alpha + \beta$) field or you are on a boundary that touches two phases total. Here, $F = 1$. This means that if you specify *one* of $\{T, W_\alpha, W_\beta\}$, then everything else is determined. (A tie line construction passing through W_α, W_β, or T fixes the unknown temperature or phase composition(s).)
- If $\varphi = 3$, then your system is located on a boundary between two two-phase fields with three distinct phases between them, e.g., a eutectic. Since the temperature of this boundary is fixed and the compositions $\{W_\alpha, W_\beta, X_L\}$ come from a tie line at that temperature, we have $F = 0$.

Note that F cannot be less than zero, so it is not possible to have $\varphi > 3$ at any location on a binary phase diagram. It is also not possible for any phase field to have more than two phases within it. These restrictions (themselves a consequence of the phase rule) constrain the variety that is possible in phase diagrams.

Example 4.9:

Identify the number of phases and degrees of freedom at each location A, B, and C in the Pb-Bi phase diagram below.

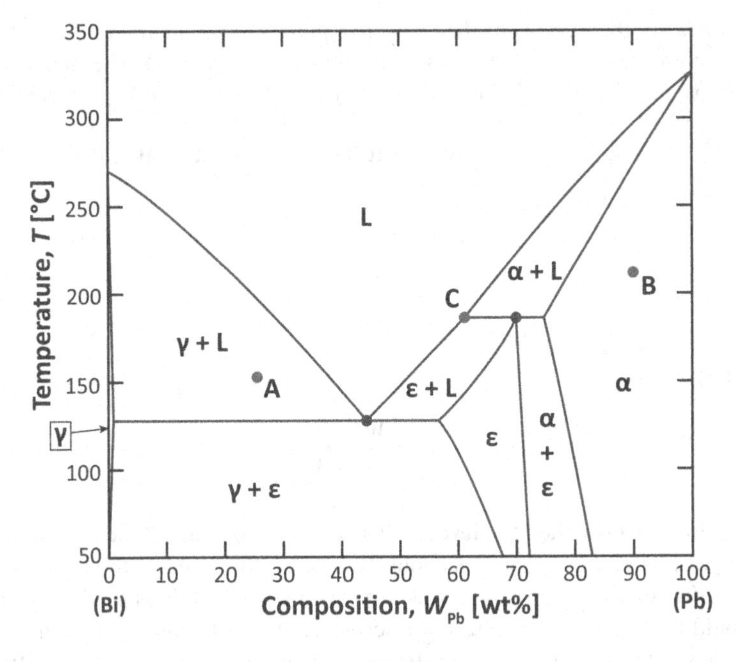

SOLUTION

a. This point lies entirely within the γ + L two-phase field, so $\underline{\boldsymbol{\varphi} = \mathbf{2}}$. The phase rule then indicates $\underline{\boldsymbol{F} = \mathbf{1}}$. What does this mean? Just knowing that you are in the γ + L field does not provide enough information to establish any crucial information about the system state. If, however, you specify *one* additional piece of information, the rest of the system parameters are established. For example, choosing a value for W_L ($0\% < W_L < 44\%$) fixes the height of the tie line (i.e., the temperature T) and the intersection of the tie line with the γ phase field (i.e., the composition W_γ).
b. This point lies within the single-phase α field, and therefore $\underline{\boldsymbol{\varphi} = \mathbf{1}}$ and $\underline{\boldsymbol{F} = \mathbf{2}}$. You must specify both T and W_α in this case (there is no tie line in a one-phase field).
c. Point **C** lies on a boundary between multiple phase fields: L, α + L, and ε + L. Since all of these phases may coexist simultaneously, we count $\underline{\boldsymbol{\varphi} = \mathbf{3}}$ at this location. The temperature is fixed (at the peritectic temperature T_p = 185°C) and the tie line at this temperature establishes the compositions of the various phases.

In a single-phase/α field, the overall composition W of the alloy is the same as the composition of the α phase: $W = W_\alpha$. Consider now the behavior of a system in a state inside of a two-phase/($\alpha + \beta$) field. At a given temperature T, the tie line establishes the compositions W_α and W_β of the two phases. What is not fixed in this case is the overall system composition W, which we know lies somewhere in between W_α and W_β. As it turns out, the overall system composition depends on *how much* of each phase is present. These quantities are all related by

$$W = W_\alpha f_\alpha + W_\beta f_\beta$$

where the fs are the weights of the respective phases, taken as fractions of the entire system weight. (Since there are only two phases, $f_\alpha + f_\beta = 1$.) Alternatively, if we know W, we can find the weight fractions f_α and f_β. Wince $W_\alpha + W_\beta = 1$, we obtain

$$W = W_\alpha f_\alpha + W_\beta f_\beta = W_\alpha f_\alpha + W_\beta \left(1 - f_\alpha\right) = f_\alpha \left(W_\alpha - W_\beta\right) + W_\beta$$

or

$$f_\alpha = \frac{W - W_\beta}{W_\alpha - W_\beta} \tag{4.10(a)}$$

Correspondingly

$$f_\beta = \frac{W_\alpha - W}{W_\alpha - W_\beta} \tag{4.10(b)}$$

Equation 4.10 is called the **lever rule** because it "balances" the contributions of the components in the two phases about the central "fulcrum" composition W.

The value of the lever rule is that it provides the fractions of the two phases you would expect to observe in the microstructure in terms of the compositions of the phases (along with the overall system composition). The compositions W_α and W_β are present in the phase diagram itself and may be read off directly (see **Example 4.9**).

Example 4.10:

What are the phase fractions f_α and f_L in the problem of **Example 4.8** ($T = 1022°C$)? If you have a sample with a mass of 150 g, what is the weight of the solid fraction?

SOLUTION

In the solution to **Example 4.8**, we determined that the overall system composition is $W = 55$ wt%, the liquid composition is $W_L = 54.5$ wt%, and the solid composition is $W_\alpha = 56.5$ wt%. From just these three values, we can determine

$$f_\alpha = \frac{W - W_L}{W_\alpha - W_L} = \frac{55 - 54.5}{56.5 - 54.5} = 0.25 = \underline{\mathbf{25\%}}$$

The liquid phase fraction can be found in the same way

$$f_\beta = \frac{W_\alpha - W}{W_\alpha - W_\beta} = \frac{56.5 - 55}{56.5 - 54.5} = 0.75 = \underline{\mathbf{75\%}}$$

(Note that this result can also be obtained by taking $f_L = 1 - f_\alpha = 1 - 0.25 = 0.75$.) The solid fraction is 25%, so in a sample with weight $mg = 0.150(9.81) = 1.47$ N. Therefore

$$w_\alpha = mgf_\alpha = 1.47(0.25) = \underline{\mathbf{0.368\ N}}$$

Example 4.11:

Suppose you have a sample of alloy Al20Cu (wt%) at a temperature of 500°C. What are the fractions of the κ phase and the fraction of the θ phase?

SOLUTION

Consider the diagram below. For an overall system composition W = 20 wt% at the given temperature, we identify W_κ = 4.5 wt% and W_θ = 52.5 wt%.

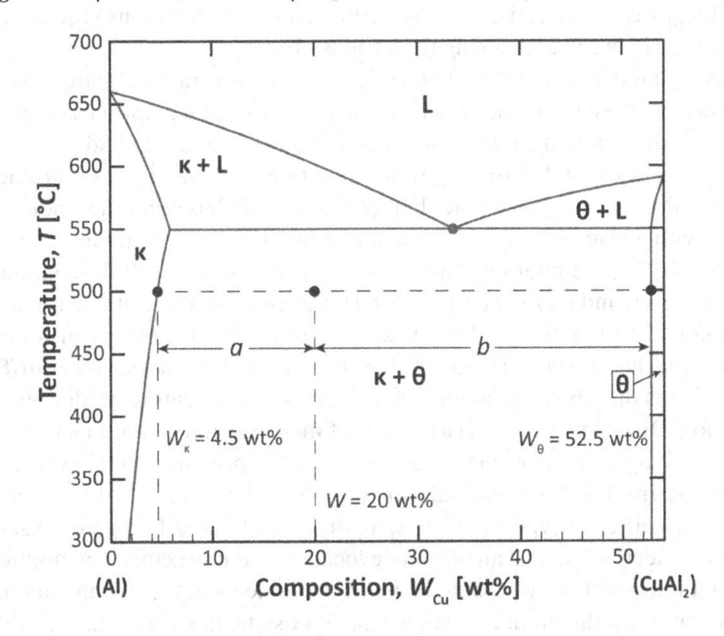

Note that the "distance" $a = W - W_\kappa$ = 15.5 wt% and $b = W_\theta - W$ = 32.5 wt%. Also, $a + b = W_\theta - W_\kappa$ = 48.0 wt%. We therefore find

$$f_\kappa = \frac{b}{a+b} = \frac{32.5}{48.0} = 0.68 = \underline{\mathbf{68\%}}$$

and then

$$f_\theta = \frac{a}{a+b} = \frac{15.5}{48.0} = 0.32 = \underline{\mathbf{32\%}}$$

4.5 DIFFUSION

We have introduced several different types of transformations that are possible in materials systems. Some notable examples include

- Solidification – a transformation from a liquid phase into a solid phase.
- Solid-state transformation – a solid phase transforms into a different solid phase (different crystal structure).

- Eutectic solidification – when a liquid solidifies into two distinct solid phases.
- Precipitation – a new, different solid phase emerges from a solid parent phase.

The phase diagram tells you about the conditions under which you would expect a transformation to occur. However, it does not tell you *how* or *how long* the transformation takes. This is important information. Any attempt to alter the properties of a material requires an alteration of its structure, and such alterations require reallocation of matter between and among the various phases.

To understand the "how" and "how long" of phase transformations and other treatments, we must study the underlying **kinetics**. The assumption used in assembling the information in a phase diagram is that the system has had adequate time to achieve its most stable or "equilibrium" state, and the diagram provides the conditions that produce that state. For example, if a phase diagram indicates that a system with a given (T, X_1, and X_2) will contain β-phase precipitates at a fraction f_β, then during a precipitation transformation the structure will develop until this state is achieved and *no further*. In contrast, the kinetics associated with the transformation is the physics that describes the "motion" of the system from its prior state to the terminal state. The most important kinetic phenomenon is **diffusion**.

Diffusion is the wholesale motion of molecular constituents to produce mixing or segregation. Diffusion occurs via a number of **mechanisms** within a material, any or all of which may be in operation simultaneously. The primary mechanisms are illustrated in **Figure 4.6**. It is important to recognize that these different mechanisms are related to different types of defects, as shown in **Figure 4.6(a)**. For example, an interstitial defect, situated at an off-lattice location in the structure, can migrate from one location to another by passing between the lattice atoms. Motion from one lattice location to another lattice location is also possible; this is the vacancy diffusion mechanism. A lattice atom may move from its accustomed lattice site into an adjacent site only if the adjacent site is empty (i.e., there is a vacancy in the structure). Finally, the disordered region surrounding a grain boundary (or phase boundary) is also a pathway for diffusional transport. The grain boundary has an open, fluidlike structure that atoms can traverse easily.

Figure 4.6(b) shows a schematic representation of diffusion via the interstitial mechanism. In order to jump from one interstitial location to another, the atom must push past the atoms in the lattice that confine it. That is, the diffusing atom must work against the intermolecular forces that hold the crystal together (represented here by a spring). The potential-energy picture shows that the amount of work required to move the atom from one location over to the next stable spot is ΔU. This is the **activation energy** for the mechanism and is similar to the activation energy required for a chemical reaction. What the presence of this activation-energy requirement (or "activation barrier") means is that some energy input is required in order for diffusional "jumping" of the type depicted to occur. This energy comes from the thermal energy contained in the crystal. If the crystal is at a temperature T, then we expect that a typical energy E possessed by an atom to be $E \sim k_B T$. ($k_B \approx 1.381 \times 10^{-23}$ J/K is Boltzmann's constant). The frequency ω

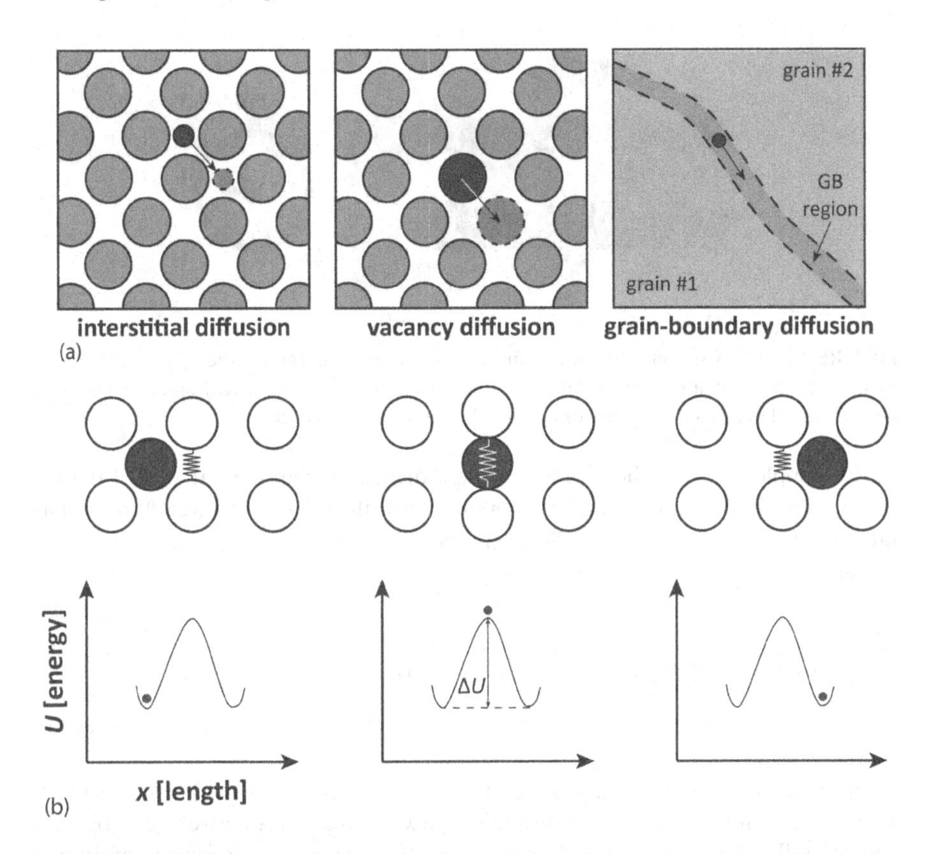

FIGURE 4.6 Diffusion mechanisms. The interstitial, vacancy, and grain-boundary mechanisms are illustrated in (a). These mechanisms correspond to different types of diffusive motion via defects in a crystal. The thermal-activation picture for the interstitial mechanism is in (b). By overcoming intermolecular forces (represented by the potential difference ΔU), the atom can "hop" into the next unoccupied interstitial location over. The atom obtains the energy to do this from the thermal environment.

with which we expect atomic jumps to occur depends on how large ΔU is compared to E:

$$\omega \sim \exp\left(-\frac{\Delta U}{E}\right) = \exp\left(-\frac{\Delta U}{k_B T}\right) \tag{4.11}$$

The smaller the ratio $\Delta U/E$, the higher the value of ω, and the higher the frequency ω, the more rapidly atoms diffuse through the crystal. Based on what we know about the variation in the strength of intermolecular forces among materials, ΔU will vary considerably, and hence diffusion rates. It is also important to recognize that heat energy plays an important role in the diffusion process.

To account for the kinetic behavior of atoms during phase transformations, we require a way to count the diffusing species as they move so that we know how rapidly a transformation is progressing. The counting setup is illustrated in **Figure 4.7**.

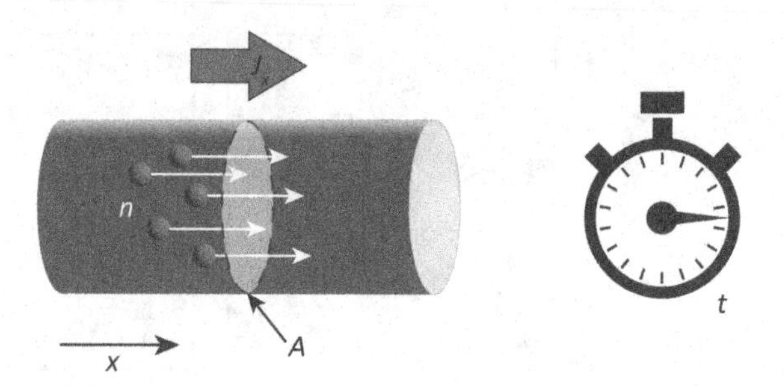

FIGURE 4.7 Illustration of the flux concept. During an interval of time t, some number of atoms n (or, equivalently, some total mass of atoms) moving in the x direction cross the plane with area A. This establishes an atomic flux $Jx = n/At$ in the cylinder.

In this simple 1D case, the solute atoms are diffusing from the left side of the rod to the right. By counting the atoms (= n) that cross the plane (with area A) during an interval of time t, we obtain the **flux** J_x in the x direction:

$$J_x = \frac{n}{At} \tag{4.12}$$

Equivalently, we may compute the flux in terms of the mass to get

$$J_x = \frac{m}{At} \tag{4.12'}$$

We recognize that the units of flux are $[J]$ = [number/area/time] = #/m²/s or $[J]$ = [mass/area/time] = kg/m²/s. (It is straightforward to convert between number flux and mass flux, since $m = n \times A$, where A is the atomic mass.) Flux is an important quantity in describing transport kinetics of materials since it provides a way to account for "how much" and "how long". Also, note that it is possible that atomic motion occurs in both the leftward (negative x) and rightward (positive x) directions. These combined motions combine into a "net" flux ΔJ given by

$$\Delta J = J_x - J_{-x} \tag{4.13}$$

i.e., the leftward and rightward motions cancel each other.

Example 4.12:

Suppose you have a thin, flat, rectangular filter membrane that is 10. cm × 10. cm. If the membrane can filter 5.3 mol of alcohol molecules per half hour, estimate the average flux across the membrane assuming steady-state conditions.

SOLUTION

The membrane is square, so we compute its area A = 10. cm × 10. cm = $1\underline{0}0$ cm². Now, by direct computation, we find

$$J = \frac{n}{At} = \frac{5}{100(0.5)} = 1\,\frac{\text{mol}}{\text{cm}^2\text{h}} \approx \underline{\mathbf{2.8}\,\frac{\textbf{mol}}{\textbf{m}^2\textbf{s}}}$$

The flux J is a convenient way of representing internal molecular motion physically, and it is something that can be measured in an experiment. There are a number of factors that determine the value of flux. These are: how steep the **concentration gradient** is and intrinsic rapidity that the atoms are capable of within the material. A concentration gradient is required for any net atomic motion. The motion of atoms in the material is essentially random, a behavior that can be observed in the phenomenon of "Brownian motion". This being the case, if there are equal numbers of atoms on either side of the plane in **Figure 4.7**, then the jumping of atoms is *equally likely* in the leftward and the rightward directions. The flux then would be $\Delta J = 0$, and so no net motion would be detected. However, if there are unequal numbers of atoms on either side of the plane, i.e., a concentration gradient, the net motion will be biased in the direction away from the more highly concentrated side.

The situation is illustrated in **Figure 4.8**. The number of solute atoms is higher on the left side of the rod than on the right, i.e., the composition is not uniform across the rod. In this situation, it makes sense to represent the concentration C of the diffusing species "A" within the rod by computing the number of atoms n_A per unit volume:

$$C_A = \frac{n_A}{V} \tag{4.14}$$

or the mass of atoms per unit volume[4]

$$C_A = \frac{m_A}{V} \tag{4.14'}$$

For example, in **Example 4.7**, the concentration of Al in the Si is $C_{Al} = 2.2\ \text{mg/m}^3$. In the diagram of **Figure 4.8**, we then have a situation in which the composition is a decreasing function of x. The concentration gradient C' is then the *slope* of this function:

$$C' = \frac{dC}{dx} \approx \frac{\Delta C}{\Delta x} \tag{4.15}$$

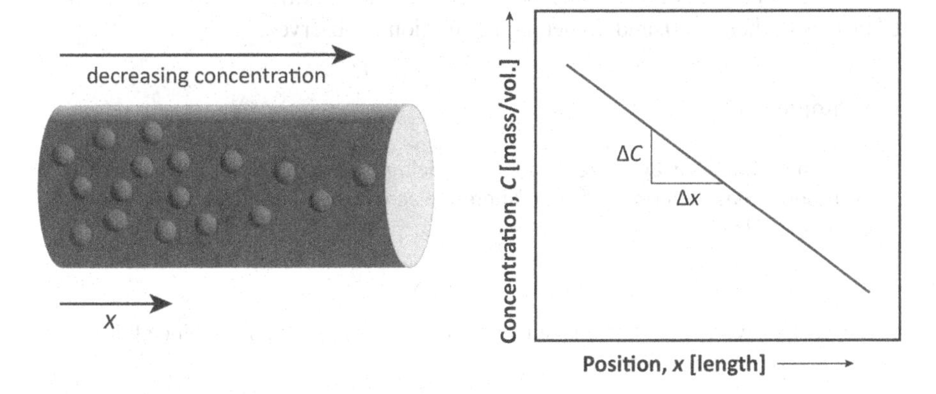

FIGURE 4.8 Concentration gradient in a 1D system. A higher concentration C_L on the left and a lower concentration C_R on the right give rise to a gradient in concentration $\Delta C/\Delta x$.

Note that $[\Delta C]$ = [number/volume] or [mass/volume] and $[\Delta x]$ = [length], so $[C']$ = [number/length^4] or [mass/length^4]. Consider what this means for diffusion. The flux to the right will be greater than the flux to the left, so the net flux will be to the right. That is, atoms tend to move "down" the concentration gradient. Also, the steeper the gradient, the greater the flux will be.

As it turns out, the relationship between the concentration gradient and the flux is one of proportionality: $J \propto C'$. We call this relationship **Fick's First Law** and typically write it

$$J = -D \times C' = -D\frac{dC}{dx} \tag{4.16}$$

where D is the coefficient of proportionality. As the concentration gradient increases, so does the flux. Also, the flux is in the positive x direction when the gradient C' is negative, hence the (–) sign in **Equation 4.16**. We know from the discussion above that the flux depends on the concentration gradient *and* the intrinsic mobility of the atoms. For this reason, the coefficient D must be indicative of how fast the atomic motion is overall. We call D the **diffusivity** of the system and find that D is related to the jump frequency ω by

$$D \propto \omega$$

and so we find

$$D = D_0 \exp\left(-\frac{\Delta U}{k_B T}\right) \tag{4.17}$$

where D_0 is a constant for a particular material system. **Equation 4.17** establishes that the intrinsic rate of atomic motion, and hence the diffusion flux, is a "thermally activated" phenomenon. Practically, this means that the rate of diffusion increases as the temperature increases. Also note that, because flux is the product of two independent factors, both of these must be present in order for diffusion to occur. If either C' or D is 0, then $J = 0$, and no net atomic motion is observed.

Example 4.13:

If the membrane of **Example 4.12** has a thickness of d = 1.3 mm, what would you estimate as the concentration difference across the filter? Take the diffusivity parameter $D = 10^{-2}$ m^2/s.

SOLUTION

We wish now to apply Fick's First Law to the circumstances above. Since this law is, roughly

$$J \approx -D\frac{\Delta C}{\Delta x}$$

So we can write

$$\Delta C \approx -\frac{J}{D}\Delta x$$

The flux is $J = 2.8$ mol/m²s, from above. Since the width of the concentration gradient is effectively the thickness of the membrane, we take $\Delta x = d$ and get

$$\Delta C \approx -\frac{2.8\ \text{mol/m}^2\text{s}}{10^{-2}\ \text{m}^2\text{/s}}(0.0013\ \text{m}) = \underline{\mathbf{-0.4\ \frac{mol}{m^3}}}$$

(Note that the sign doesn't mean much to the answer here, since there is no x direction assigned.) The molar mass of ethanol is $M = 46.07$ g/mol. Note that this composition is equivalent to

$$\Delta C \times M = 0.4\ \frac{\text{mol}}{\text{m}^3} \times 46.07\ \frac{\text{g}}{\text{mol}} = 17\ \frac{\text{g}}{\text{m}^3}$$

A few points regarding **Equation 4.17**. First, D and D_0 must have units $[D] = [D_0] = [\text{length}^2/\text{time}] = \text{m}^2/\text{s}$. This may make the quantity D seem a little strange, physically, but consider an analogy. The velocity V of an object in rigid-body motion has units $[V] = [\text{length/time}] = \text{m/s}$. The distance d an object travels during an interval of time t is then

$$d = Vt$$

From basic physics, we can estimate the distance traveled by a diffusing particle executing a 3D "random walk" via diffusive jumps. By considering the statistics for a large sample of random walks, we obtain an "expected" displacement of

$$d = \sqrt{2Dt} \tag{4.18}$$

That is, D is like a "diffusive velocity" that can be used to relate how far a typical particle goes to how long it moves. Another point: the activation energy ΔU is the energy of the barrier in **Figure 4.6(b)**. This is the energy requirement for jumping for a single atom. Frequently, this energy requirement is provided not for a single atom but for 1 mol of them. In this case, **Equation 4.17** can be modified to read

$$D = D_0 \exp\left(-\frac{Q}{RT}\right) \tag{4.18'}$$

where Q is the energy input required for N_A atoms: $[Q] = [\text{energy/number}] = \text{J/mol}$. The atomic energy ($= k_B T$) is correspondingly transformed into that of a *molar* energy ($= RT$). It is also important when working with expression like **Equation 4.17** that the temperature T be given in Kelvin rather than a scale that is not absolute (e.g.,°C or °F). **Table 4.2** provides diffusivity data for some important materials systems.

TABLE 4.2
Diffusivity Data for Selected Material Systems

Material System	D_0 [cm²/s]	Q [kJ/mol]
C in α-Fe	2.2	123
C in γ-Fe	0.15	142
Cu in Al	0.647	135.0
Sn in Cu	0.11	188
Ag in Au	0.072	176.8
V in β-Ti	0.00031	135
Al in Si	8.0	338
As in Si	0.32	338
P in Ge	2.5	241

Example 4.14:

The diffusivity D of **Equation 4.18′** is a function of T: $D = D(T)$. If the diffusion coefficient D_1 for carbon in nickel is 5.5×10^{-14} m²/s at $T_1 = 600°C$. Additionally, $D_2 = 3.9 \times 10^{-13}$ m²/s at $T_2 = 700°C$. What are D_0 and Q?

SOLUTION

Using the formula for the diffusivity at temperature T:

$$D = D_0 \exp\left(-\frac{Q}{RT}\right)$$

and the facts that 600°C = 873 K and 700°C = 973 K, we can represent the two pieces of data we have as

$$D_1 = D(873) = D_0 \exp\left(-\frac{Q}{RT_1}\right) = 5.5 \times 10^{-14} \ \frac{\text{m}^2}{\text{s}}$$

and

$$D_2 = D(973) = D_0 \exp\left(-\frac{Q}{RT_2}\right) = 3.9 \times 10^{-14} \ \frac{\text{m}^2}{\text{s}}$$

These are *two* equations in *two* unknowns: Q (in J/mol) and D_0 in (m²/s). These can be solved using the standard techniques. For example, since

$$D_0 = D_1 \exp\left(\frac{Q}{RT_1}\right)$$

This gives

$$D_2 = D_1 \exp\left(\frac{Q}{RT_1}\right) \exp\left(-\frac{Q}{RT_2}\right) = D_1 \exp\left[\frac{Q}{R}\left(\frac{1}{T_1} - \frac{1}{T_2}\right)\right]$$

or

$$Q = R\left(\frac{1}{T_1} - \frac{1}{T_2}\right)^{-1} \ln\left(\frac{D_2}{D_1}\right) = \left(8.314 \frac{\text{J}}{\text{mol} \times \text{K}}\right)\left(\frac{1}{873\text{ K}} - \frac{1}{973\text{ K}}\right)^{-1} \ln\left(\frac{5.5 \times 10^{-14}}{3.9 \times 10^{-14}}\right)$$

$$= \mathbf{140}\,\frac{\mathbf{kJ}}{\mathbf{mol}}$$

The calculation of D_0 follows directly:

$$D_0 = D_1 \exp\left(\frac{Q}{RT_1}\right) = \left(5.5\times10^{-14}\right)\exp\left[\frac{140{,}000\text{ J/mol}}{(8.314\text{ J/mol/K})(873\text{ K})}\right] = 0.000010\,\frac{\text{m}^2}{\text{s}}$$

or, $\mathbf{D_0 = 0.10\ cm^2/s}$.

Astute readers will notice that the description of diffusion encoded in Fick's First Law and described by the diffusivity and concentration gradient is an incomplete description of atomic motion in materials. The flux produced according to Fick's First Law will redistribute atoms, altering the concentration gradient over time. For this reason, we say that Fick's First Law is a *steady-state* representation of atomic motion. It applies at either a single instant in time or in cases where the concentration gradient is held fixed by externally applied conditions (e.g., atoms are being added into/removed from the system). For systems that are not at the steady state, we require a rule that describes how the accounts for how the concentration $C(x)$ changes over time, i.e., a rule for finding $C(x, t)$. This rule is called **Fick's Second Law** and it relates the time rate of change of the concentration gradient $C(x, t)$ to the net flux $\Delta J = J_{in} - J_{out}$. From the diagram in **Figure 4.9**, we can estimate the change in concentration ΔC in the thin strip Δx as

$$\Delta C = \frac{\Delta J}{\Delta x}\Delta t$$

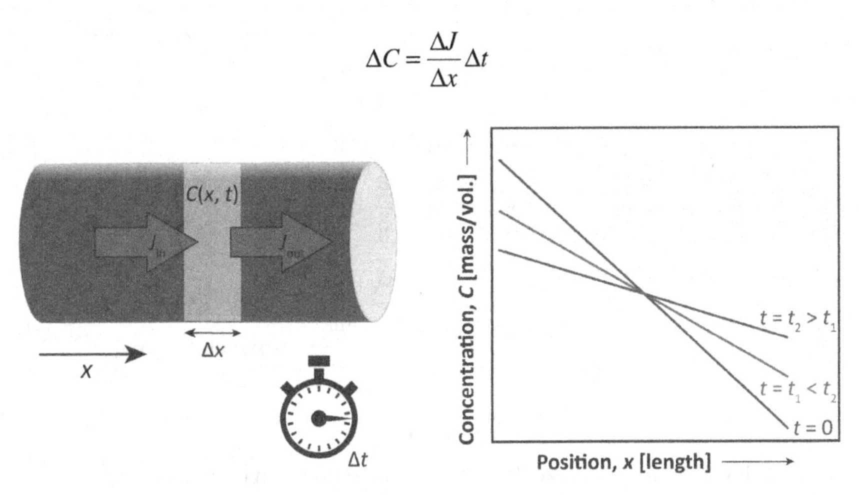

FIGURE 4.9 The meaning of Fick's Second Law. Within the small strip of material Δx, the net flux $\Delta J = J_{out} - J_{in}$. Under non-steady-state conditions, material will either accumulate or deplete in the region Δx, altering the concentration profile over time, as shown.

When the times involved become very brief and the distances become very short, we obtain the relationship for *continuous* change throughout the material:

$$\frac{\partial C}{\partial t} = \frac{\partial J}{\partial x}$$

In cases where the diffusivity D is constant throughout the material, this becomes

$$\frac{\partial C}{\partial t} = D\frac{\partial^2 C}{\partial x^2} \tag{4.19}$$

(In the context of the diagram, ΔJ is a negative quantity, so the (–) sign of **Equation 4.16** is unnecessary.)

We recognize **Equation 4.18** as a 1D second-order partial differential equation in the independent variables x and t. When Fick's second law is combined with the geometry of the system and a set of initial and boundary conditions, we obtain a solution $C(x, t)$ that gives the composition of the system.

Example 4.15:

Consider two large pieces of electrum that have been joined together. The piece on the left has a composition of W_1 = 10.0 wt% Ag, and the piece on the right has a composition W_2 = 5.0 wt% Ag. This "diffusion couple" is shown below, alongside the initial concentration profile (C vs. x). If the couple is heated to 975°C, compute and sketch the composition $C(x, t)$ after 100 h. What is the composition at x = +1.0 mm? (Assume a constant diffusivity D.)

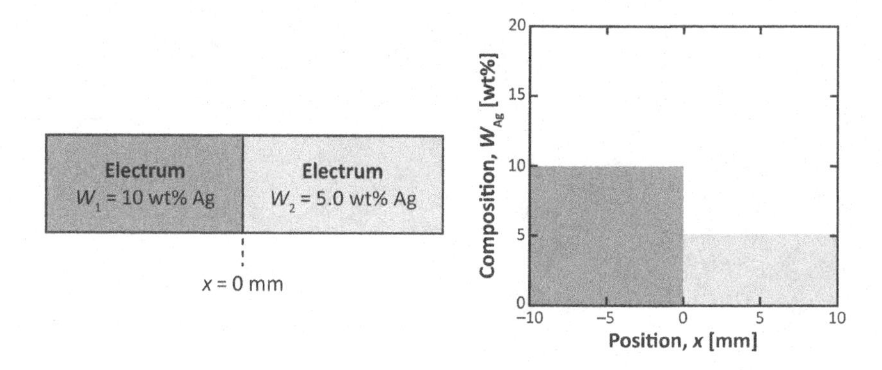

SOLUTION

From **Table 4.3**, we recognize the solution to the diffusion couple geometry as

$$C(x,t) = C_2 + \frac{C_1 - C_2}{2}\left[1 - \text{erf}\left(\frac{x}{\sqrt{4\pi Dt}}\right)\right]$$

TABLE 4.3
Important Solutions to Fick's Second Law

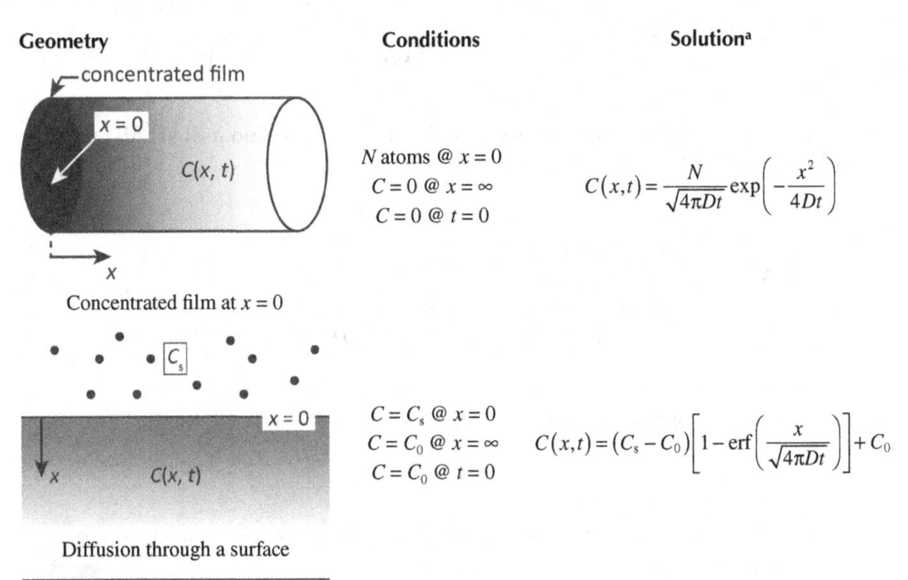

Geometry	Conditions	Solution[a]
Concentrated film at $x = 0$	N atoms @ $x = 0$; $C = 0$ @ $x = \infty$; $C = 0$ @ $t = 0$	$C(x,t) = \frac{N}{\sqrt{4\pi Dt}} \exp\left(-\frac{x^2}{4Dt}\right)$
Diffusion through a surface	$C = C_s$ @ $x = 0$; $C = C_0$ @ $x = \infty$; $C = C_0$ @ $t = 0$	$C(x,t) = (C_s - C_0)\left[1 - \text{erf}\left(\frac{x}{\sqrt{4\pi Dt}}\right)\right] + C_0$
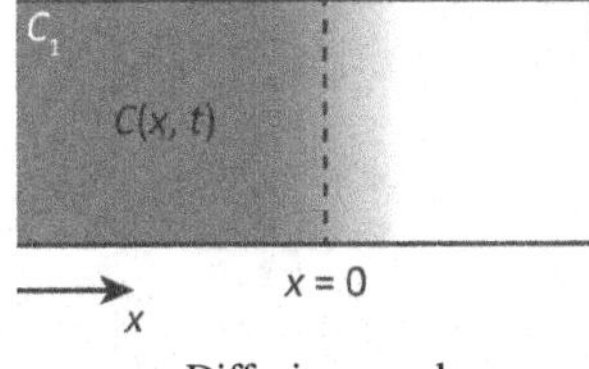 Diffusion couple	$C = C_1$ @ $x = -\infty$; $C = C_2$ @ $x = +\infty$; $C = C_1$ @ $t = 0, x < 0$; $C = C_2$ @ $t = 0, x > 0$	$C(x,t) = C_2 + \frac{C_1 - C_2}{2}\left[1 - \text{erf}\left(\frac{x}{\sqrt{4\pi Dt}}\right)\right]$

[a] In these solutions, "erf" is the "Gaussian error function". Since erf(z) cannot be broken down into simpler, more familiar functions, computing it typically requires software or tables of values.

Since W_1 = 80 wt% Ag, ρ_{Ag} = 10.49 g/cm³, and ρ_{Au} = 19.3 g/cm³, we compute

$$C_1 = \frac{W_{Ag}}{W_{Ag}/\rho_{Ag} + W_{Au}/\rho_{Au}} = \frac{0.10}{0.10/\left(10.49 \text{ g/cm}^3\right) + 0.050/\left(19.3 \text{ g/cm}^3\right)} = 1.8\frac{\text{g}}{\text{cm}^3}$$

of Ag and

$$C_2 = 0.93 \text{ g/cm}^3$$

of Ag. Next, we determine the diffusivity of Ag in Au (using **Table 4.2**) as

$$D = D_0 \exp\left(-\frac{Q}{RT}\right) = \left(0.072\frac{\text{cm}^2}{\text{s}}\right)\exp\left[-\frac{176{,}800 \text{ J/mol}}{(8.314 \text{ J/mol/K})(1248 \text{ K})}\right] = 2.9 \times 10^{-9} \frac{\text{cm}^2}{\text{s}}$$

(Note that we have used a temperature of 1248 K = 975°C to find *D*.) We now have

$$C(x,t) = 0.93 + \frac{1.8 - 0.93}{2}\left[1 - \text{erf}\left(\frac{x}{\sqrt{4\pi\left(2.9\times10^{-9}\ \text{cm}^2/\text{s}\right)t}}\right)\right]\frac{\text{g}}{\text{cm}^3}$$

To visualize this and facilitate future calculations, we set up a MATLAB® function utilizing the built-in `erf()` function:

```
function comp = C(x, t)
    % x in cm; t in s
    comp = 0.93 + ((1.8 - 0.93)/2)*...
              (1 - erf(x/sqrt(4*pi*2.9*10^-9*t)));
end
```

We can now plot as normal over a range of, say, –5 mm < x < 5 mm:

```
>> x = -5:0.001:5;
>> y = C(x/10, 100*3600); % convert x to cm and t to s
>> plot(x, y);
>>
```

This produces

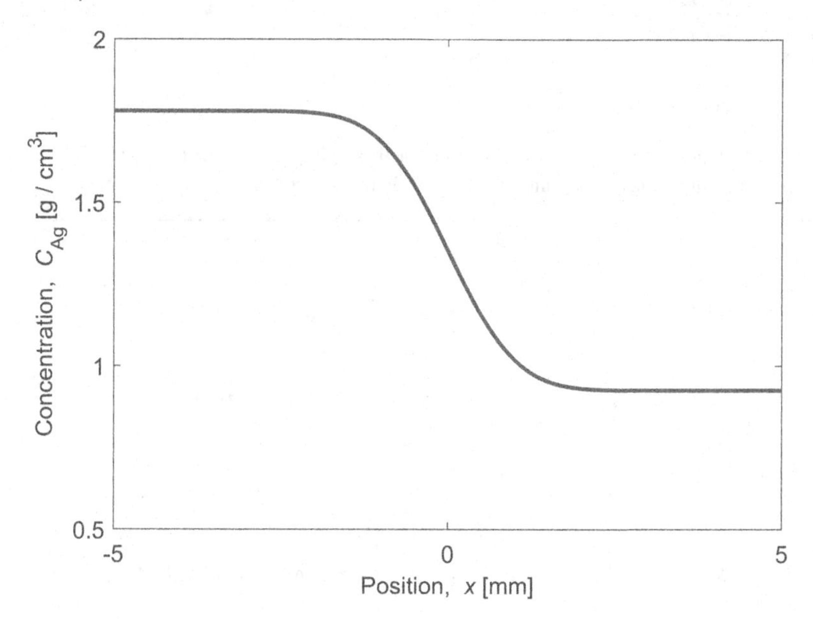

The concentration profile has been altered by the flux of Ag atoms away from the (higher-concentration) left side toward the (lower-concentration) right side.

If we wish to find C for $x = 1$ mm, we can read the value directly off of the graph or request it from our MATLAB function:

```
>> C(1/10, 100*3600)
ans =
1.0179
>>
```

in g/cm^3. That is, the concentration of Ag at a distance of 1 mm from the interface has increased by ≈10%.

4.6 PHASE TRANSFORMATIONS

The phase diagram tells us what the preferred state of a system is under a given set of thermodynamic conditions, and the diffusion characteristics of a system tell us something about how rapidly the components can rearrange themselves to change state. We wish to understand phase transformations in this context. Typically, the way we do this is to represent the rate of transformation λ as the combination of a "driving force" ΔG and a "mobility" μ:

$$\lambda = \lambda(\Delta G, \mu) \tag{4.20}$$

If either of these factors is absent, then $\lambda \approx 0$ and the transformation will not proceed. Understanding the possible conditions under which both of these factors are present requires a closer physical analysis.

Consider the case of the solidification of a pure material with melting temperature T_m. At temperatures above T_m, the material tends to melt; below T_m; the material tends to solidify. Suppose that you were to take a sample of material at temperature $T = T_m$ and **quench** it so that the temperature is reduced rapidly to $T < T_m$. The difference between the melting temperature and the quenched temperature is the **undercooling** ΔT:

$$\Delta T = T_m - T \tag{4.21}$$

We expect that the greater the quench ΔT, the greater the driving force ΔG will be. In fact

$$\Delta G = \frac{\Delta T}{T_m} \Delta H_f \tag{4.22}$$

where ΔH_f is the "heat of fusion" of the material with units $[\Delta H_f]$ = [energy/volume] = J/m^3. The heat of fusion essentially represents the difference in chemical/bonding energy between the system in its liquid state and solid state, and we interpret this difference in energy as a "chemical force" (per unit area) that pushes the system toward a new, more stable state. In typical materials systems, the mobility parameter μ is closely associated with the diffusivity D: $\mu \propto D/T$ or

$$\mu \propto \frac{1}{T}\exp\left(-\frac{Q}{RT}\right)$$

High values of D (and hence, high μ values) are associated with rapid transformation rates.

The solidification transformation proceeds along essentially two pathways:

1. **Nucleation** – During nucleation, small, nanoscale "islands" of the new (solid) phase form in the parent (liquid) phase. The formation of these island nuclei is a necessary precursor to the remainder of the transformation, and their emergence is a statistical phenomenon not unlike the barrier-crossing behavior depicted in **Figure 4.6**. For this reason, the rate of nucleus formation $\dot{n}$ obeys a relationship like

$$\dot{n} = k_1 \omega n^* \exp\left(-\frac{Q}{RT}\right) \tag{4.23}$$

   where k_1 is a constant and $n^* = n^*(\Delta T)$ is the number concentration of nuclei that are statistically *possible* at a given level of undercooling. Note that $[\dot{n}]$ = [number/volume/time] = #/m^3/s. Since the greater the value of ΔT, the greater the value of n^*, nucleation is favored at large degrees of undercooling. (Typical undercoolings required for metals to exhibit significant nucleation are ~200°C.)
2. **Growth** – Nucleation is not the only process driving transformation of a parent phase into a new phase. The nuclei formed during the transformation will also undergo growth. Once stable nuclei have developed, these islands of the solid phase can grow by the sequential addition of new material from the liquid. The island growth rate g is

$$g = k(T)\exp\left(-\frac{Q}{RT}\right) \approx k_2 \exp\left(-\frac{Q}{RT}\right) \tag{4.24}$$

   where $k(T)$ is a "weak" function of the temperature near T_m (i.e., approximately a constant k_2). For this reason, growth is favored at higher temperatures because of the high mobility factor. g is, physically, the velocity of the boundary of the expanding island: $[g]$ = [length/time] = m/s.

The two pathways are illustrated in **Figure 4.10**.

Because both of the processes in **Figure 4.10** are essentially distinct, both of them can happen simultaneously during a transformation (although nucleation must occur *somewhere* before growth begins). The overall transformation rate can therefore be recast as

$$\lambda = \lambda(\dot{n}, g) \tag{4.24'}$$

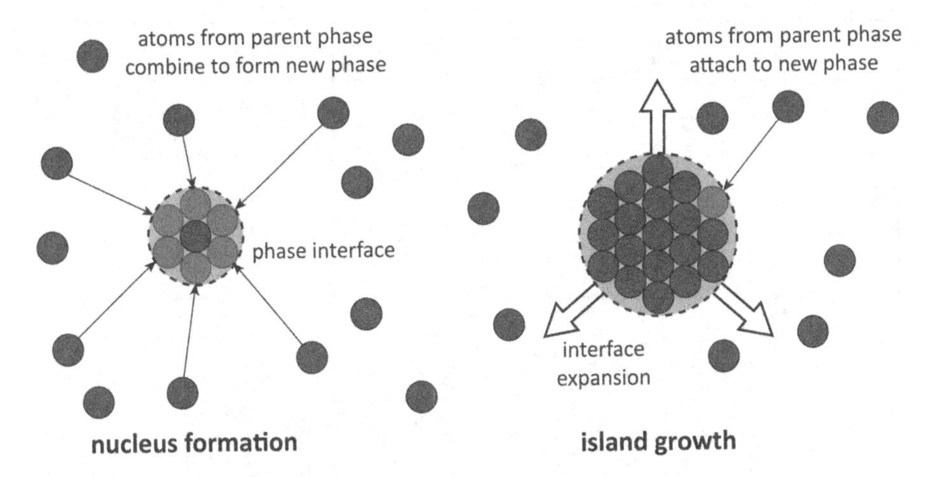

FIGURE 4.10 Kinetic processes of nucleation and growth. During nucleation, molecules from the parent phase reorganize themselves into the structure of the emerging phase. Once these small regions of the new phase have been established, they can grow by recruiting more molecules from the parent phase into their structure.

The overall transformation process and the expected temperature dependence of λ are depicted in **Figure 4.11**. **Figure 4.11(a)** shows the emergence of regions of the new phase and their subsequent expansion. At the start of the transformation ($t = 0$), the material is mostly the parent phase, but nuclei of the new phase emerge at the rate $\dot{n}$ expected at the system temperature T. At a later time ($t_1 > 0$), the nuclei initially formed have grown (at a rate g) while more nuclei continue to emerge. The transformation progresses in this way until the islands of the new phase have grown in size and number to the point where they begin to contact each other and their growth is impeded (t_3). The final transformed structure is one where the islands have consolidated and can grow no further ($t = \infty$).

The expected rates as a function of temperature are depicted schematically in **Figure 4.11(b)**. Curves illustrating the nominal contributions associated with nucleation and growth are presented alongside the overall rate. The nucleation process proceeds most rapidly at low temperatures, but the nuclei (being individually small) contribute little to the developing phase unless they can grow. Growth proceeds rapidly at higher temperatures, but without nucleation, there is nothing to grow *from*. The tension between these two factors predicts that there is a range of intermediate temperatures where the overall transformation rate is the fastest possible balance of nucleation and growth. For a given system, we can then represent the overall transformation rate as a purely temperature-dependent quantity:

$$\lambda = \lambda(T) \tag{4.24''}$$

Hence, the transformation rate can be controlled by controlling the temperature.

It is now a good idea to consider the development of the new phase from the parent in terms of the volume fraction f of the material that the new phase occupies.

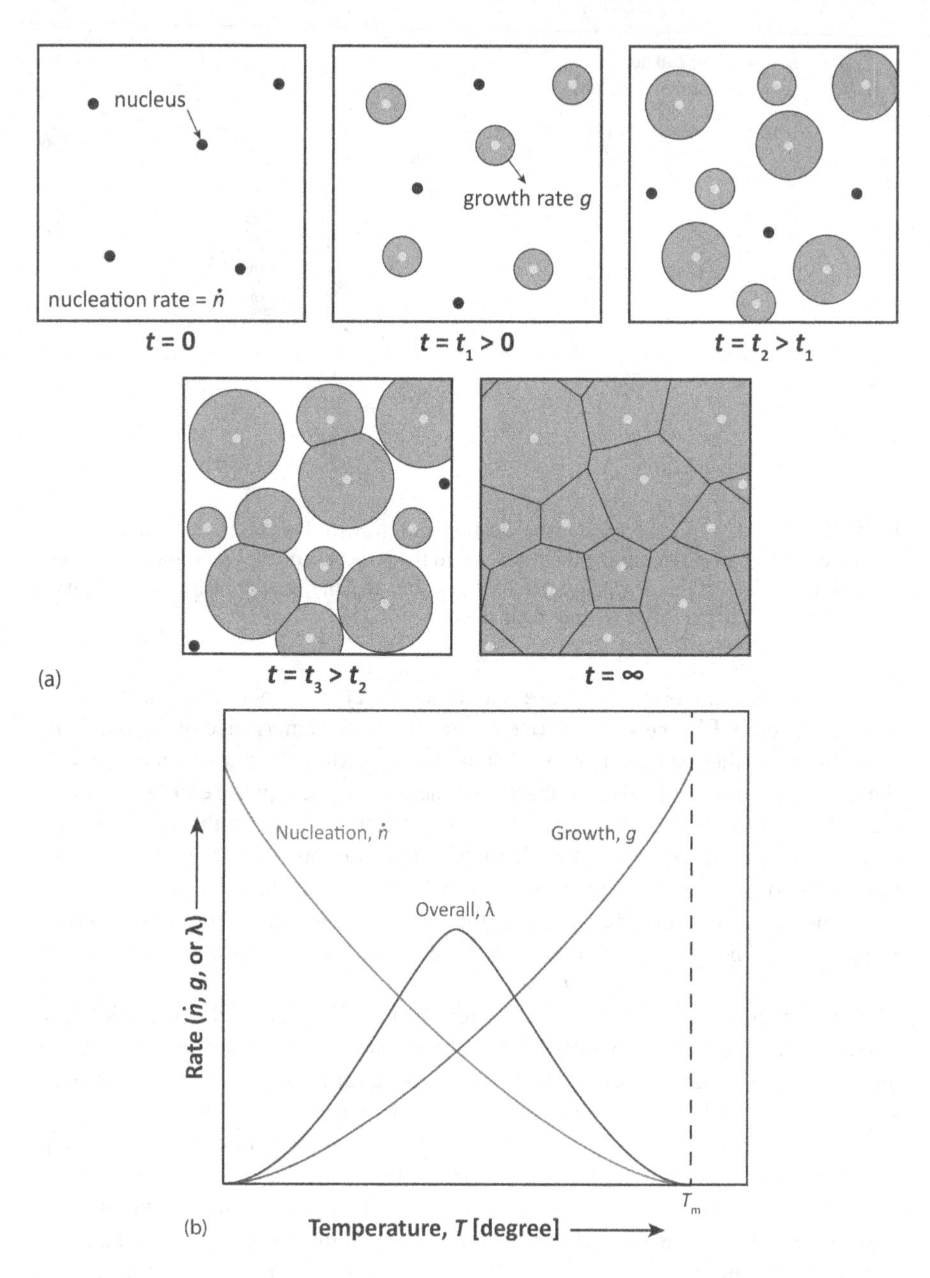

FIGURE 4.11 The transformation process is dependent on temperature. In (a), the development of the new phase from the parent involves the appearance of discrete nuclei that begin to appear at time $t = 0$. Nuclei appear continuously over time in the parent phase, while simultaneously, the new phase grows outward from the nuclei that appeared during earlier epochs. As nucleation and growth proceed, the parent phase is replaced by the new phase, until eventually the new phase reaches its equilibrium fraction (100% in this case). The rates of nucleation and growth depend on the temperature, as shown in (b), so the progress of the transformation in (a) will depend on the temperature.

At the start of the transformation, $f = 0$. f then increases according to the value of $\lambda(T)$ until it reaches its maximum possible value f_{max}. This maximum possible value might be $f_{max} = 1 = 100\%$ for complete transformations or some value less than 100% for transformations like precipitation or solidification in a freezing range where the emerging phases do completely replace the parent. Generally

$$f = f(\lambda, t) = f(T, t) \tag{4.25}$$

Consider a growing island i of the new phase in the parent. If the island grows as a sphere, then at a fixed temperature T its size (i.e., volume V_i) at time t is

$$V_i(t) = \frac{4}{3}\pi r_i^3$$

where $r_i = r_i(t - \tau_i)$ is the island's radius and τ_i is the time at which the island's nucleus initially formed. If the island radius is given by

$$r_i = g \times (t - \tau_i)$$

then

$$V_i(t) = \frac{4}{3}\pi g^3 (t - \tau_i)^3$$

The total transformed volume of material V at time t is then

$$V(t) = \sum_i V_i(t) = \frac{4}{3}\pi g^3 \sum_i (t - \tau_i)^3$$

For a large system of n islands in a volume V_0, $dn = \dot{n}V_0 dt = \dot{n}V_0 d\tau$ and we replace summation with integration:

$$V(t) = \frac{4}{3}\pi g^3 \int_0^t (t - \tau)^3 \, dn = \frac{4}{3}\pi g^3 \dot{n} V_0 \int_0^t (t - \tau)^3 \, d\tau = \frac{\pi}{3} g^3 \dot{n} V_0 t^4$$

as the transformed volume.

From V, we can find

$$f(t) = \frac{V(t)}{V_0} = \frac{\pi}{3} g^3 \dot{n} t^4 \tag{4.26}$$

as the transformed fraction. **Equation 4.26** works well for small values of f. However, as f increases

1. the volume of untransformed parent material will decrease, producing a decrease in nucleation, and
2. island growth will be inhibited by the presence of other islands, producing a decrease in effective growth rate.

For these reasons, there will be "diminishing returns" in the transformation as time goes on. A more appropriate expression that captures this effect would be

$$f(t) = 1 - \exp\left(-\frac{\pi}{3} g^3 \dot{n} t^4\right) \tag{4.26'}$$

Equation 4.26′ describes a highly idealized version of what happens during a transformation. Additionally, the temperature will affect the progress of the transformation through the parameters g and $\dot{n}$, as described above. A more general expression that can describe the behavior of a wide range of real materials would look like

$$f(t,T) = 1 - \exp\left[-K(T)(t - t_0)^a\right] \tag{4.27}$$

where the parameter $K(T)$ reflects the (temperature-dependent, thermally activated) contributions of g and $\dot{n}$ and t_0 indicates a delay in the onset of transformation associated with the development of the initial nuclei. The value of the parameter a ($1 < a \leq 4$, typically) is chosen to best fit the observed behavior. **Equation 2.27** is sometimes called the Avrami equation. Some examples of the schematic f curves are shown in **Figure 4.12**. As the temperature T increases, so does K, and it follows that the overall transformation will be shifted to shorter intervals of time.

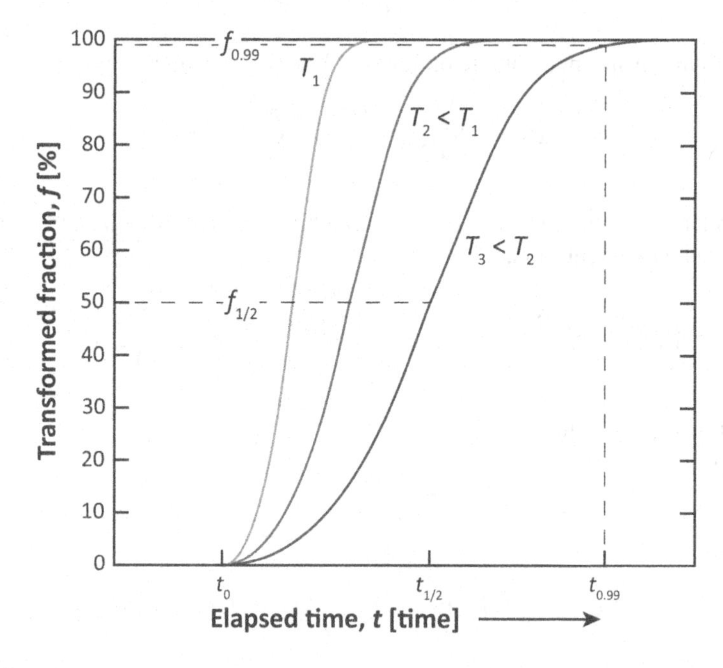

FIGURE 4.12 Transformation progress in terms of the transformed fraction f. The transformation increases from a transformed fraction of $f_0 = 0\%$ up to a transformed fraction of $f_{max} = 100\%$ (completely transformed). Decreasing the temperature at which the transformation occurs leads to lower transformation rates and generally longer transformation times. The maximum transformation rate occurs at a transformed fraction $f_{1/2} = 50\%$.

Example 4.16:

You have an alloy system that solidifies according to the Avrami equation with $a = 3$, $K = 10.0\ \text{s}^{-3}$, and $t_0 = 1.0$ s at a particular temperature. How long does it take the system to reach a transformed fraction $f = 50\%$? How much longer does it take to reach $f = 99\%$? What is the maximum transformation rate df/dt?

SOLUTION

For the given set of transformation parameters $\{K, t_0, a\}$ and taking f as a percentage, we have

$$f(t) = \left\{1 - \exp\left[-K(t - t_0)^a\right]\right\} \times 100$$

or

$$t = t_0 + \sqrt[3]{-\frac{1}{k}\ln\left[1 - \frac{f(t)}{100}\right]}$$

For the transformation of $f = 50\%$ and the given values, we obtain

$$t_{50\%} = 1.0 + \sqrt[3]{\frac{1}{10.}\ln\left[1 - \frac{50}{100}\right]} = \underline{\mathbf{1.4\ s}}$$

It is easy to determine the transformation rate at this time. Since

$$\frac{df}{dt} = 100\frac{d}{dt}\left\{1 - \exp\left[-K(t - t_0)^a\right]\right\} = 100aK(t - t_0)^{a-1}\exp\left[-K(t - t_0)^a\right]$$

we get

$$\frac{df}{dt} = 100(3)(10.\ \text{s}^{-3})(1.4 - 1.0)^{a-1}\exp\left[-(10.\ \text{s}^{-3})(1.4\ \text{s} - 1.0\ \text{s})^3\right] = \underline{\mathbf{250}\frac{\%}{\mathbf{s}}}$$

Similar results can be obtained easily in MATLAB:

```
>> syms t
>> a = 3; K = 10; t0 = 1;
>> f(t) = 100 - 100*exp(-K*(t - t0)^a);
>> t50 = double(solve(f(t) == 50))
T50 =

    1.4108 + 0.0000i
    0.7946 + 0.3557i
    0.7946 - 0.3557i

>>
```

(We recognize the first result as the only useful one `t50(1)`.) Also

```
>> df(t) = diff(f(t));
>> df(t50(1))
ans =
   253.1094
>>
```

Furthermore, if we wish to know how much longer the transformation takes after the halfway point $t_{1/2}$, we calculate

```
>> t99 = double(solve(f(t) == 99));
>> t99(1) - t50(1)
ans =
   0.3615
>>
```

We can also assess transformation processes using a modified diagram of the type shown in **Figure 4.13**. This type of diagram is called a **time-temperature-transformation diagram** or "TTT diagram". In **Figure 4.13(a)**, the transformation curves of **Figure 4.12** have been projected onto a set of axes "time" and "temperature" so that the initial time t_0, the ½-completion time $t_{1/2}$, and the time to "completion" $t_{0.99}$ coincide with the leftmost, middlemost, and rightmost points in a transformation "band" at a given temperature (e.g., 400°C). The pathway in (t, T)-space by which a material traverses the transformation band influences not only how long the transformation takes but also the microstructure that results. For example, the treatment pathway specified in **Figure 4.13(b)** corresponds to an **age-hardening** process in which Al_2Cu precipitates are "ripened" over a number of days to obtain the optimal dispersion and size for a given application.

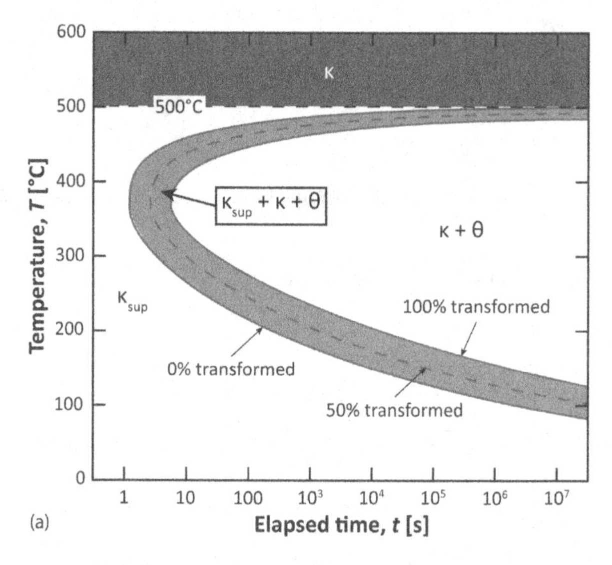

FIGURE 4.13 Example TTT diagram and heat-treatment design. The TTT diagram for a typical Al-Cu alloy in the precipitation range ($W_{Cu} < 6$ wt%) is provided in (a). *(Continued)*

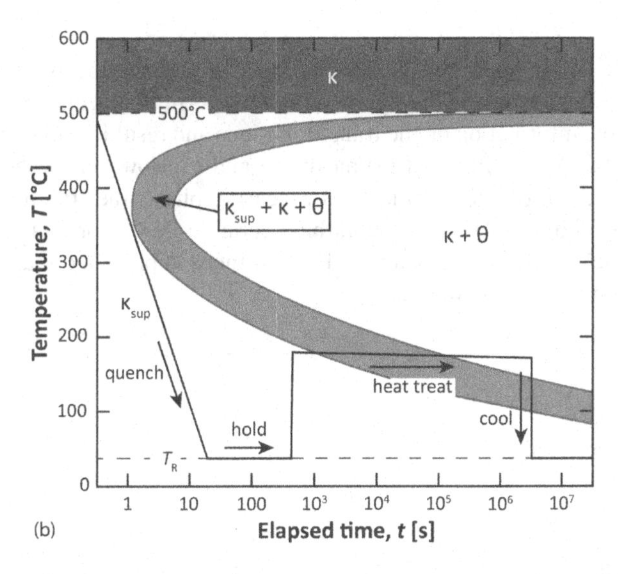

FIGURE 4.13 *(Continued)* The transformation from a supersaturated solid solution κ_{sup} into a heterogeneous, two-phase mixture of θ precipitates in a κ matrix proceeds as the transformation bounds (0% → 100%) are traversed. A schematic heat treatment is depicted in (b). The alloy is brought to room temperature T_R via a quench that evades the "nose" of the precipitation transformation, retaining its supersaturated (metastable) structure. Subsequent reheating at a later time permits a controlled transformation at a lower temperature (i.e., age hardening).

APPLICATION NOTE – IRON AND STEEL METALLURGY

One of the most important phase diagrams in engineering is the Fe-Fe_3C binary diagram, shown in the figure below. The form of this diagram provides considerable information on the transformations and resulting microstructures in steels. Steels are alloys of primarily Fe and C. However, because of the kinetics of the formation of the carbon-bearing phase, the structure of most steels is a mixture of Fe and the carbide ceramic Fe_3C. For this reason, the "metastable" phase diagram for Fe-Fe_3C is more appropriate than the true-equilibrium Fe-C phase diagram.

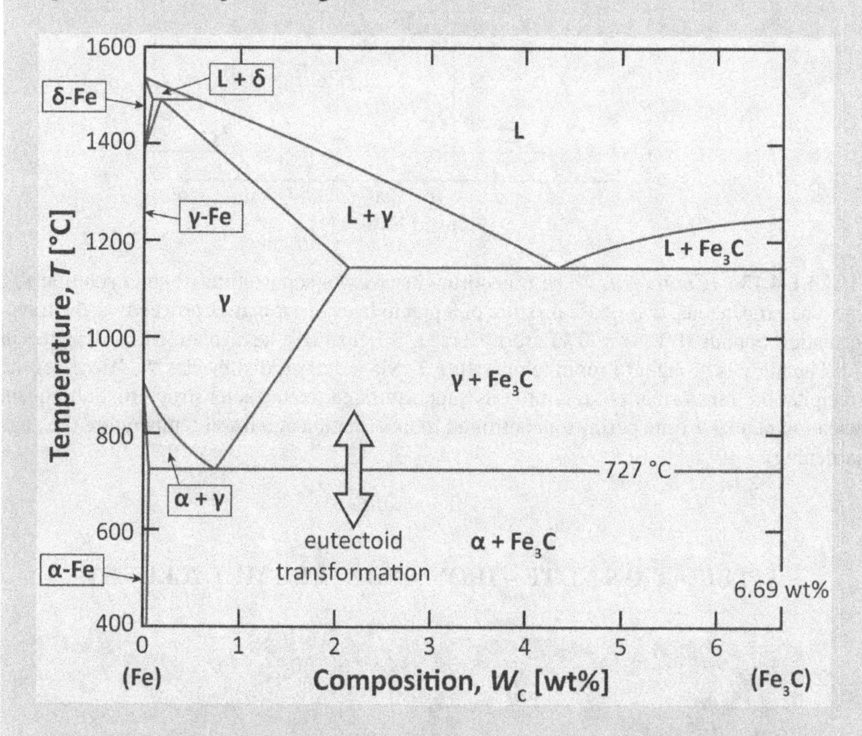

Many other metallic elements (e.g., Mn, Cr, and Ni) can be added to steel to adjust the metal's composition, forming different solid solutions, altering carbide formation, and for other reasons. (Sometimes, elements such as Si are included, as well.) The ability to substantially modify the properties with various alloying elements makes steel a versatile material that can be produced in quantity and to order. The Fe-Fe_3C diagram is only set up with two components but is useful for understanding the basic transformations at play. There are also, in fact, many steels that contain only F and C for which the diagram is entirely sufficient. Since the combination of 3 Fe atoms and 1 C atom produces a material that is 6.69 wt% C, the composition of the axis is labeled by wt% C and terminates at 6.69 wt% (i.e., pure Fe_3C).

The most common types of steels have 0–2 wt% C, so most study of the Fe-Fe_3C diagram is confined to the left-hand side. (Compositions > 2% C are typical of cast irons.) Along the left (100 wt% Fe) axis, we recognize the allotropes of Fe: α-Fe, γ-Fe, and δ-Fe and the respective solid-state transformation temperatures. The α and γ gamma phases are solid solutions of C in Fe, where the structure of the Fe is either BCC (for α) or FCC (for γ). The α phase is sometimes called "ferrite", and the gamma phase has the alternative name "austenite". Fe_3C is called "cementite".

There is another solid-state transformation of note in this diagram; this is the **eutectoid transformation** at $T_{e'}$ = 727°C. It is represented like

$$\gamma \underset{\alpha+Fe3C}{\longleftrightarrow} \alpha + Fe_3C$$

When the temperature of the alloy drops below the eutectoid temperature $T_{e'}$, any of the solid-solution γ phase that is present transforms into a mixture of two phases: α and Fe_3C. The two phases in the mixture will have a form similar to the structure of **Figure 4.4(c)**: closely spaced, alternating layers. This layered mixture is called "pearlite" and is shown below.[5] The pearlite (dark) exists as "islands" in the ferrite (white) matrix. A close inspection of the pearlitic islands reveals closely spaced layers of α and Fe_3C. At low-C concentrations (< 0.77 wt% C), the microstructure exhibits both ferrite and pearlite; at higher C concentrations, the structure is cementite and pearlite. The relative amount of each of the α or Fe_3C phases (including the quantity of each bound up in the pearlite) can be determined by applying the lever rule to compositions/temperatures within the α + Fe_3C two-phase field.

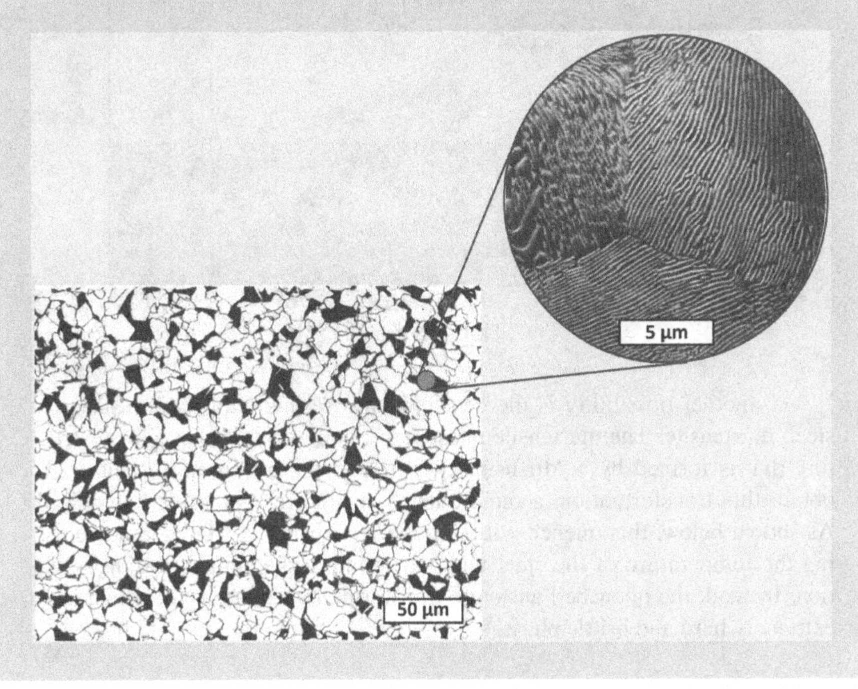

The properties of steel result from the combined properties of the individual phases. Ferrite is a phase with the characteristics of a BCC metal and Fe_3C has the properties of a ceramic. For example, we expect that the metal will have some hardness/brittleness to it because of the ceramic cementite component, but steel alloys can still be tough and flexible by virtue of the metallic component and the presence of large amounts of interface in the eutectoid mixture. The TTT diagram for steel is shown below. This diagram indicates the possibility of obtaining a number of different phase arrangements, including coarse pearlite and fine pearlite. Coarse pearlite has the lamellar structure of pearlite but with a wide lamellar spacing; fine pearlite has a small lamellar spacing. These two configurations (and the possibilities that span them) produce steels with different properties. (Bainite is yet another interwoven phase configuration composed of ferrite and cementite that is produced by a non-eutectic transformation.) Both pearlite and bainite consist of the equilibrium phases (α and Fe_3C) inferred from the Fe-Fe_3C phase diagram.

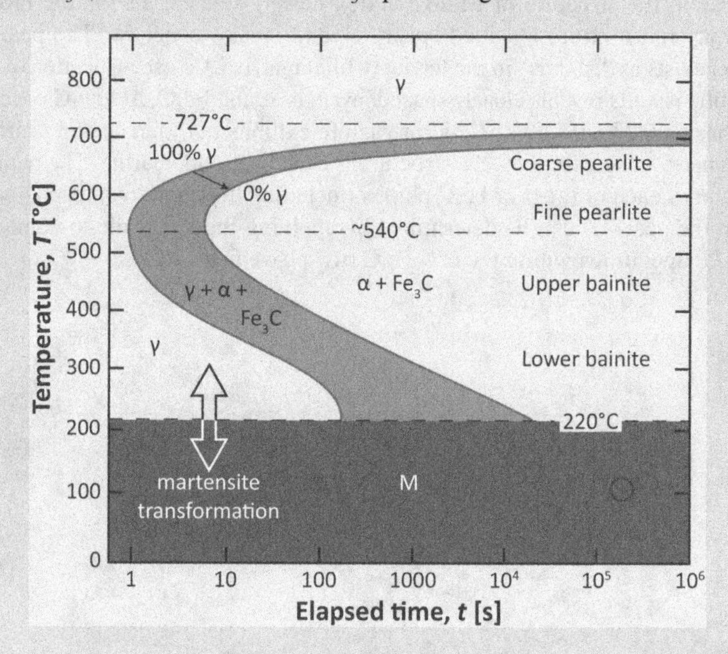

Yet another possibility is the development of a non-equilibrium phase in steel: martensite. The martensite phase has a body-centered tetragonal structure that is formed by a "diffusionless" transformation from austenite/γ. To obtain this transformation, a rapid quench at high temperatures is required. As shown below, this quench will skip the nose of the TTT diagram, reducing the temperature of the steel without initiating the eutectoid transformation. Instead, the quenched austenite transforms into martensite, which is an extremely hard and brittle phase.[6]

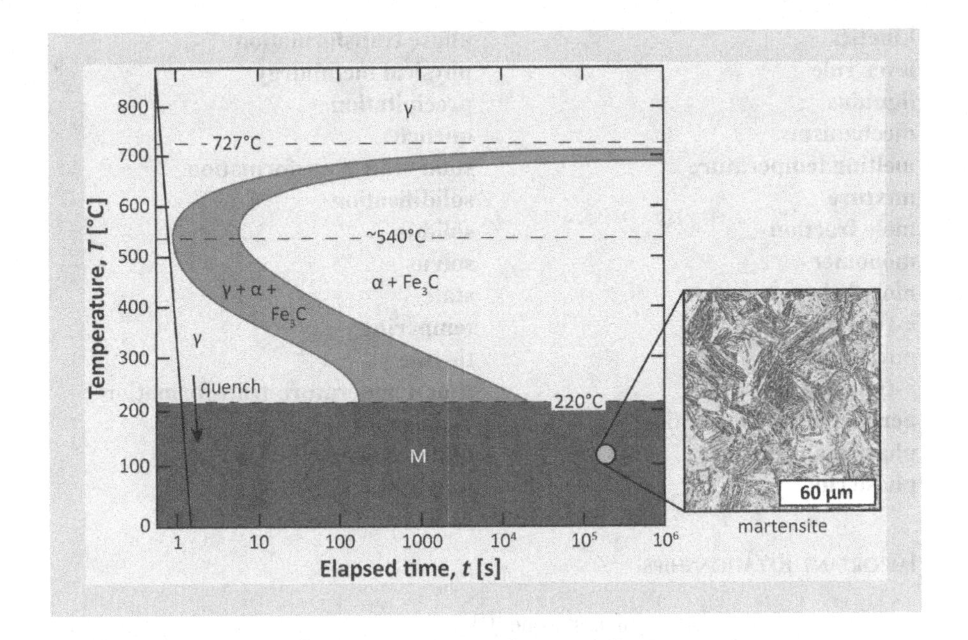

4.7 CLOSING

We have seen in this chapter some important examples of how materials are formed from their "raw" constituents. The constituents/raw materials are typically blended, and the chemical forces at work within the blended materials drive the material toward a particular structural configuration via a discrete number of transformations. A phase diagram predicts the end (or "equilibrium") state of these transformations based on thermodynamic principles. The progress of these transformations also has an impact on the morphology of the phases, i.e., their organization, size, and dispersion. The principles of transformation are those of the kinetics of mass transport (i.e., diffusion) and phase formation (nucleation and growth). These principles have considerable application in the production of many different materials, including important metallurgical systems like steel.

4.8 CHAPTER SUMMARY

Key Terms

activation energy
age-hardening
allotropy
binary phase diagram
component
concentration gradient
copolymers
degree of polymerization
diffusion
diffusion flux
diffusivity
eutectic solidification
eutectic temperature
Fick's First Law
Fick's Second Law
freezing range
growth
intermetallic

kinetics
lever rule
liquidus
mechanisms
melting temperature
mixture
mole fraction
monomer
morphology
***n*-type doping**
nucleation
***p*-type doping**
peritectic transformation
phase diagram
phase rule
phase transformation
physical metallurgy
precipitation
quench
solid-state transformation
solidification
solidus
solvus
state
tempering
tie line
time-temperature-transformation diagram
undercooling
weight fraction

Important Relationships

$$X_1 = \frac{\text{Moles of component 1}}{\text{Total moles of all components}} \quad \text{(mole fraction)}$$

$$W_1 = \frac{\text{Weight of component 1}}{\text{Total weight of all components}} \quad \text{(weight fraction)}$$

$$W_1 = \frac{X_1 A_1}{X_1 A_1 + X_2 A_2} \quad (X \text{ to } W \text{ conversion})$$

$$X_1 = \frac{W_1 A_2}{W_1 A_2 + W_2 A_1} \quad (W \text{ to } X \text{ conversion})$$

$$M = A \times n \quad \text{(molecular weight)}$$

$$\bar{M}_\text{n} = \frac{\sum_i N_i M_i}{\sum_i N_i} \quad \text{(number-average molecular weight)}$$

$$\bar{A} = A_1 X_1 + A_2 X_2 + \cdots + A_n X_n \quad \text{(average monomer weight)}$$

$$M = \bar{A} \times n \quad \text{(copolymer molecular weight)}$$

$$P = P_1 \phi_1 + P_2 \phi_2 \quad \text{(rule of mixtures)}$$

$$P = \frac{P_1 P_2}{\phi_1 P_2 + \phi_1 P_2} \quad \text{(inverse rule of mixtures)}$$

$$\varphi = C + 1 - F \quad \text{(phase rule)}$$

$$f_\alpha = \frac{W - W_\beta}{W_\alpha - W_\beta} \text{ and } f_\beta = \frac{W_\alpha - W}{W_\alpha - W_\beta} \quad \text{(lever rule)}$$

$$J_x = \frac{n}{At} \text{ or } J_x = \frac{m}{At} \quad \text{(flux definition)}$$

$$C_A = \frac{n_A}{V} \text{ or } C_A = \frac{m_A}{V} \quad \text{(concentration)}$$

$$J = -D\frac{dC}{dx} \quad \text{(Fick's First Law)}$$

$$d = \sqrt{2Dt} \quad \text{(typical diffusion distance)}$$

$$D = D_0 \exp\left(-\frac{\Delta U}{k_B T}\right) \text{ or } D = D_0 \exp\left(-\frac{Q}{RT}\right) \quad \text{(diffusivity)}$$

$$\frac{\partial C}{\partial t} = D\frac{\partial^2 C}{\partial x^2} \quad \text{(Fick's Second Law)}$$

$$\Delta T = T_m - T \quad \text{(undercooling)}$$

$$\Delta G = \frac{\Delta T}{T_m}\Delta H_f \quad \text{(driving force for transformation)}$$

$$\dot{n} = k_1 \omega n^* \exp\left(-\frac{Q}{RT}\right) \quad \text{(nucleation rate)}$$

$$g = k_2 \exp\left(-\frac{Q}{RT}\right) \quad \text{(growth rate)}$$

$$f(t) = 1 - \exp\left(-\frac{\pi}{3} g^3 \dot{n} t^4\right) \quad \text{(Avrami equation)}$$

$$f(t,T) = 1 - \exp\left[-K(T)(t - t_0)^a\right] \quad \text{(general transformation)}$$

4.9 QUESTIONS AND EXERCISES

Concept Review

C4.1 Describe the difference between *interstitial* diffusion and *vacancy* diffusion. Why would you say interstitial diffusion is typically more rapid than vacancy diffusion?

C4.2 Explain the importance of controlling the temperature during materials processing.

Discussion-forum Prompt

D4.1 Using campus library resources, find a phase diagram for a binary system. You may wish to search the technical literature directly or utilize the ASM Handbooks online (https://dl.asminternational.org/handbooks/). Post your phase diagram and briefly describe the engineering importance of the binary system.

PROBLEMS

P4.1 Suppose you have a Cu-Zn alloy with $X_{Zn} = 0.30$. What is the composition of the alloy in wt%? What if you have an alloy of $Cu_{60}Zn_{30}Sn_{10}$?

P4.2 Calculate the fraction of styrene monomer units (in mole %) for a styrene-butadiene copolymer with 35 wt% styrene. The butadiene monomer has a formula $-(C_4H_6)-$.

P4.3 The specific volume v' of a polymer that is part crystalline ("c") and part amorphous ("a") is

$$V' = W_a V_a' + W_c V_c' = W_a V_a' + (1 - W_a) V_a'$$

where $[V'] = $ [volume/mass] and the W_i are the weight fractions of the amorphous and crystalline components. Calculate the crystallinity (as a weight fraction W_c) of a polyethylene (PE) sample with density $\rho = 930$ kg/m³. The specific volume of crystalline PE is $v_c' = 0.99 \times 10^{-3}$ m³/kg, and the specific volume of amorphous PE is $v_a' = 1.2 \times 10^{-3}$ m³/kg.

P4.4 Label all of the phase fields 1–10 on the phase diagram below. The labels will be "L", "α", "β", "γ", "δ", and "ε" and any necessary combinations.

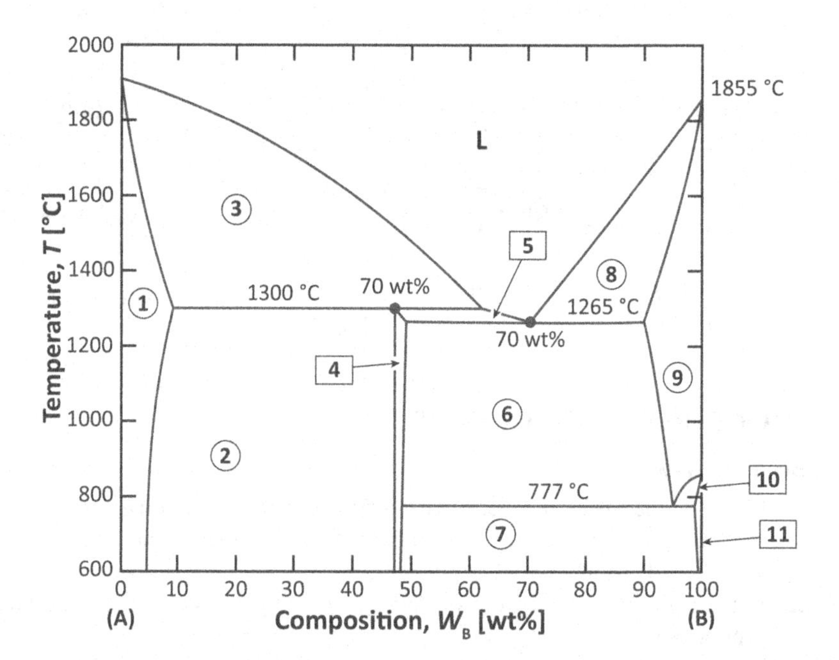

P4.5 For a system composition $W_B = 30\%$ and a temperature of $T = 1100°C$ in the phase diagram of **Exercise P4.4** above, what is the number of degrees of freedom of the system?

P4.6 The diffusivity of copper in brass is $D_{Cu} = 10^{-20}$ m²/s at 400°C. If the activation energy for diffusion in this system is 195 kJ/mol, what is the diffusivity at 600°C?

P4.7 An engineer wishes to apply a nitridization treatment to improve the surface properties of a steel component. The treatment involves diffusing nitrogen atoms to a depth of 0.50 mm into the Fe surface. Approximately how long will nitridization take at 1,200°C? **Data:** $D_0 = 5.0 \times 10^{-7}$ m²/s, $Q_d = 77{,}000$ J/mol.

P4.8 Suppose that a solidification transformation has an n value of 2 and that $k(T) = k_0 T^m$, where $m = 1.2$. If the transformation takes 10 s to complete (i.e., $f = 99.9\%$) at 400 K, how long does it take to complete at 700 K?

MATLAB Exercises

M4.1 Write a MATLAB function that performs the lever-rule calculations for a given two-phase material system at a fixed temperature: `leverrule(W, W1, and W2)`. The function will have three inputs: the weight fraction of phase #1 (`W1`) at temperature, the weight fraction of phase #2 (`W2`), and the composition of the system overall (`W`).

M4.2 Using the "diffusion through a surface" solution of **Table 4.3**, plot the composition profile $C(x)$ of a system at three time intervals $t = \{30$ min, 60 min, 90 min$\}$. The system has $D = 3 \times 10^{-12}$ m²/s, exterior composition $C_s = 1.0$ g/m³, and initial interior composition $C_0 = 0$ g/m³.

NOTES

1. E. van der Lingen et al. "Metallography and Microstructures of Precious Metals and Precious Metal Alloys". In: *ASM Handbook, Volume 9: Metallography and Microstructures*. Ed. by G. F. Vander Voort. Materials Park, OH, USA: ASM International, Dec. 2004, pp. 860–876.
2. D. A. Porter, K. E. Easterling, and M. Y. Sherif. *Phase Transformations in Metals and Alloys* (revised reprint). Third edition. Boca Raton, FL, USA: CRC Press, Feb. 2009.
3. Doru M. Stefanescu and Roxana Ruxanda. "Metallography and Microstructures of Precious Metals and Precious Metal Alloys". In: *ASM Handbook, Volume 9: Metallography and Microstructures*. Ed. by G. F. Vander Voort. Materials Park, OH, USA: ASM International, Dec. 2004, pp. 71–92.
4. Note that C_A may be obtained from W_A according to $C_A = W_A/[(W_A/\rho_A) + (W_B/\rho_B)]$.
5. Adapted with permission from B. L. Bramfitt and S. J. Lawrence. "Metallography and Microstructures of Carbon and Low-Alloy Steels". In: *ASM Handbook, Volume 9: Metallography and Microstructures*. Ed. by G. F. Vander Voort. ASM International, Dec. 2004, pp. 608–626 and G. Kraus. "Microstructures, Processing, and Properties of Steels". In: *ASM Handbook, Volume 1: Properties and Selection: Irons, Steels, and High-Performance Alloys*. Ed. by ASM Handbook Committee. ASM International, Jan. 1990, pp. 126–139.
6. Adapted with permission from M. E. Finn. "Machining of Carbon and Alloy Steels". In: *ASM Handbook, Volume 16: Machining*. Ed. by ASM Handbook Committee. ASM International, Jan. 1989, pp. 666–680.

5 Designing with Materials
From Imagination to Reality

LEARNING OBJECTIVES

After completing this chapter, you should be able to:

1. List and describe the steps in the engineering design process.
2. Explain why the design process is iterative.
3. Differentiate between an objective and a constraint.
4. Describe how materials understanding is incorporated into the design process.
5. Derive a material index describing the objective of a given design.
6. Apply the materials-selection process to identify the best material for a given application.
7. Describe the components of the cost of a material, the factors influencing its lifespan, and the long-term implications for its use.

5.1 DESIGNING FOR USE

Engineers are frequently tasked with creating manufactured goods (and individual components of manufactured goods) that do not currently exist. After the need for the product has been recognized, but before it has entered production, the product exists only as a **design**. The method by which the design is conceived and developed so as to bring it closer to a product is the **engineering design process** (here, just "the design process"). Typically, this process involves the identification, explication, and solution of several engineering problems. The problems encountered in the design process can be simple and resolvable on their own, or they can be complicated and interrelated. Since the latter, complex arrangement of problems is typically the rule, there is unlikely to be a straightforward "one size fits all" strategy for obtaining an adequate (much less perfect) design. Rather, the goal of the design process is to determine the *best* solution out of a menu of possibilities. These design problems we refer to are therefore "open-ended", and unlike the exercises you find yourself tasked with in most of your engineering textbooks.

Consider a few possible ways to approach and solve the interrelated problems involved in producing something new and satisfactory:

- **Trial and error.** Since there is no obvious, direct path to obtaining a solution to a complex problem, you might just "take a stab at it", proposing a solution that you guess is appropriate in a number of critical aspects, trying it out, considering the successes and shortcomings you observe, and making a new guess. This approach is not intrinsically wrong, but if the guesses are not well-informed by prior results or well-defined requirements, then

DOI: 10.1201/9781003214403-6

the trial-and-error process may be unsatisfactory because of the excessive time and effort required.

- **Designer experience.** A designer may have considerable experience in the application area that the product is meant for and may leverage that experience in the design process through customary (sometimes unwritten) rules or practices or other "heuristics". Knowledgeable designers might synthesize aspects of existing designs into a new design in an intuitive and potentially effective manner. This kind of experience is invaluable but will be limited in applicability when the design aspects or materials involved are new to the designer.
- **Computer-assisted methods.** Computers are capable of balancing a large number of variables within given constraints and, therefore, can be useful tools in engineering design. Trial-and-error–based design strategies can be automated in some aspects to be made more efficient and effective. Artificial intelligence routines can access design possibilities that might never result from a human-directed decision-making process via algorithms that leverage "evolutionary" or "neural network" principles and that lack the biases of human designers. These computer contributions might inform various stages in the development of the solution.
- **Formal process.** Formal processes are employed to ensure that the overall process remains grounded with respect to the original goals, the problems involved are well-defined, and the resolution of the problems is evidence-based (i.e., the process is *rational*, overall). These processes are typically organized as a series of steps that involve ordered, distinct activities.

The engineering design process is such a formal process, and we introduce it in more detail below. Since the design process for a physical product necessarily involves decisions about what materials to make it out of, we will dedicate considerable attention to this aspect of the design process. We call this materials component of the process **materials selection** and introduce it in **Section 5.3**.

5.2 THE ENGINEERING DESIGN PROCESS

Learning and applying the design process is an important component of an engineering education. You may have been exposed to this process in your other courses, but we will review it here and go on to discuss the important ways that material understanding plays a role. The design process is depicted in **Figure 5.1**. The design process proceeds in ordered stages, or "phases", each associated with unique activities. The beginning of the process is the recognition of a need or the identification of a problem to be solved, and the process concludes with a full specification for the device or solution. Additionally, the design process is *iterative*, meaning it involves full or partial repetition under circumstances when it is deemed necessary.

Phase 1. Identify the need or define the problem. This phase may be either self-evident or may require some additional thought on the part of the designer. The designer may have been provided with a detailed description

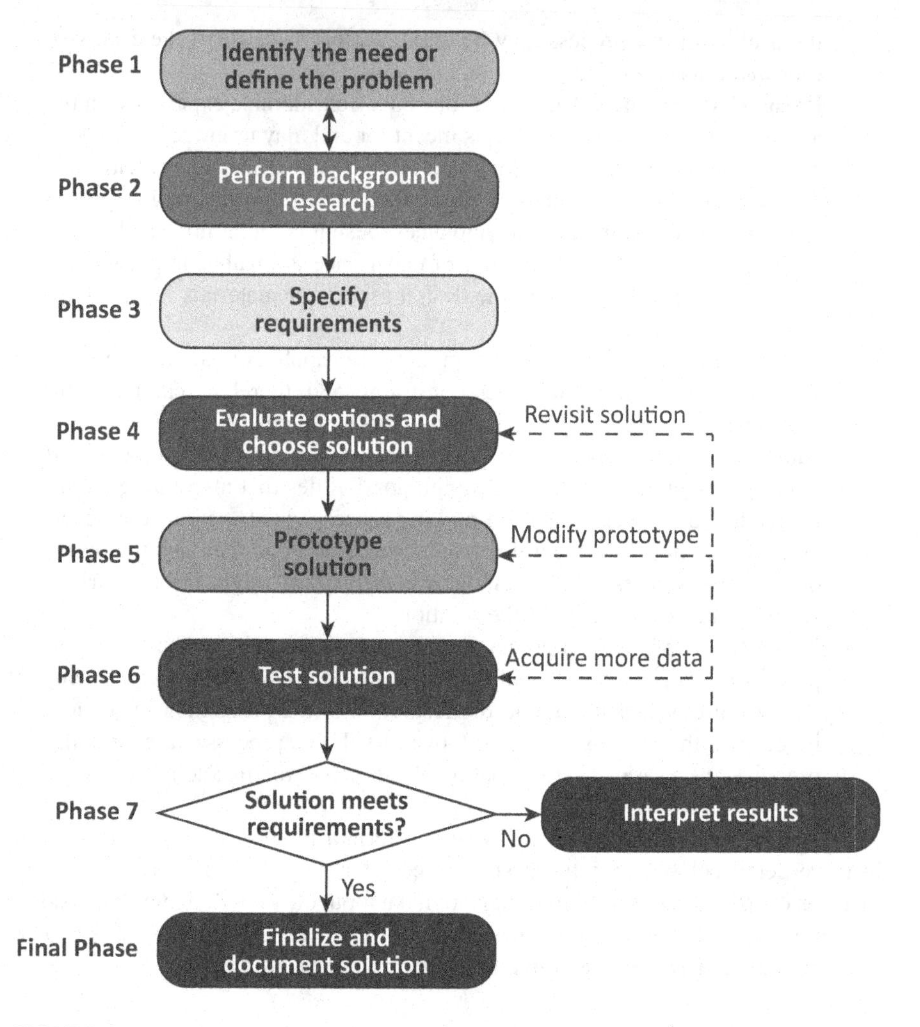

FIGURE 5.1 The engineering design process. The design process consists of seven phases, plus a final "wrap-up" phase.

of the purpose of the manufactured component, or the need for a solution generally may be recognized and discussed in less precise terms. In the former case, the designer should ensure they understand how the requested design "fits" with the larger project or otherwise satisfies the end user, and in the latter case, the designer should carefully consider the context that produces the need and how to satisfy the end user. Ask yourself: what is it that the product should do? This is its **function**. In addition to satisfying the function required, what qualities or attributes would a successful design possess? These become **objectives** for your design work.

Phase 2. Perform background research. Some questions will likely have occurred to you, the designer, in formulating and refining the problem in

Phase 1: Are there existing designs that satisfy the need? What is important about the function, either as a standalone device or a subcomponent of a larger assembly? *Who* will use it? When we consider the qualities of a successful design, which are strictly required, and which are merely desired? Performing research on questions like these will help refine your initial thoughts and organize your efforts in the next phase.

Phase 3. Specify requirements. The specification phase requires a deepening of your thinking about the engineering principles and the way you talk about the design. The requirements must be expressed as **constraints** along with the established objectives in suitable engineering terms. A "hard" constraint is a nonnegotiable condition that *must* be met, and a "soft" constraint is a negotiable (but desirable) condition. During the specification phase, objectives frequently assume the form of properties to be minimized or maximized. The end result of Phase 3 could include drawings, schemes, and plans that meet the requirements within the constraints. There may be multiple possible solutions in the offering at this time; all of these are carried over into the next phase.

Phase 4. Evaluate options and choose solution. The results of the specification phase need to be sorted and evaluated. Out of all the proposed designs that satisfy the constraints, which one meets the objectives *best*? This solution becomes the "working" design for Phase 5.

Phase 5. Prototype solution. The working design is typically converted from a sketch into a complete, dimensioned drawing and fabricated at this time. The important property of the prototype is that *it must be capable of being evaluated under conditions resembling those of service*. For example, if the prototype is physical, it must have dimensions that conform to the requirements closely enough so that it can be assessed for a good fit and/or has adequate range of motion. If the prototype has been made from the material selected, the prototype can be tested for the appropriate strength, stiffness, weight, conductivity, etc. If the prototype is not physical but instead a computer rendering [i.e., produced using computer-aided design (CAD) software], it can be evaluated for fit and possibly in numerous other ways using software that can simulate service conditions.

Phase 6. Test solution. When the prototype that is sufficient for the task of testing has been fabricated, the tests are carried out. It is important that these tests are not of the "let's try it and see what happens" variety but should be organized so that definite measurements performed during the tests can be compared to predetermined metrics that assess how well the objectives are met. The results of testing should be thoroughly documented and retained throughout the design process.

Phase 7. Establish success/Interpret results. Following the test(s) of Phase 6, you must decide if the prototype meets all the metrics required for success. If the answer is "yes", move on to the final phase. If the answer is "no", then an additional investment of effort is required before moving on. The successes and shortcomings of the prototype are recorded, and these data are fed back into the process as supplementary information in another iteration of the design process. Typically, the process doesn't restart from Phase 1 but reenters it at the nearest previous phase that the analysis of the data indicates. Adjusting experimental parameters (and *not* metrics) may be sufficient to establish a different answer to the questions of Phase 7 in cases where the results were borderline (i.e., lack precision) or lack sufficient accuracy. Or, if there were aspects of the prototype that were clearly insufficient, a new prototype may be made. This prototype may be more accurate in its geometry, be made of better-suited materials, etc. If the prototype cannot satisfy all metrics and cannot be modified to do so, the design must be replaced by choosing from one of the options discarded in Phase 7.

Final Phase. Finalize and document solution. To conclude the design process, all of the design decisions and experimental results must be documented in sufficient detail. "Sufficient detail" means that, for example,

a. a supervisor, engineer-in-charge, or chief engineer can assess the design solution and results and authorize (or not) the production of the design, or
b. the manufacturer or anyone with the required expertise can produce the design according to the specifications.

It should be obvious that good record-keeping and transparent decision-making practices throughout the process will pay off when it comes time to document the solution.

Be aware that descriptions of the design process can take different forms according to the engineering subject area that it is intended for, or it may vary based on the considerations of the one presenting it. The process that we present here captures the essential activities. This version also possesses the iterative structure that makes the process robust. Using an iterative process, the designer can "hone in" on the best available solution through repeated refinements.

5.3 THE MATERIALS-SELECTION PROCESS

The design process outlined above is a general-purpose methodology for finding the solution to an engineering problem. Being general, it has little to say about the specific mechanical aspects of the design, the aerodynamic aspects of design, the systems aspect of design, the materials aspects of design, etc. These distinct (yet interrelated) aspects become the focus of designers in those respective engineering areas. The *materials* portion of the design process is what is important to us in our course. This component is what we have called the materials-selection process. Like the design process in general, we wish for a process that is formal and rational.

Typically, a manufactured product will have several components. The overall product is a **technical system**, and the overall technical system may have assemblies, sub-assemblies, sub-sub-assemblies, and so on, as shown in **Figure 5.2**. At the

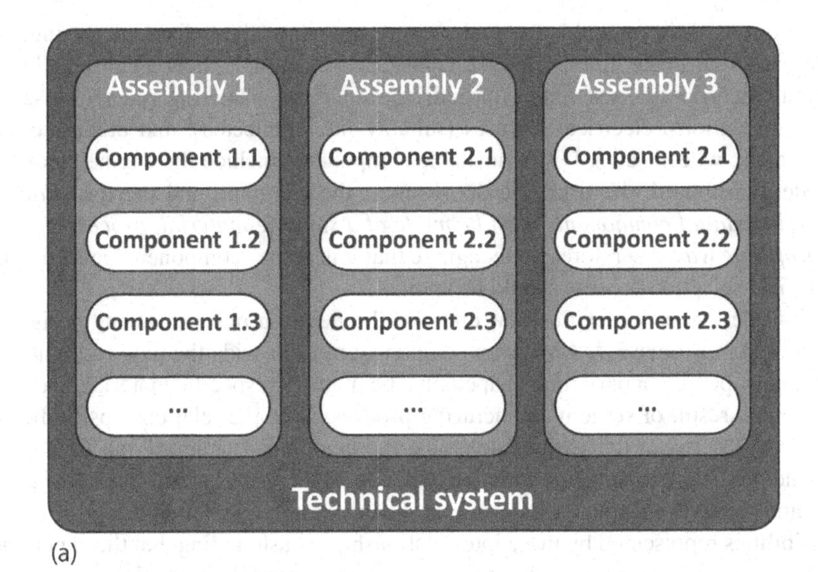

(a)

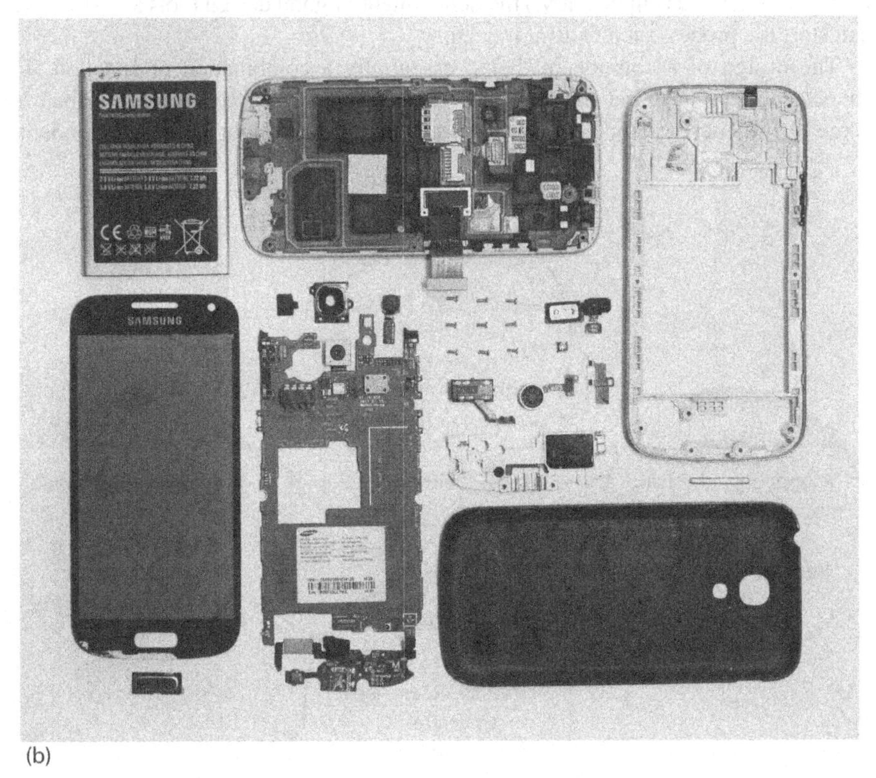

(b)

FIGURE 5.2 The breakdown of a technical system. The overall product is likely made from collections of parts that perform focused functions; these collections are the assemblies. The individual parts in the collections are the components, and these components, as individual and nonreducible pieces of the system, must be made of a particular material.

very bottom of the assembly hierarchy are components: the individual, distinguishable parts that make up the product. Consider, for example, how a car (a technical system) incorporates assemblies like a drivetrain, control/steering system, passenger accommodations, electrical features (suitably interconnected), and other accessories (trunk, wheels, etc.). Distinguishable components include the engine block, the molded dashboard, the tires, the doorframes, the gas tank, and the rest. *Each of these individual components must be made of a suitable material, processed in the appropriate way.* Therefore, we recognize that it is at the "component" level that the materials-selection process should be applied.

Consider the complexities that underlie the decision-making during design, as illustrated in **Figure 5.3**. Any given component must provide the required functionality, must possess a particular shape, must be made of a specific material, and must be the end result of some manufacturing process. Note the relationships (indicated by arrows) between the various domains: the component material and final shape depend on the processing, the function depends on the material and shape, the material and shape are mutually determined during design, etc. The variety of design possibilities represented by these interrelationships is astounding, but the experience of the designer and (importantly!) the deployment of good design tools and strategies can start the process on a fruitful trajectory.

The design of a component, being essentially a combination of function (F), shape/geometry (G), and material (M) considerations, is frequently described by some overall performance measure P. For example, P might be the component's

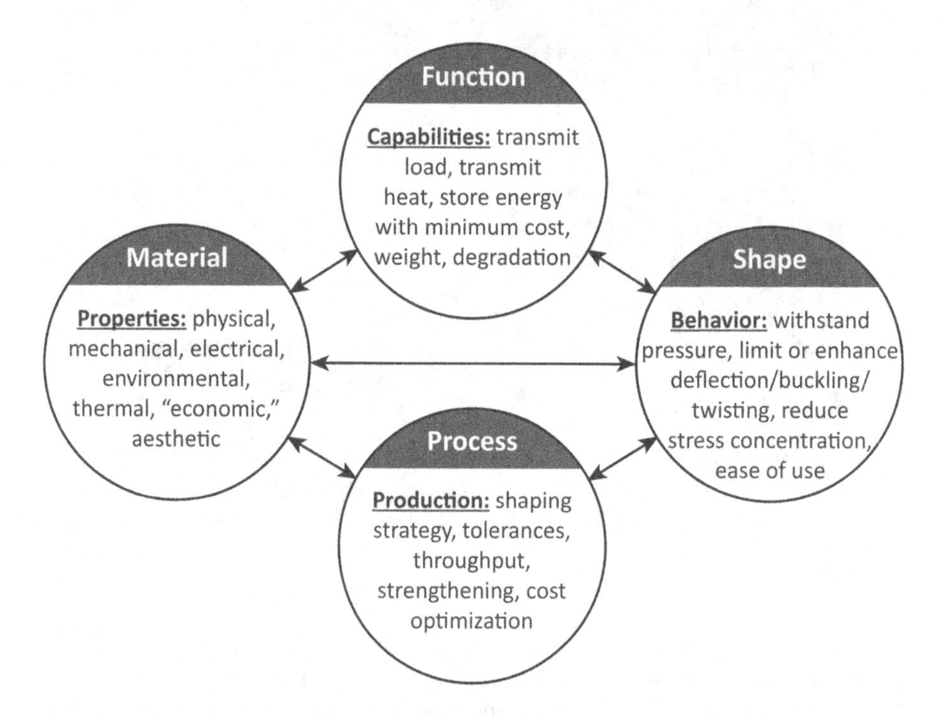

FIGURE 5.3 The interrelationship between various aspects of the materials-selection process.

weight, cost, or lifespan. The performance can sometimes be summarized as a general relationship:

$$P = P(\text{function, shape, material}) = P(F, G, M) \tag{5.1}$$

In **Equation 5.1**, the form of P is a representation of the way that the various factors F, G, and M (given as arguments to P) interact and combine (whether in a simple or complicated way). There may be maximum or minimum requirements on P (i.e., to satisfy a constraint), or it may be of interest to make P as high (or low) as possible (i.e., to pursue an objective). Frequently, the impact of the various factors can be isolated as factors:

$$P(F, G, M) = f(F) \times g(G) \times h(M) \tag{5.1'}$$

where f, g, and h are the separate functions. The materials-dependent factor h plays an important role and is called the **material index** of p. The remainder factor, $f \times g$, is called the "structural index".

Example 5.1:

Consider a simply supported beam of length L loaded by a force F in the center, like that shown below. Some of the important mechanical characteristics of this structure are provided in **Table 2.3**. The stiffness S of the beam is determined by how much force is required to produce some amount of deflection δ (taken at the center of the beam):

$$S = \frac{F}{\delta_{max}} = \frac{48EI}{L^3}$$

The beam has a square cross section with side length b and area $A = b^2$. The volume $V = AL$. From **Table 2.2**, the parameter I is

$$I = \frac{b^4}{12} = \frac{A^2}{12}$$

Suppose that the beam design includes a specific length $L = L_0$ and a minimum stiffness $S \geq S_0$ as requirements. Furthermore, the weight of the beam is to be as small as possible. What is the material index h that captures these requirements? How do you identify the materials properties that minimize the weight of the beam?

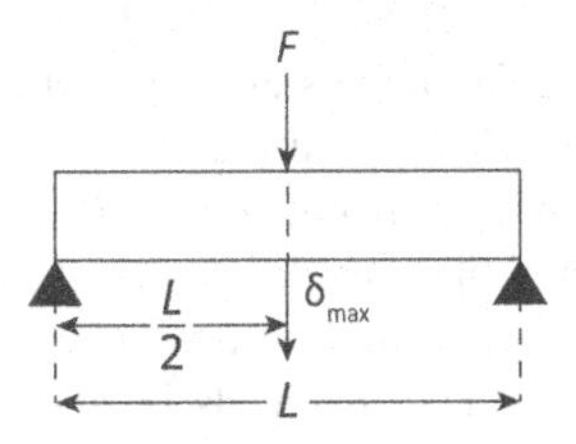

SOLUTION

The parameters L and S are specified by the requirements on length (= L_0) and stiffness (= S_0). The side length b (and hence A) is unspecified and is therefore a "free" parameter that may be chosen for convenience. How do we identify the crucial materials properties for the design and their roles? Consider a performance parameter P of the design: the weight of the beam. Since the mass of the beam $m = \rho V$, we find

$$P = mg = \rho g V = \rho g A L$$

where g is the gravitational acceleration on Earth. We can eliminate parameter A (since it is not related to any requirements) from the expression for P using the expression for I:

$$P = \rho g A L = \rho g L \sqrt{12 I}$$

Finally, incorporating the stiffness S gives

$$P = \rho g L \sqrt{\frac{L^3 S}{4E}}$$

Since $S \geq S_0$ and $L = L_0$, we must have

$$P \geq g \sqrt{\frac{L^5 S_0}{4}} \times \frac{\rho}{\sqrt{E}}$$

This requirement incorporates the constraints on L and S and gives us a basis for determining how well the design meets the minimum-weight objective. The performance parameter/weight P is composed of two independent factors. The first factor is the structural index (which includes only *fixed, known* values) and the second is the material index

$$\boldsymbol{h} = \left(\frac{\sqrt{E}}{\rho}\right)^{-1}$$

The material index depends *only* on the relevant materials properties of the beam (in this case, Young's modulus and density). We can minimize the weight of the beam by choosing a material that has a value of h as small as possible. Alternately, we can minimize weight by choosing $\sqrt{E}/\rho$ as large as possible. (The latter approach is quite useful to us; see below.)

Materials selection enters at each stage of the design process. At the beginning of the process (say, Phases 1 and 2), the designer should not have ruled out very many materials for the component. (Without any formalized requirements in place, who is to say what kind of material is or is not suitable?) During the following phases (3 and 4), it should become apparent that some materials will *not* be able to satisfy the requirements. Any material that cannot be used is excluded, and any material that *might* work is placed on a list. The materials in this list are then ranked, and the

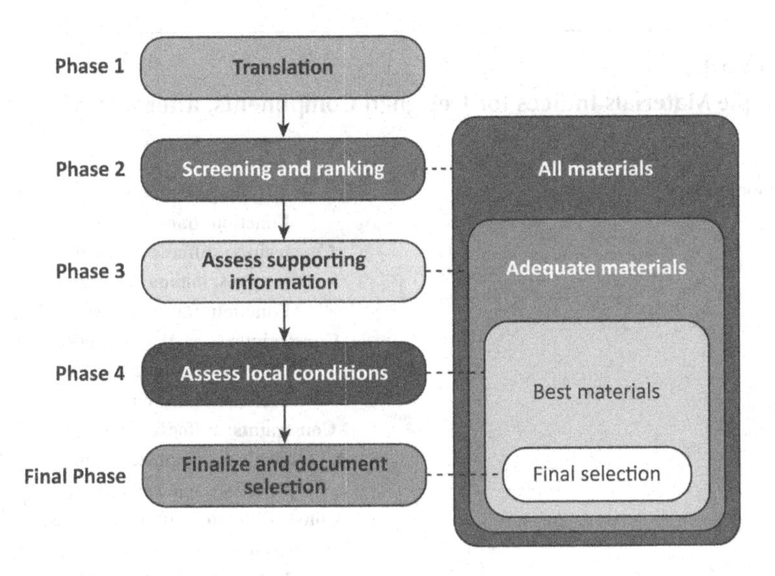

FIGURE 5.4 The materials-selection process. As the process proceeds, the materials that are under consideration at a given stage are winnowed from those in the previous stage.

top-ranked material is incorporated into the prospective design. The selected material (or a sufficiently close analog of it) is then used in the prototype (Phase 5).

There are a number of formal strategies for identifying the best material for a given component. The materials-selection process we present and discuss here is the general-purpose strategy of Ashby.[1] The process is illustrated in **Figure 5.4**. The various phases of the process are given on the left side of the diagram; the "space" of materials is on the right. At various phases, certain materials will be included (or not) in the considerations of that phase. For instance, when "assessing supporting information" in Phase 3, only those materials that were judged "adequate" in the previous phase are under consideration.

Phase 1. Translation. Before any information on candidate materials is assembled and analyzed, the design's requirements must be *translated* (or "converted") into a prescription for selection. Carefully consider the function(s) of the component. Identify the constraints that must be satisfied (both hard and soft) and the objectives the design should attempt to accommodate. Consider if there are any free parameters that might be subject to modification by the designer (e.g., shape changes). These constraints, objectives, and free parameters become inputs into the next phase.

In the simplest possible case, the constraints may come in the form of minimum or maximum property values that a candidate must have. For instance, if the component performs its function at a "service" temperature somewhere up to 300°C, the melting temperature T_m of the material selected must satisfy the simple criterion $T_m \geq 300°C$. In other cases, the use of a material index h may be required. Some examples of materials indices are given in **Table 5.1**. (**Example 5.1** illustrates how

TABLE 5.1
Example Materials Indices for Designed Components, after Ashby[1]

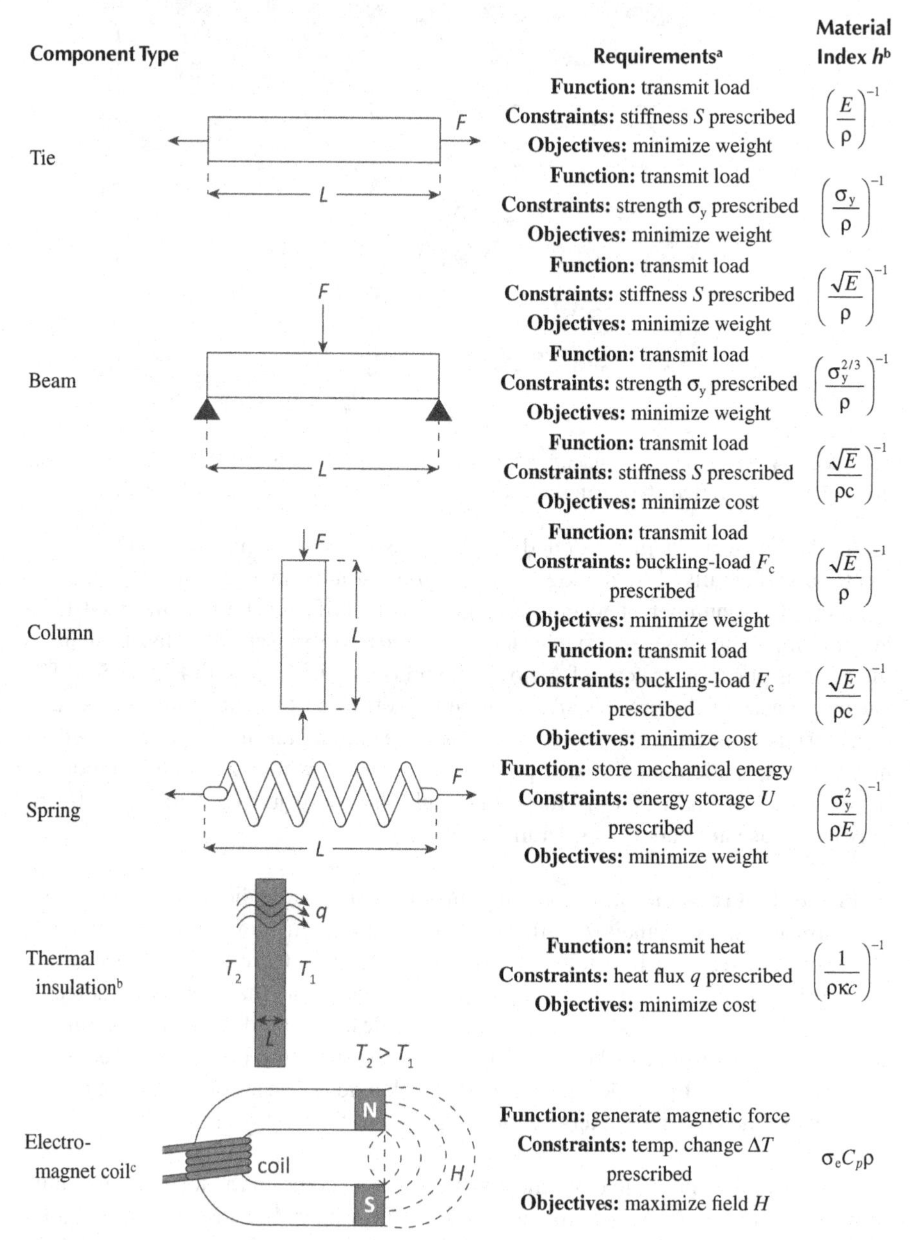

Component Type	Requirements[a]	Material Index h[b]
Tie	**Function:** transmit load **Constraints:** stiffness S prescribed **Objectives:** minimize weight	$\left(\frac{E}{\rho}\right)^{-1}$
	Function: transmit load **Constraints:** strength σ_y prescribed **Objectives:** minimize weight	$\left(\frac{\sigma_y}{\rho}\right)^{-1}$
Beam	**Function:** transmit load **Constraints:** stiffness S prescribed **Objectives:** minimize weight	$\left(\frac{\sqrt{E}}{\rho}\right)^{-1}$
	Function: transmit load **Constraints:** strength σ_y prescribed **Objectives:** minimize weight	$\left(\frac{\sigma_y^{2/3}}{\rho}\right)^{-1}$
	Function: transmit load **Constraints:** stiffness S prescribed **Objectives:** minimize cost	$\left(\frac{\sqrt{E}}{\rho c}\right)^{-1}$
Column	**Function:** transmit load **Constraints:** buckling-load F_c prescribed **Objectives:** minimize weight	$\left(\frac{\sqrt{E}}{\rho}\right)^{-1}$
	Function: transmit load **Constraints:** buckling-load F_c prescribed **Objectives:** minimize cost	$\left(\frac{\sqrt{E}}{\rho c}\right)^{-1}$
Spring	**Function:** store mechanical energy **Constraints:** energy storage U prescribed **Objectives:** minimize weight	$\left(\frac{\sigma_y^2}{\rho E}\right)^{-1}$
Thermal insulation[b]	**Function:** transmit heat **Constraints:** heat flux q prescribed **Objectives:** minimize cost	$\left(\frac{1}{\rho \kappa c}\right)^{-1}$
Electro-magnet coil[c]	**Function:** generate magnetic force **Constraints:** temp. change ΔT prescribed **Objectives:** maximize field H	$\sigma_e C_p \rho$

[a] c = cost per weight, κ = thermal conductivity, σe = electrical conductivity, Cp = specific heat. [b] See **Topic 2** for a discussion of thermal properties. [c] See **Topic 3** for a discussion of electrical and magnetic properties. [b] Frequently, the index h is written without the "−1" exponent, so that *maximizing* (rather than minimizing) the index gives the best performance.

to analyze a given situation to determine the appropriate index.) A constraint is satisfied (or not) by a material whose index has the requisite value.

Phase 2. Screening and ranking. Screening is the task that places the translated requirements alongside the properties of the materials available. Those materials that do not meet the minimum required properties (constraints) are excluded from further consideration. When the requirements are formulated as exact property values, the suitability of various materials can be assessed using a chart like that shown in **Figure 5.5**. This bar chart provides the ranges of a specific property for a variety of material types and some particular materials within those types. It is important to remember that within a given material type, there is variation in the properties according to the composition of that alloy, the formulation of the polymer, the fabrication method of the ceramic, etc. These differences give rise to differences in properties. For example, the modulus of Zn alloys can vary with different additions, treatments, etc., so its modulus is represented by a range of values. When compared against some hard constraint (say, 300 MPa indicated in the chart), the alloys that are suitable, unsuitable, and potentially suitable can be resolved.

In cases where the requirements are formulated as required material indices, candidate materials may be identified using charts like those in **Figure 5.6**. These **materials-property charts** show where the different materials lie in the "space" of their properties. For example, **Figure 5.6(a)** (reprised from **Chapter I**) shows how much strength σ_f a material has alongside its density ρ. The ratio between the quantities on the two axes $(\rho/\sigma_f)^{-1}$ is a material index. A line of constant σ_f/ρ (i.e., constant "strength-to-weight

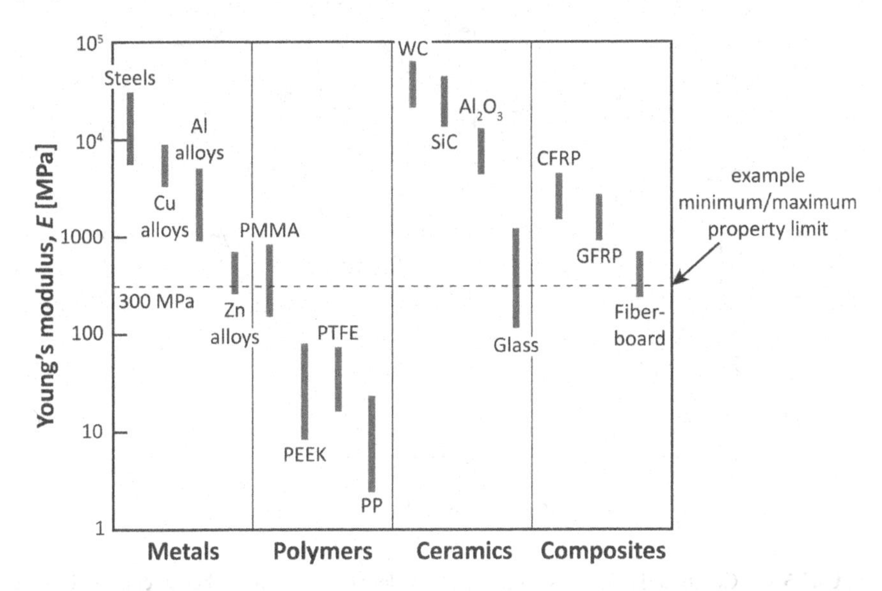

FIGURE 5.5 Materials bar chart. The chart indicates the property ranges for various materials types. As an example, a hard constraint of $E > 300$ MPa eliminates many polymers from consideration but indicates most metals are suitable.

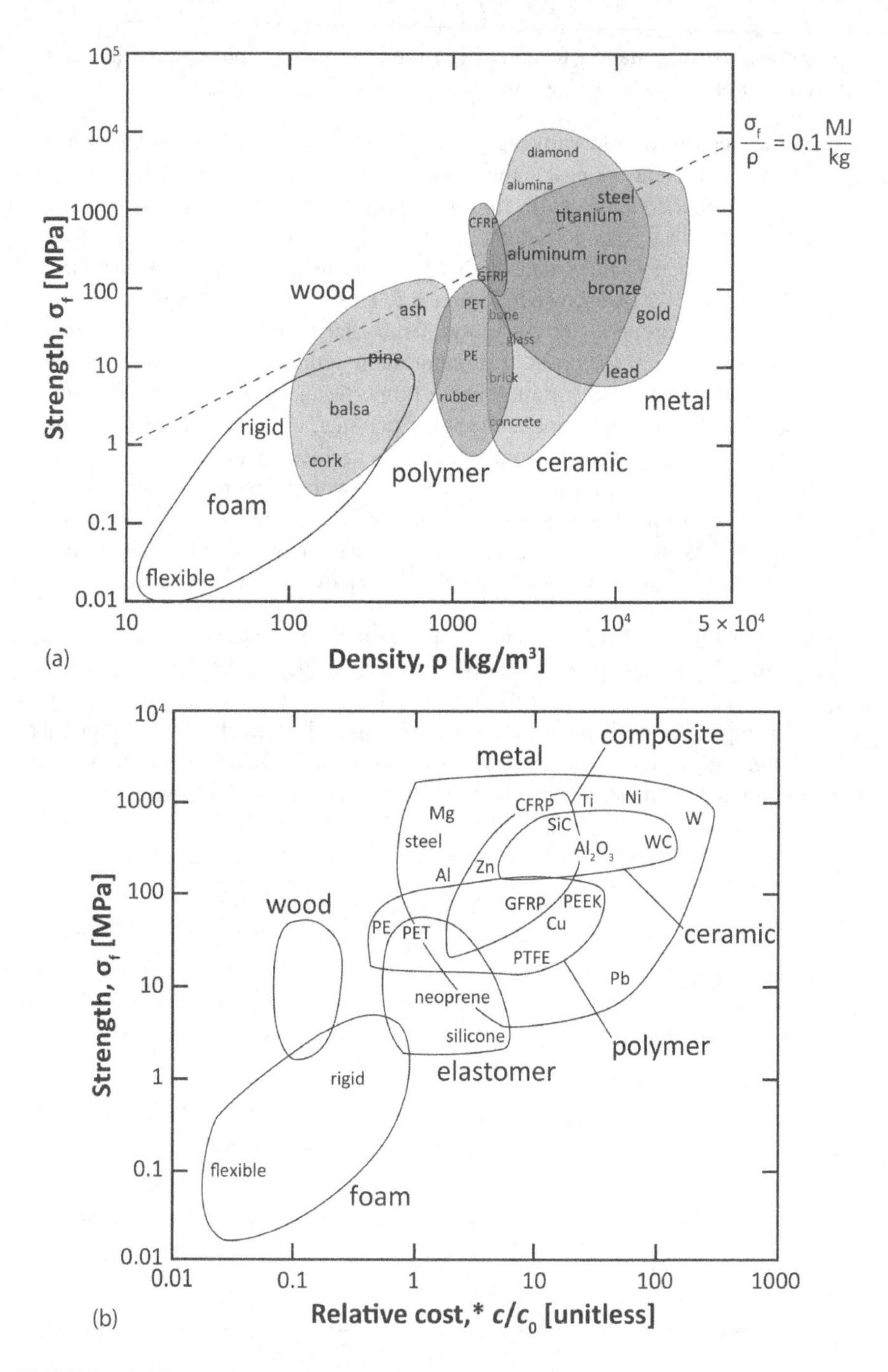

FIGURE 5.6 Combined materials property ("bubble") charts. Unlike the bar chart in **Figure 5.5**, the combined property charts permit comparison across multiple properties. Strength (σ_f) vs. mass density (ρ) is in (a). Lines of constant σ_f/ρ can be used to indicate requirements. Strength (σ_f) vs. normalized cost (c/c_0) is in (b).

ratio") like that shown on the chart can be used to identify materials sufficient for the requirements in a manner like that of **Figure 5.5**. The dashed line separates the space into two regions. Above the line, materials have a sufficient strength-to-weight ratio to satisfy the 0.1 MPa × m³/kg = MJ/kg requirement. Below the line, they do not. **Figure 5.6(b)** provides similar strength data compared to the material's cost. Please note that the cost c of a material, unlike its strength, is not fixed at a single numerical value by physics but is instead determined to a large degree by historical and social factors and so may change over time. Additionally, the cost of a material is only partially determined by the price requested by the supplier. Lifetime and disposal considerations also influence the overall cost associated with the material (see below).

Materials properties and materials indices are used to screen out those materials that cannot do the job as defined by the constraints. But of the materials that remain, these constraints have little to say about which material is the *best* fit for the application. The adequate materials (obtained by screening) can be *objectively* ranked from best to worst by considering how well the properties of the candidate materials align with the objectives. Consider the component of **Example 5.1**. The objective – to minimize the weight of the component – is embodied quantitatively in the material index $h = (\sqrt{E}/\rho)^{-1}$. The candidate materials with large values of $\sqrt{E}/\rho$ will be ranked higher than those with small values of $\sqrt{E}/\rho$.

Example 5.2:

Suppose that you are selecting a material for a small, efficient spring like that described in **Table 5.1**. The following materials with the given properties satisfy the constraints: What is the ranking of these materials from best to worst according to the objective given?

Material	E [MPa]	ρ [kg/m³]	σ_y [MPa]
"1095" spring steel	207,000	7840	570
"4983" Ti alloy	110,000	4650	900
"6/6" nylon	2400	1140	72.4

SOLUTION

The governing parameter is the material index $h = (\sigma_y^2/\rho E)^{-1}$, so we expect that materials with higher $\sigma_y^2/\rho E$ values will be better suited to the role. We represent the calculated values using the units $[\sigma_y]^2/[\rho][E]$ = [force/area]/[mass/volume] = [energy]/[mass] = J/kg. The ranking is then

Rank	Material	$\sigma_y^2/\rho E$ [J/kg]
1	6/6 nylon	1920
2	4983 Ti alloy	1580
3	1095 spring steel	200

If weight is indeed the only objective that we are interested in, we conclude that the nylon material is most appropriate.

Phase 3. Assess supporting information. The "supporting information" is information about the shortlisted/best materials that goes beyond the property values necessary for screening and ranking. Information on the candidate materials' appearance, availability, cost, lifecycle, recyclability, etc. needs to be collected and assessed as part of the selection process. Since this phase requires considerable effort for each material, only the top-ranked candidates should be subjected to assessment.

Phase 4. Assess local conditions. After the supporting information is accumulated and considered, the final decision must be made. This final choice will likely depend strongly on "local conditions", meaning any concerns that might be present at the point of manufacture that could limit the deployment of any of the materials. These include availability at the point of manufacture, the level of expertise that technical staff have with the material, and so on. The impact of these considerations, like that of the supporting information, has a certain subjectivity to it. But overall, the selection process hones in on the best material. Even if you choose to go with a type of material that is not ranked in first place after Phase 2 because of such local considerations, you still retain all the information utilized in the selection process for reference.

Final Phase. Finalize and document selection. As with the engineering design process, communicating the reasoning you employed throughout the process to another interested party is crucial. The final report should include the details of the translation phase, indicate clearly the importance of the various material indices employed, provide copies of the data sets utilized in the screening and ranking process, and highlight the crucial supporting and local information.

Now that you have been introduced to the process of material selection, please consider the following case study, which puts all of the phases together.

CASE STUDY – MATERIAL SELECTION FOR A HIGH-THROUGHPUT FAN BLADE

Many engineered systems rely on fans for cooling: automobile engines, electronics, materials processed at elevated temperatures, etc. For performance's sake, the fan must operate at significant speeds, and for safety's sake, it must not fail and scatter debris that could harm equipment or personnel.

A schematic of a fan blade is illustrated below. The fan blades and the central hub they are attached to comprise the *impeller*. The impeller is contained in a housing and rotates with an angular velocity ω. The housing has a radius R. We identify the requirements for this design as[1]

- **Function.** Move large volumes of air for cooling purposes.
- **Constraints.** The impeller size is fixed at the value $\approx R$ so that the fan fits in the housing (with a small clearance). The blades have a cross-sectional area of A. The stress in the blades σ must be less than the yield or failure stress of the material σ_f *including a margin of safety.*

- **Objectives.** Maximize angular velocity (this maximizes airflow and hence cooling).

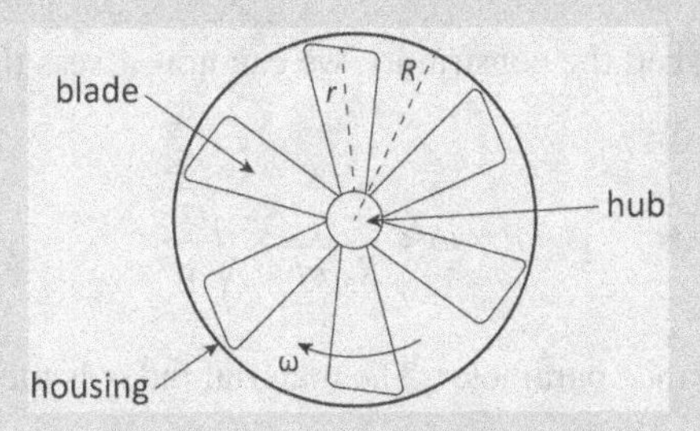

The force F acting on any one of the blades is given by

$$F = ma$$

where m is the mass of the blade and a is its acceleration during rotation. Assuming that the blade length r is a simple fraction b of the impeller size ($r = bR$), we determine that the volume of the blade is then $V = Ar = bAR$. This gives

$$F = ma = \rho Va = \rho bARa$$

Now, according to the physics of rotating bodies, the acceleration $a = \omega^2 R$, and so

$$F = \rho bA\omega^2 R^2$$

Assuming that the force is distributed uniformly over the cross section, we get

$$\sigma = \frac{F}{A} = \rho b\omega^2 R^2$$

The impeller size R is fixed by constraint, and the blade's cross-sectional area A is a free parameter.

The other constraint requires that

$$\sigma < \frac{\sigma_f}{f}$$

where $f > 1$ is the **factor of safety**. When f is close to 1, the design stresses can approach the limiting material strength. *This is not safe*, since small fluctuations in stress may push the material past its limit. (The material used in the component might also not be as strong as advertised.) Rather than designing a component that is intended to operate near the critical limit, we choose a value of f at some convenient value > 1. For example, if we know the value of σ_f can vary in the material selected by as much as 5%, a choice $f \approx 1.05$ can place the component in an operating range that is nominally safe but that still carries a statistical risk. If an additional guarantee is required, we can choose *f* higher. ($f \approx 3$ is typical in many designs.) Note that the choice of *f* modifies the constraint quantitatively and that going to higher values of *f* will screen more materials.

Having established the constraints, we can now assess the objective: maximum ω. We have

$$P = \omega < \frac{1}{R\sqrt{bf}} \times \sqrt{\frac{\sigma_f}{\rho}}$$

as our performance parameter. The material index is then

$$h = \sqrt{\frac{\sigma_f}{\rho}} \quad \text{or} \quad h = \frac{\sigma_f}{\rho}$$

This means that choosing a material with the largest values of σ_f/ρ will give the highest ω. The appropriate data to assess the constraints and objectives is contained in **Figure 5.6(a)** (reproduced and annotated below). If, for example, the minimum failure stress $\sigma_f = 10$ MPa (safety factor already included), then the materials in the dark gray region are excluded from further consideration. Of the materials that remain, we can assess how well they meet the objective using the diagonal guideline. The guideline has a slope = 1 and is situated at a convenient example location corresponding to $\sigma_f/\rho = 3$ kJ/kg. The higher the position of the material above this line, the more suitable it is in terms of the objective (high ω). The materials below the line (in the light gray region) meet the constraint but only satisfy the objective poorly. The materials above this line may be added to the "shortlist" of candidates.

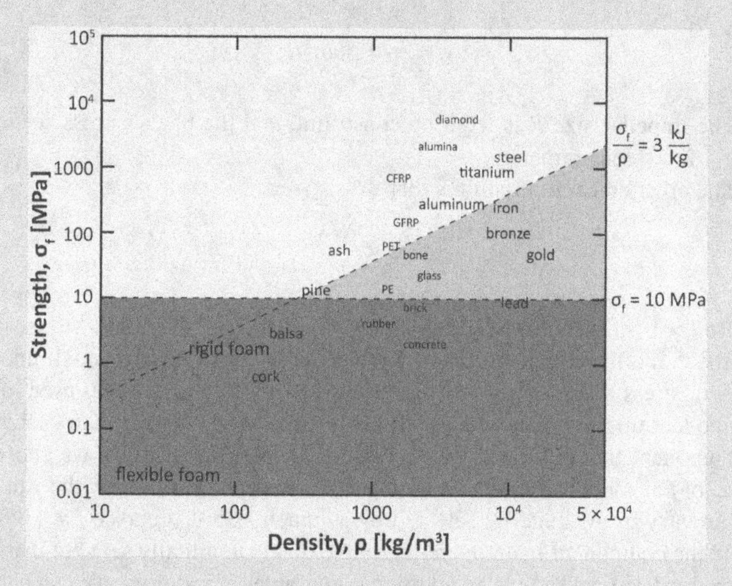

Following the creation of the shortlist, we can look more closely at the candidates supporting information to assess other factors that could influence the final selection. For example, consider the following

- **Steel.** Steel is a metal available in a wide variety of compositions, in quantity, and in many different pre-made forms (sheet, billet, etc.). Steels that do not contain large amounts of difficult-to-obtain alloying elements are also inexpensive. The finished component can also be treated to improve some of the important properties and sometimes coated to alter the finish. The failure mode of steel is typically ductile, so steel has some resistance to shattering during service.
- **PET.** PET is poly(ethylene terephthalate), a type of polymer called a polyester. It can be obtained as a resin that can include a large fraction of recycled material. As a moldable resin, it is readily formed into a wide variety of complex shapes at low cost. PET also does not shatter readily during failure.
- **Alumina.** Alumina is a technical ceramic that is strong in a limited sense: it is resistant to failure in compression. In tension, it ruptures readily and can shatter; this makes it mostly unsuitable in many applications like the one considered here. Ceramics are also not as readily processed into the required shape as metals or polymers, so such processing will be expensive.
- **GFRP.** Glass-fiber–reinforced polymer (GFRP or "fiberglass") is a composite of glassy ceramic fibers in a polymer matrix (typically epoxy). The materials are relatively easy to obtain and mold into a wide variety of shapes ("layup"). Fiberglass is resistant to shattering, despite the ceramic component. However, the layup process can be time-consuming and labor-intensive, so fiberglass might not be viable if many components are needed for production.

All properties of the various candidate materials being equal, it is most economical to go with the cheapest. However, safety is a significant consideration in this application and the failure modes of the various candidates. Though alumina gives the highest value of σ_f/ρ, the chances of catastrophic failure (leading to additional expenses) are too great to ignore. Depending on the environment and the steel's composition, the steel might suffer from a short lifespan as a result of corrosion. Fiberglass and polyester do not suffer similar limitations. If a metallic component is still desired, an aluminum alloy could make a suitable replacement for steel without some of the lifespan limitations.

5.4 IDENTIFYING MATERIALS

The number of materials that are available for the designer to consider during the selection process is staggering. Some materials are wholly dissimilar, and some have similar pedigrees and properties. What we can do is create a model for how the

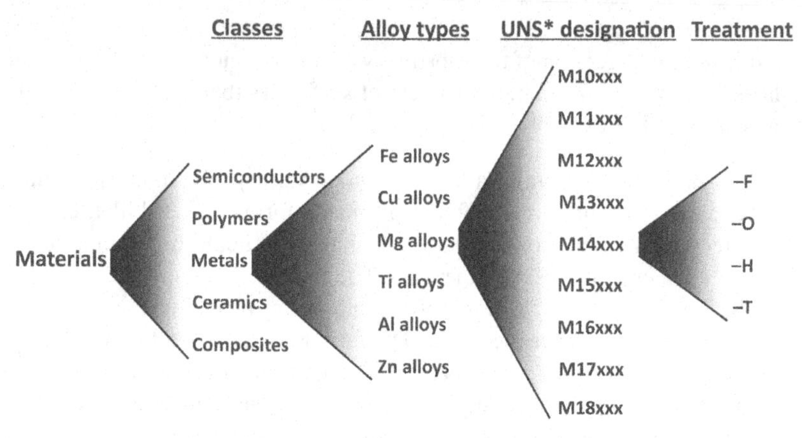

FIGURE 5.7 Example of locating a particular material in a hierarchy. Out of all the materials that exist, we look at the class of "metals", then the various alloy systems, then the alloy designation numbers, and then the codes that indicate different treatments (F = as fabricated; O = annealed; H = work hardened; T = thermally treated).

materials are organized into types, subtypes, sub-subtypes, etc. Consider the classification (or "taxonomy") of a specific material shown in **Figure 5.7**. First, the material is a metal (i.e., an alloy); metallic elements make up the vast majority of the atoms in the material. Second, the alloy is primarily Mg atoms, so we identify it as an "Mg alloy". Out of all the possible Mg alloys, there are a large number of particular compositions that have been determined to be useful in applications and are in common production. These alloy compositions are recorded and are referred to by designation numbers. For instance, the UNS designation system (see below) uses an alphanumeric code to organize the alloys. Since the full composition of the alloy may be lengthy and awkward to recite over and over again, the designation number becomes shorthand for the composition of an alloy. It is important to recognize that the composition of an alloy is typically not enough to indicate everything about the material. The physical and chemical processing that the material has undergone (heat treatment, coating, mechanical working, etc.) can influence the properties as well. Typical treatments are given letter designations that accompany the designation number.

The diagram of **Figure 5.8** expands the classification of metallic alloys so that the differences between them are emphasized by their separation distance in the diagram. Because of the high degree of development of ferrous metallurgy, many important Fe-based alloys occur throughout the various engineering disciplines. For this reason, the primary branch in the taxonomy distinguishes between ferrous and nonferrous alloys. Within the ferrous main branch, the alloys are segregated into steel and cast irons. In the nonferrous main branch, separation occurs between the engineering alloys in common use and the more highly specialized ones. Most of the alloys are separated by their majority metallic constituent. Within the common engineering alloys, they are most easily distinguished by their majority constituents (Al, Cu, etc.). When you reach this level in the taxonomy, the possible alloy matches up with the series in the UNS Designation System.

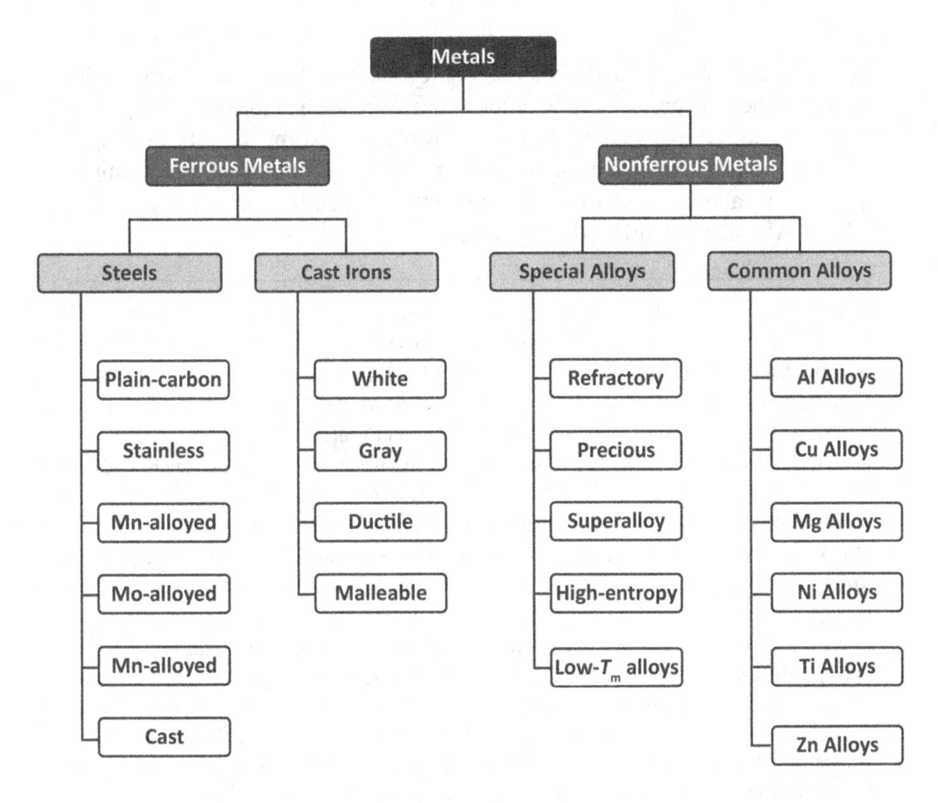

FIGURE 5.8 Proposed classification of metals. This diagram is a model for how to organize metal alloys under consideration during design.

APPLICATION NOTE – THE UNS ALLOY DESIGNATION SYSTEM

The UNS numbering system (established in ASTM Designation E527–16) organizes metallic alloys using a six-digit alphanumeric code. Each UNS code/designation/"number" consists of a single-letter prefix followed by five digits:

$$\left[\text{letter prefix “A”} - \text{“Z”}^{a}\right]\left[\text{five} - \text{digit numerical code } 00001 - 99999\right]$$

The letter prefix frequently indicates the majority metal comprising the alloy, such as in the case of Al, Cu, Zn, etc. alloys (the "A/B", "C", and "Z" series, respectively). Sometimes the prefix has a suggestive association with the alloy covered by that series (e.g., "A" is used for predominantly Al alloys), though this is not always the case (some Al alloys are classified using series "B").

The important thing to remember about the UNS codes is that they indicate the nominal composition[b] of an alloy that currently has established use in industry. The numerical part of the code organizes the alloys by composition within the series. In cases where the series includes multiple majority-component alloys (for instance, series "E" and "P"), the first few digits indicate the majority component. When the prefix is for an alloy series with a single majority component (like "A" and "C" and all of the ferrous-alloy prefixes except "D"), the first digits typically indicate the next most predominant alloying component. For example, a "G13xxx" alloy is steel with predominantly Mn additions, and an "A02xxx" alloy is Al with predominantly Cu additions.

Series	Description
	Nonferrous Alloys
A00001–A99999	Al and Al alloys
B00001–B99999	Al and Al alloys
C00001–C99999	Cu and Cu alloys
E00001–E99999	15 rare earth and rare-earth–like metals and alloys [e.g., lanthanoids (see **Figure 1.7**)]
L00001–L99999	Low–melting-temperature metals and alloys (e.g., Pb)
M00001–M99999	12 miscellaneous nonferrous metals and alloys (e.g., Mg)
N00001–N99999	Ni and Ni alloys
P00001–P99999	8 precious metals and alloys (e.g., Au)
R00001–R99999	14 reactive and refractory metals and alloys (e.g., W)
Z00001–Z99999	Zn and Zn alloys
	Ferrous Alloys
D00001–D99999	Steels with specified mechanical properties
F00001–F99999	Cast irons
G00001–G99999	Carbon and alloy steels (except tool steels)
H00001–H99999	H-steels
J00001–J99999	Cast steels (except tool steels)
K00001–K99999	Miscellaneous steels and ferrous alloys
S00001–S99999	Heat- and corrosion-resistant (i.e., "stainless") steels
T00001–T99999	Tool steels
W00001–W99999	Welding-filler metals, covered and tubular electrodes, classified by weld deposit composition

The assignment of alloy designations is administered by several trade associations/professional organizations, such as The Aluminum Association, Inc. and the Society of Automotive Engineers (SAE). In many cases, the numerical portion of a UNS designation is inherited from a preexisting designation system. For example, carbon steels assigned numbers by the American Iron and Steel Institute (AISI) like "1045" have UNS designations like "G10450". Or the SAE "standard 304 grade" stainless steel has a UNS designation "S30400".

[a]Some letters, "I", "O", "Q", "U/V", "Y", and "X" are not currently in use.

[b]Except when the mechanical properties are the distinguishing feature like in the case of the "D" series.

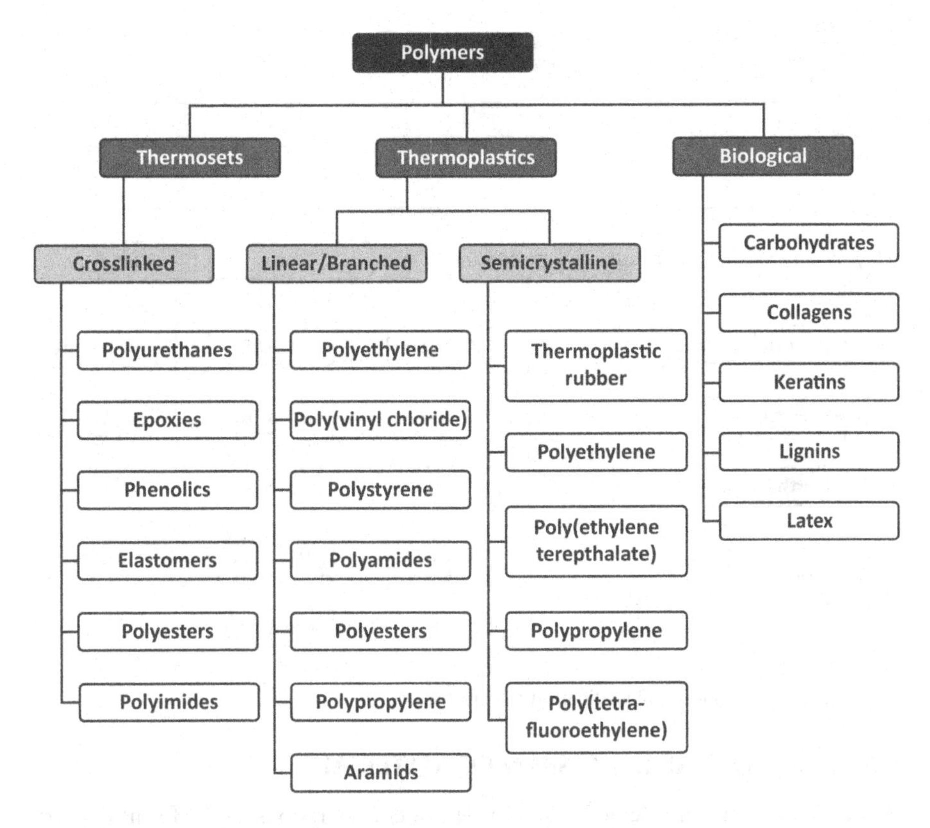

FIGURE 5.9 Proposed classification of polymers.

What are the nonmetallic materials? There are many in common use, and it is important to consider models that not only reflect the chemistry of the various materials types but also something more about their structural organization. Consider the proposed classification of polymers in **Figure 5.9**. Aside from the materials of natural origin (like latex rubber), the synthetic materials are organized by their macromolecular arrangement. The crosslinked polymers, the linear or branched polymers, and the polymers with crystallinity represent distinct materials, though the macromolecules involved may have identical compositions. (For example, polyethylene can be processed as a linear material or a crystalline one.) A classification for ceramics is provided in **Figure 5.10**. At the topmost level, they are categorized by their *use*. The "traditional ceramics" are identified as materials for domestic use and as basic building supplies (bricks, cement, and glass). The "refractory ceramics" are required for applications where heat resistance (see **Topic 2**) is the primary requirement. Finally, "technical ceramics" are used in applications where high-temperature wear resistance and chemical and dimensional stability are necessary. Some of these materials are also prized for their insulating properties (see **Topic 3**).

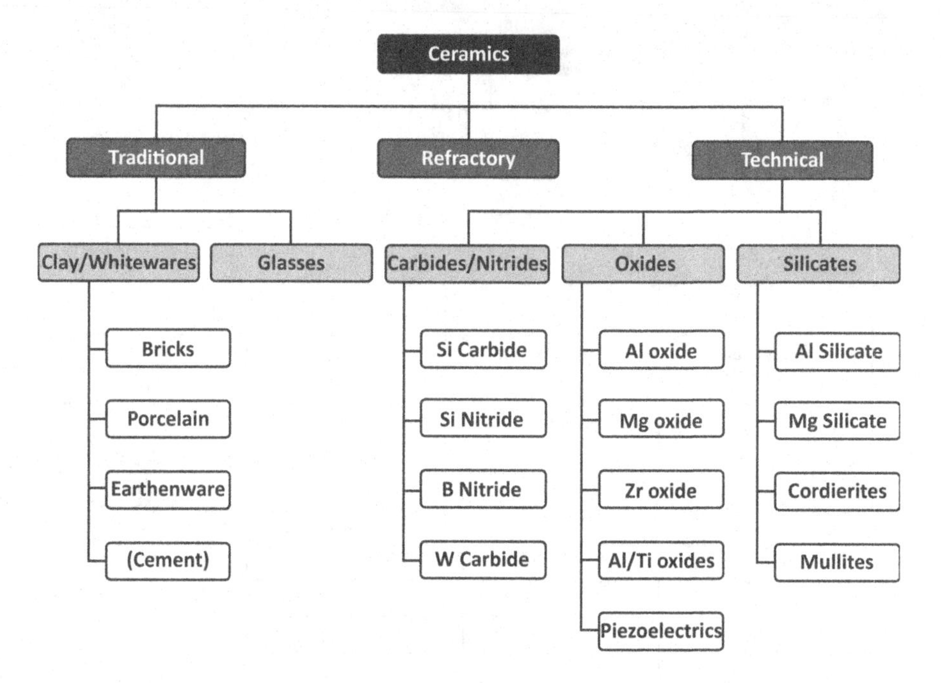

FIGURE 5.10 Proposed classification of ceramics.

5.5 THE COST AND LIFESPAN OF MATERIALS

The cost c of a material enters into our design considerations in the form of a cost per weight (or a cost per volume). This cost is most frequently interpreted in terms of currency: $[c]$ = [cost/weight] = \$/kg or $[c]$ = [cost/volume] = £/m^3, and so on. Sometimes, we encounter design/selection problems in which "cost is no object". That is, the cost of the component is considered unimportant. This outlook typically arises in circumstances where the health and safety of the end user are paramount to expense. Ordinarily, however, the cost must be taken into account since components that are too costly can impair the economic viability of the manufacturer or discourage purchase by the end user. Consider the origins of the cost of a particular material as a finished component in a design:

- **Raw material.** The material in its unfinished form. It may yet require shaping or treatment. Part of this cost is the cost of transporting the raw material to the manufacturing facility, so the cost of the raw material is getting it to your doorstep.
- **Fabrication.** The expense related to shaping the component from the raw material. Machinery/tooling, workspace, labor, and utilities are typical factors.
- **Treatment.** If the material requires additional treatment to alter the bulk or surface properties, that expense is added on top of the rest of the costs. If the treatment is to be performed at a different location, more transport costs are incurred.

When considering the cost of materials during Phase 3 and Phase 4 of the selection process, it is worthwhile to compare across these subfactors to identify costs that might fluctuate during production.

There are other, sometimes unconsidered, components to the cost of a material. These costs are environmental in origin. The production of a material involves an investment of energy q, e.g., combustion of fuel during the reduction of iron oxides in a blast furnace, consumption of electricity in the electrolysis of alumina ore, heat fed to autoclaves during the synthesis of composites, etc. The accounting of this energy q' is typically on a per-mass basis: $[q'] = [\text{energy/mass}] = \text{J/kg}$. The energy required is not the only cost associated with the production of different materials. Depending on the origin of the energy invested in the process and the byproducts of the process itself, there may be negative environmental impacts downstream of the production process. These negative effects include toxic byproducts, unrecyclable components, and nonbiodegradable waste. **Table 5.2** lists some of the eco-factors involved in making metallic Al. These factors all have more or less unrelated origins, possess different and incompatible units, and represent different degrees of negative impact. In order to establish some kind of equivalence between the impacts of the production of different materials, the environmental load factors are bundled into an **eco-indicator** parameter that collects, weights, and normalizes the various contributions. These indicators are provided alongside the strictly energy-related inputs for various materials in **Table 5.3**.

Information derived for ecological assessments of the production of different materials can be incorporated into materials-property charts like those in **Section 5.3**. Some examples are shown in **Figure 5.11**. **Figure 5.11(a)** shows the stiffness (i.e., Young's modulus E) plotted against the energy input cost q'. Typically, a fraction of the energy input q' is associated with the release of CO_2 or other greenhouse gases into the atmosphere. (The other fraction of these energy expenditures comes from "clean" or carbon-neutral/-negative sources, but globally, these contributions remain a relatively small part.) Additionally, the production and use of materials are associated

TABLE 5.2
Ecological Impact Profile for the Production of Al

Environmental Load	Unit
Energy inputs	[energy] = J
Raw materials	[mass] = kg
Greenhouse emissions	["global-warming potential" vs. mass of CO_2] = GWP
Ozone depletion	["ozone depletion potential" vs. mass of CFC[a]=111] = ODP
Acidification	["acidification potential" vs. mass of SO_2] = AP
Oxygen depletion	["nutrification potential" vs. mass of PO_4] = NP
Heavy metals	[toxicity vs. mass of Pb ions] = Pb equivalents
Carcinogenicity	[carcinogenicity vs. 1 unit PAH[b]] = PAH equivalents
Solid waste	[mass] = kg

[a] Chlorofluorocarbon; [b]Polycyclic aromatic hydrocarbons

TABLE 5.3
Energy Content and Eco-indicator Values of Some Common Engineering Materials

Class	Material	Specific Energy Content [MJ/kg]	Eco-indicator [#/kg]
Metals	Ti alloys	555–565	80–100
	Mg alloys	410–420	20–30
	Cast irons	60–260	3–10
	Al alloys	290–305	10–18
	Steels	50–60	4.0–4.3
Polymers	6/6 nylon	170–180	12–14
	Polypropylene	108–113	3.2–3.4
	Low-density polyethylene	80–104	3.7–3.9
	High-density polyethylene	103–120	2.8–3.0
	Synthetic rubber	120–140	13–15
	Natural rubber	5.5–6.5	14–16
Ceramics	Glasses	13–23	2.0–2.2
	Refractories	1–50	10–20
	Pottery	6–15	0.5–1.5
	Concrete	3–6	0.6–1.0
Composites	GFRP	90–120	~12
	CFRP	130–300	20–25
	Woods	1.8–4.0	0.6–0.8

with waste that carries additional costs in terms of disposal. These disposal costs are somewhat mitigated by recycling, which serves to preserve some of the energy expended (and reduce the wasted material fraction) in the next generation of material produced. **Table 5.4** contains some data on the atmospheric emissions related to the production of the polymer polyethylene. Much like the strategy for the overall eco-indicator above, we can collapse all of these disparate quantities into a single index (air-pollution index or API) and use that index to compare materials. **Figure 5.11(b)** shows the materials-property chart that compares the API for common materials.

Example 5.3:

In the aftermath of natural disasters, there is frequently a need for temporary shelter in large quantities. Suppose that you are selecting the material for the internal framework of these temporary structures and that the selection criteria call for a material that possesses a good balance of stiffness and low production-energy content. What materials do you think are best? What materials are to be avoided?

SOLUTION

Following **Table 5.1**, the governing material index h for a stiff structural member vs. currency cost c is $h = (\sqrt{E}/\rho c)^{-1}$, where E is the Young's modulus and ρ is the

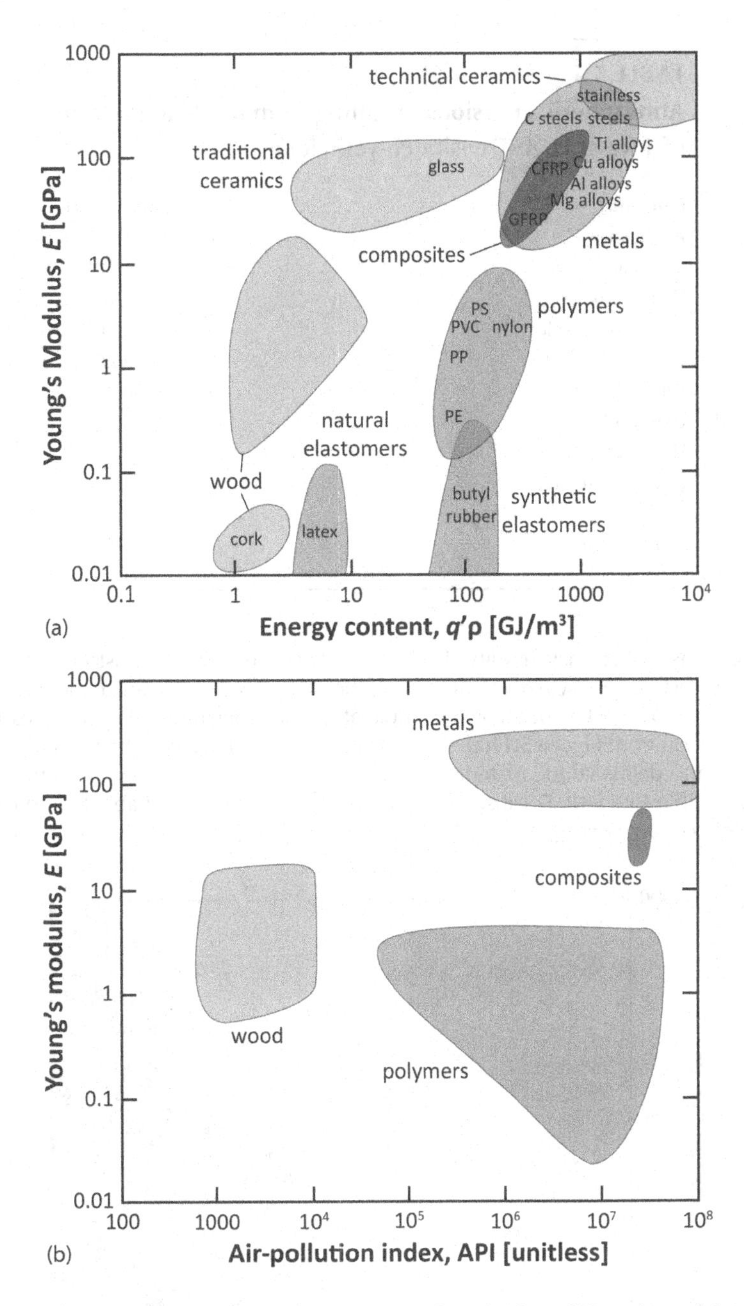

FIGURE 5.11 Combined materials property chart for selection of Young's modulus against energy "cost" $q'\rho$ and against greenhouse gas production. (a) Like the strictly currency-based cost c in **Figure 5.10**, the energy cost of a material varies with the process type and process conditions, so it is not exactly intrinsic to the material but depends on a number of other factors. (b) Additionally, the energy cost and materials processing strategy might be associated with a certain release of pollutants. These pollutants are represented in an effective air pollution index, or API.

TABLE 5.4
Atmospheric Emissions Resulting from the Manufacture of 1 kg of High-Density Polyethylene[2]

Emission	Quantity [mg]
Acidic ions	100
Ammonium ions	10
Carbon dioxide	9.4×10^5
Carbon monoxide	600
Chloride ions	800
Dust	2000
Hydrocarbons	50
Hydrogen chloride	1
Hydrogen fluoride	1
Metals	1000
Nitrogen oxides	5
Other	6000

mass density. We may modify this by replacing the currency cost $c\rho$ with the energy cost $q'\rho$: $h = (\sqrt{E}/q'\rho)^{-1}$. Therefore, the materials that maximize the quantity $1/h = \sqrt{E}/q'\rho$ are the best for our application. The diagram below reprises the information from **Figure 5.11(a)**. A line of constant $1/h$ has been drawn across the diagram to delineate the materials that have "high" suitability from those that are less suitable (shaded). For instance, certain metals are more suitable than others, as well as polymers.

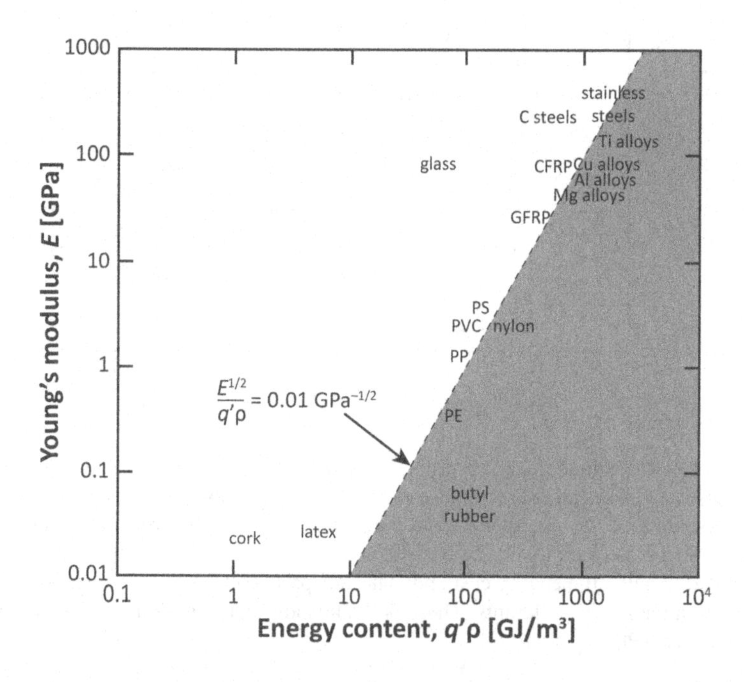

The rankings of some of the most suitable materials are provided in the table below, along with approximate values of $1/h$. Again, materials with higher values of $\sqrt{E}/q'\rho$ give better performance, as defined by the index. (Note that glass, while ranked highly, is unlikely to be suitable for this structural application given its tendency to *fracture*; see **Topic 1**.)

Rank	Material	$\sqrt{E}/(q'\rho)$ [$Pa^{-1/2}$]
1	Glass	3800
2	Carbon steel	640
3	Polyvinyl chloride (PVC)	450

It is also necessary to consider how the costs associated with the production of materials may not be shared in a perfectly socially equitable way. The *currency* cost of materials is advanced by the producers and collected from the consumers and is determined by the market principles that influence the cost of fuel, labor, raw materials, etc. However, the *other* costs associated with production, e.g., atmospheric and water-based emissions, affect specific populations that might not otherwise be a party to the currency exchange. Communities that are adjacent to extraction, production, transportation, or disposal facilities (i.e., stations on the "supply chain") are disproportionately impacted by any negative economic, health, environmental, and sometimes legal factors. It is difficult to describe these overlapping geographical and social aspects of materials production in the explicit way that our selection procedure requires. Nevertheless, these considerations cannot be ignored and should be reflected on in, e.g., the "Assess Local Conditions" step.

All materials currently in use have finite **lifespans**. By lifespan (sometimes called "service life"), we mean the effective duration over which the material can fulfill its engineering function in a design. For instance, materials used in structures can deteriorate to the point that they are incapable of sustaining the loads that they were designed for. This deterioration can come in the form of changes in the material's properties ("degradation") or changes in the geometry of the component it is incorporated in ("erosion"). It is worthwhile (and frequently quite necessary) to consider how the deterioration of a material introduces additional costs related to inspection, repair, and replacement.

The primary effects that contribute to a material's lifespan include environmental incompatibility, **corrosion**, **fatigue**, and **wear**.

- **Environmental incompatibility.** The environment that the material is deployed in might introduce factors that produce deterioration. These factors might include exposure to extreme temperatures, large variations in temperature, atmospheric humidity, ultraviolet light (or other electromagnetic) exposure, and chemical attack.
- **Corrosion.** The collection of phenomena we call "corrosion" is a form of environmental incompatibility primarily related to chemical modification of a material through **oxidation** (e.g., "rusting"). The replacement of the native material by the oxide product will negatively influence the overall

properties of the material, and in cases where the oxide is mechanically incompatible, the material may erode.

- **Fatigue.** Fatigue is a process by which the mechanical integrity of a component degrades via the expansion of (initially) small cracks or fissures in the material subject to repetitive loading. The component's lifespan is determined by the material's intrinsic resistance and the magnitude of the stresses involved. Fatigue is discussed in detail in **Topic 1**.
- **Wear.** Wear is an erosive process related to the mechanical interaction of materials in contact with one another. This interaction can cause localized yielding of the material on the respective surfaces, as well as scratching (during purely frictional sliding) and "galling" (in cases where there is some adhesion between the materials). Wear is an essentially plastic-deformation–related phenomenon since the strength properties of the materials (like the yield strength and the ultimate strength) dictate how much deformation/scratching/galling occurs given the (high) stresses that are present at the interface between the materials.

An additional factor influencing the service life of a component is its exposure to mechanical fatigue, the principles of which are introduced in **Topic 1**.

The most important of these degradation processes is perhaps **oxidation**. During oxidation, metals generally deteriorate via mass loss due to unwanted chemical reactions. A generic oxidation reaction for a metal ("M") looks like

$$M \rightarrow M^{+n} + ne^- \quad (5.2)$$

The oxidation state of the metal changes during this reaction from neutral ("0") to "+2", and so the electrons required for charge balance appear on the "product" side. **Equation 5.2** is a partial "electrochemical" reaction, involving the generation and transfer of electrons (the n e^-, in this case). This reaction is only half of the picture, as the electrons produced must be transported to and absorbed by a different substance Q in the associated **reduction** reaction:

$$Q + ne^- \rightarrow Q^{-n} \quad (5.3)$$

The Q in this case may be oxygen gas, protons, or some other metal undergoing reduction. Since the metal M^{+n} ions are typically easily removed from the bulk material and dissolved in the surrounding medium (water, etc.), some of the mass of the metal will thereby be lost to the environment.

As an example, consider a typical oxidation-reduction reaction for Fe rusting in the presence of (aerated, neutral) water:

$$2Fe + O_2 + 2H_2O \rightarrow 2Fe^{+2}(aq) + 4OH^-(aq) \quad (5.4a)$$

That is, electrons have been removed from the metallic Fe to drive production of Fe^{+2} and OH^{-} ions in solution ("aq"). The associated "half-cell" reactions are

$$2Fe \rightarrow 2Fe^{+2}(aq) + 4e^{-} \tag{5.4b}$$

akin to **Equation 5.2** and

$$O_2 + 2H_2O + 4e^{-} \rightarrow 4OH^{-}(aq) \tag{5.4c}$$

like **Equation 5.3**. You will recall from chemistry class that **Equation 5.4b** is associated with the *anode* of an idealized electrochemical cell and that **Equation 5.4c** is associated with the *cathode*. Indeed, electrochemical cells are important experimental tools in studying the corrosion properties of materials.

The tendency of a given pair of anode/cathode materials to corrode is described by an **electrode potential**. Since these potentials are all relative, they are frequently collected in a series that is centered around the potential relative to the standard hydrogen gas electrode reaction:

$$2H^{+} + 2e^{-} \rightarrow H_2 \tag{5.5}$$

These ranked potentials form what is called the "standard EMF series". The standard hydrogen gas electrode reactions are located at $v_0 = 0$ V, and the magnitude of a material's tendency to oxidize can be gauged by where it lies relative to 0 V. Materials that have high v_0 values (and are therefore more cathodic, noble, or inert) tend not to lose electrons; the materials that are lower in the series (more anodic or reactive) are most likely subject to electron loss when paired with another material. The rule is: when any two materials are combined in a cell, the one with the lower v_0 becomes the anode and will tend to corrode, and the higher-v_0 material will function as the cathode.

Establishing that the lifespan of a material is limited by its tendency to corrode is important, but it raises the additional question of how rapidly that corrosion can occur. Given the electrochemical nature of corrosion – it can only proceed as fast as the anode material can lose its electrons – we can represent the instantaneous corrosion rate r as the rate of mass lost:

$$r = \frac{dm}{dt} \tag{5.6}$$

where m is the material's mass and t is the elapsed time. Suppose the mass lost in time Δt is

$$\Delta m = \frac{A}{n\mathcal{F}} \Delta Q$$

where A is the atomic mass of the corroding species, ΔQ is the total number of charges removed, the value of n is the oxidation state from **Equation 5.2**, and $\mathcal{F} \approx 96485$ is Faraday's constant. We can then compute

$$r = \frac{\Delta m}{\Delta t} = \frac{A}{n\mathcal{F}} \times \frac{\Delta Q}{\Delta t} = \frac{A}{n\mathcal{F}} i_{\text{corr}} \tag{5.7}$$

where $i_{\text{corr}} = \Delta Q / \Delta t$ is the current established during corrosion. We expect that this current will generally depend on the driving force Δv for corrosion:

$$r \propto -\Delta v$$

where $\Delta v = v_{0,\text{anode}} - v_{0,\text{cathode}}$ and the anode and cathode potentials are taken from the EMF series. The greater the magnitude of Δv, the higher the value of r, and the higher the value of r, the shorter the lifespan.

Example 5.4:

Suppose that you are performing a study on the suitability of a metal in a corrosive service environment. You make several specimens/coupons of the metal (each with mass m_0 = 85.0 g), expose them to this environment, and measure their mass m after a certain interval of time t has elapsed. If you obtain the data given below, how long would you predict before this material is at 95% of its original mass?

Time Elapsed, t [d]	Mass, m [g]
0.0	85.0
30	84.9
60	84.6
90	84.5
120	84.2
150	83.9
180	83.7
210	83.6
240	83.3

SOLUTION

In MATLAB®, we have

```
m0 = 85.0;
t = [0.0 30 60 90 120 150 180 210 240];
m = [85.0 84.9 84.6 84.5 84.2 83.9 83.7 83.6 83.3];
M = m/m0*100
```

where $M(t) = m(t)/m_0$ is the mass fraction after time t (as a percentage). Using `plot(t, m)`, we observe a linear rate of mass loss:

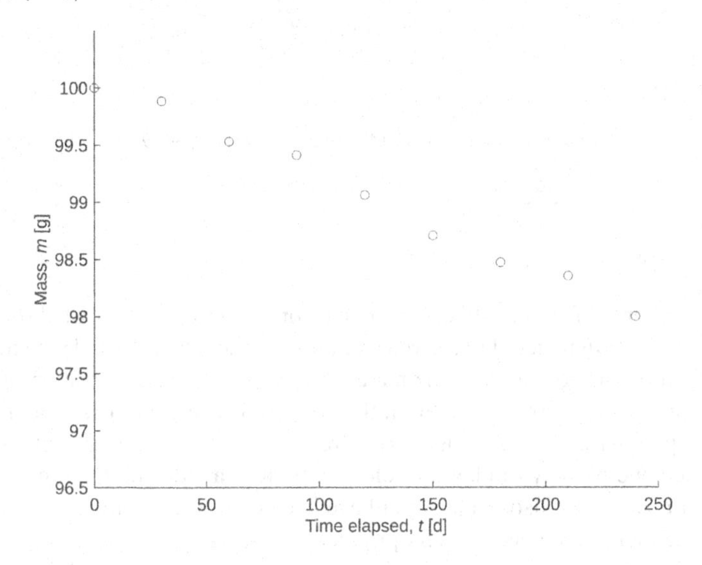

meaning $r = dm/dt$ = constant. We can obtain this rate using a linear fit. In MATLAB

```
>> fit = polyfit(t, M,  1)
fit =
    -0.0086  100.0732
```

where the first value (i.e., the slope) is the fractional mass loss rate $R = dM/dt$ = 0.0086 %/d. (The second value is the intercept b.) This fit looks like

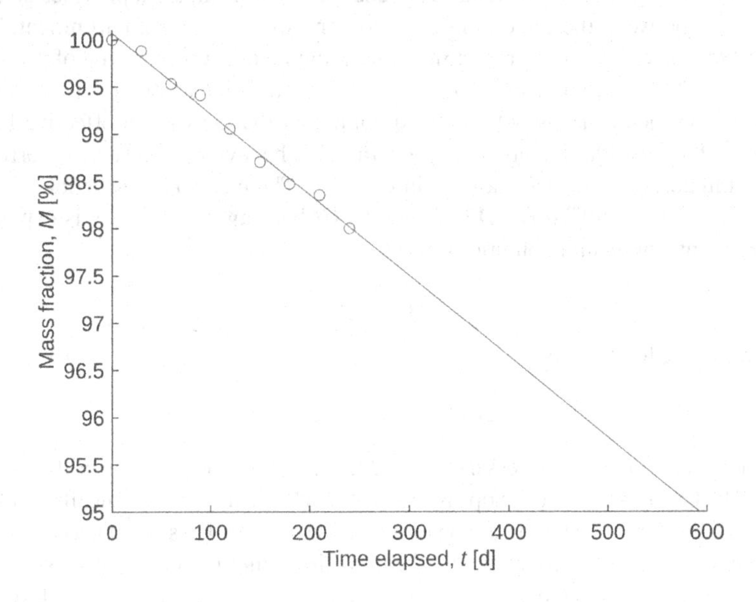

We observe that the critical time at which the mass loss is at 95% is $t_c \approx 600$ d and obtain a more exact solution by solving

$$Rt + b = 95\%$$

For example,

```
>> tc = roots([fit(1) fit(2) - 95])
tc =
  592.5191
```

or $\underline{t_c = 593 \text{ d}}$.

The number of factors influencing corrosion is large, and all of those factors might not be known in detail in a given situation. These factors include the material's microstructure, the condition of the material's surface, the concentration of various ionic species in the corrosive medium, the temperature, etc. For example, materials with multiple phases in their microstructure are typically more susceptible to corrosion since one phase will play the role of a built-in anode and the other will play the role of a cathode. Also, metals that have been cold worked are at enhanced risk because the increased defect content produced by deformation essentially catalyzes the oxidation reaction. The numerous and complicated factors underlying corrosion highlight the need to try to assess parameters like r experimentally so that the most accurate possible lifespan predictions can be obtained.

The lifespan of material subject to corrosion can also be extended by suitable modifications made to the material or component. One strategy is the application of a sacrificial coating. For example, the process of galvanization involves coating a ferrous alloy with a thin layer of Zn. Not only is Zn more anodic than Fe (see **Table 5.5**), but also the Zn coating corrodes at rates that are much slower than the typical Fe rusting process. As long as the coating remains intact, it provides an effective barrier between the underlying Fe alloy and the corrosive environment. Thus, the Zn "sacrifices" itself to protect the underlying metal. Another type of protective coating is a "passivation layer". The passivation layer is a partially or fully oxidized region of the metal that is both stable/durable and that acts as an effective barrier to the oxidation of the material underneath. This behavior is present in corrosion-resistant materials like Al, Ti, and stainless steel. The passivation principle in stainless steel is shown in **Figure 5.12**. When the Cr-bearing ferrous alloy is exposed to oxidizing environmental conditions, the reaction

$$\mathrm{Cr} \rightarrow \mathrm{Cr}^{+3} + 3\mathrm{e}^-$$

tends to occur, followed by

$$2\mathrm{Cr}^{+3} + 3\mathrm{O}^{-2} \rightarrow \mathrm{Cr_2O_3}$$

(The O_2 and O^{-2} ions necessary to produce this oxide are present in the environment.) The Cr_2O_3 film that develops is thin and mechanically and chemically stable and provides a barrier to further oxidation of metal ions in the alloy since it acts as a barrier between the exterior environment and the unoxidized interior metal. Further oxidation can only proceed by the (very slow) diffusion of O^{-2} through the passivating layer.

TABLE 5.5
Standard EMF Series (vs. Hydrogen Electrode)

Oxidation Character	Reaction	Standard Electrode Potential v_0 [V]
↑	$Au^{+3} + 3e^- \rightarrow Au$	+1.401
Cathodic (inert)	$O_2 + 4H^+ + 4e^- \rightarrow 2H_2O$	+1.229
	$Pt^{+2} + 2e^- \rightarrow Pt$	+1.118
	$Ag^+ + e^- \rightarrow Ag$	+0.7996
	$Fe^{+3} + e^- \rightarrow Fe^{+2}$	+0.771
	$O_2 + 2H_2O + 4e^- \rightarrow 4OH^-$	+0.401
	$Cu^{+2} + 2e^- \rightarrow Cu$	+0.3419
Neutral	$2H^+ + 2e^- \rightarrow H_2$	0[a]
	$Pb^{+2} + 2e^- \rightarrow Pb$	–0.1262
	$Ni^{+2} + 2e^- \rightarrow Ni$	–0.257
	$Fe^{+2} + 2e^- \rightarrow Fe$	–0.447
	$Cr^{+3} + 3e^- \rightarrow Cr$	–0.744
	$Zn^{+2} + 2e^- \rightarrow Zn$	–0.7618
	$Ti^{+2} + 2e^- \rightarrow Ti$	–1.63
	$Al^{+3} + 3e^- \rightarrow Al$	–1.662
	$H_2 + 2e^- \rightarrow 2H^-$	–2.23
	$Mg^{+2} + 2e^- \rightarrow Mg$	–2.372
Anodic (active)	$Na^+ + e^- \rightarrow Na$	–2.71
↓	$K^+ + e^- \rightarrow K$	–2.931

[a] Exact.

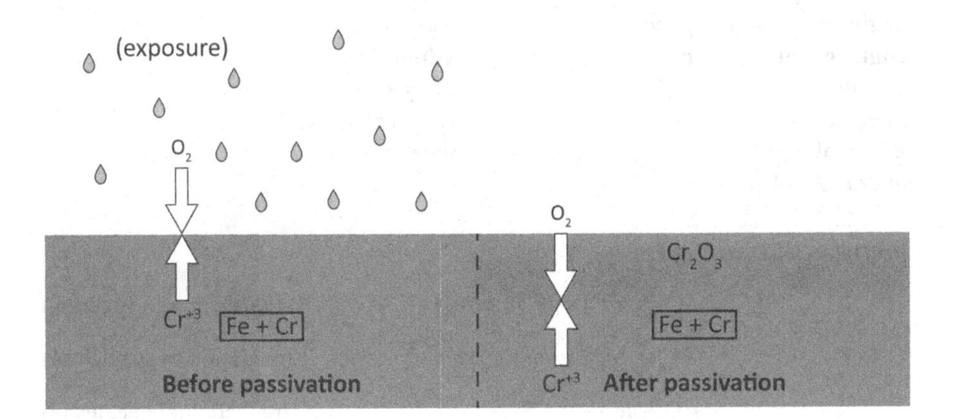

FIGURE 5.12 The passivation behavior of a stainless steel. A steel that contains significant amounts of Cr as an alloying element, when exposed to conditions that produce oxidation, will develop a thin outer layer of Cr_2O_3. As oxygen in the atmosphere reacts with Cr ions in the alloy. Since this layer/film is stable under a wide range of conditions, it becomes a semipermanent feature of the material. The alloy can only oxidize further if O^{-2} ions can be transported (via diffusion) from the exterior across the film, a process that is very slow.

5.6 CLOSING

Outside of the initial, most highly conceptual stages, engineering design must be a rational process that deals with definite questions and is oriented toward the production of concrete results. The goal of the materials-selection process (as a component of engineering design) is to identify the materials best suited to the individual components of the design. To this end, the function of the component along with its associated requirements must be considered thoroughly and quantitatively. The constraints and objectives whose satisfaction dictates the performance of the component must be translated into properties or material indices that can be matched with specific materials. The universe of materials is expansive, so the selection process is designed to pass materials through "filters" that eliminate the materials that are unsatisfactory. To this end, materials must be classified such that the selection process can navigate the classification scheme and identify and organize the candidates. From the finalists, the material that provides the best performance can be identified. While most of the inputs into the selection process are strictly physical, there might be additional economic or social considerations that strongly influence the selection. These considerations should be assessed and incorporated into the process with an eye toward the associated economic or social-equity goals.

5.7 CHAPTER SUMMARY

Key Terms

corrosion
design
eco-indicator
electrode potential
engineering design process
engineering function
fatigue
lifespan
material cost
material index
materials-property charts
materials selection
objective
oxidation
reduction
safety factor
technical system
wear

Important Relationships

$$P(F,G,M) = f(F) \times g(G) \times h(M)$$ (performance function)

$$\mathrm{M} \rightarrow \mathrm{M}^{+n} + ne-$$ (oxidation of species M)

$$\mathrm{Q} + ne- \rightarrow \mathrm{Q}^{-n}$$ (reduction of species Q)

$$r = \frac{dm}{dt}$$ (definition of corrosion rate)

$$r = \frac{A}{n\mathcal{F}} i_{\mathrm{corr}}$$ (corrosion rate vs. current)

5.8 QUESTIONS AND EXERCISES

Concept Review

C5.1 Sometimes, when evaluating different possible solutions to a design problem, a table of "figures of merit" is used. The various possible materials are ranked according to their normalized indices (i.e., the index value relative to the *maximum* value for all of the candidate materials), weighted by the presumed "importance" of that particular index (on a scale of, say, 1–10). For example,

	Material A			Material B		
Index	Value	Importance	Weighted Score	Value	Importance	Weighted Score
h_1	1	9	9	0.8	9	7.2
h_2	1	6	6	0.7	6	4.2
h_3	0.6	5	3	1	5	5
Total			T_1			T_2

From the table, we can identify the best material as "material A". Based on what you have learned about the engineering design process, describe the benefits of such a selection scheme. Compute the totals of T_1 and T_2. Which material would you judge to be the best?

C5.2 Imagine that you are designing a frame for a bicycle. How would you describe the frame's function? What would you say are a few constraints that influence the selection of the frame material? What would you say are the objectives that you want the frame to adhere to?

C5.3 Imagine that you are designing a cutting tool for a shaping machine. How would you describe the tool's function? What would you say are a few constraints that influence the selection of the tool material? What would you say are the objectives?

Discussion-forum Prompt

D5.1 Identify a material (metal, ceramic, or polymer) that is relevant to your work/studies and locate it in its respective classification tree. Post the material and your justification for its location in the classification scheme in the designated forum.

PROBLEMS

P5.1 Show that the material index for a spring designed to store elastic energy at minimum weight is given by $h = (\sigma_y^2/\rho E)^{-1}$ (see **Table 5.1**).

P5.2 Using the data in **Figure 5.5**, identify those materials that satisfy a constraint of elastic modulus $E < 800$ MPa.

P5.3 Given that the volume of a single-serving water bottle is around $v = 10$ cm^3, how would you rate the relative ecological cost of Al, HDPE, and glass as bottle materials?

MATLAB Exercises

M5.1 Suppose that you measure the mass loss of a degrading specimen over the course of 280 days in a simulated service environment. Plot these data and reflect on their characteristics.

Time Elapsed, t [d]	Mass, m [g]
0.0	100
40	92.9
80	85.7
120	81.5
160	76.9
200	75.2
240	73.6
280	72.6

M5.2 Fit the data of Exercise M5.1 using the form

$$m(t) = b_1 \exp\left(-\frac{t}{b_2}\right) + b_3$$

where $\{b_1, b_2, b_3\}$ are underdetermined coefficients. This can be accomplished using

```
nlinfit(t, m, @(b, x)(b(1)*exp(-x./b(2)) + b(3)),
b0)
```

where, e.g., `b0 = [50; 100; 50]`. What is the mass m after 320 d?

M5.3 Using the fit from exercise M5.2 above, compute the corrosion rate r as a function of time.

NOTES

1. Michael F. Ashby. *Materials Selection in Mechanical Design*. 4th edition. Oxford, England, UK: Butterworth-Heinemann, 2011.
2. L. Holloway. "Materials selection for optimal environmental impact in mechanical design," *Materials and Design* **19(4)** (Oct. 1998), pp. 133–143.

6 Materials and Your Career

From Student to Engineer

LEARNING OBJECTIVES

After completing this chapter, you should be able to:

1. Remember and describe the importance of the central materials principles presented in this book.
2. List the most common responsibilities of engineers and remember the important application areas in your field.
3. Describe the ethical considerations related to engineering decision-making.
4. Differentiate between random, systematic, and representative samples.
5. Compute the best estimate and uncertainty in a property of a given sample.
6. Utilize quantities with uncertainties to discriminate between similar results or standardized values.

6.1 APPLICATION OF MATERIALS UNDERSTANDING

Throughout our survey of topics in materials science and materials engineering, we have encountered a few broad principles. These are:

I. The development of new materials is an ongoing social process that undergirds all engineering practice. Pure scientists develop research tools and methods, then accumulate and disseminate results and interpretations. Applied scientists and engineers utilize that understanding in the production of useful materials and in the design of new components and devices. Materials science and engineering teachers reproduce the required technical understanding in the minds of new generations of scientists and engineers. All these endeavors are organized into the materials science paradigm.

1. There is a connection between the behavior and organization of matter at the smallest scales and the macroscopic, "everyday" behavior that we are familiar with. Atomic structure, and the forces that arise from it, give matter the inherent ability to organize into larger structures (crystals, glasses, macromolecules, etc.). These essential motifs, alongside the nature of the bonds involved, indicate to us how materials are sorted into different classes. To these classes, we associate properties that are (mostly) held in common with the class. We say that "the microstructure determines the properties" as an all-encompassing statement of this principle.

DOI: 10.1201/9781003214403-7

2. Recognizing materials as collections of property values permits a "high-level" description of materials that is mostly sufficient for application development. Fully understanding materials and their performance requires a recognition of the aforementioned structure/properties principles. These principles are derived from the more fundamental level of description involving atoms, interatomic forces, etc.
3. Our ability to accurately determine materials properties, as well as the materials' underlying structure, depends on our ability to design, conduct, and analyze experiments. Important inputs to our characterization efforts are experimental tools, including mechanical test platforms and microscopes, and methods of data analysis. Strategies for characterization can take on many forms, so the existence and availability of commonly agreed-upon standards are crucial to these efforts.
4. Our ability to produce materials with the required structure depends on our ability to formulate materials with the correct composition and apply transformations that lead to the correct structure. Fabricating materials to specification requires a thorough understanding of how the process variables (composition, temperature, treatment time, etc.) relate to the nature of the phases and transformations that occur in the material. The progress of these transformations is linked to the development of the material's microstructure, so producing the correct structure is dependent on the control of process conditions.
5. Finally, the application of materials in an engineering design requires a rational process that identifies the correct material for the given application. Inputs to this process include objectives and constraints inferred from the nature of the application. Materials that satisfy the constraints and best meet the objectives become preferred candidates for further investigation and testing.

All of the above are general materials principles that apply to many different materials classes, and an engineer should be aware of and prepared to apply these principles in their own day-to-day work. Focused descriptions of the materials principles most associated with particular engineering areas are provided in **Topics 1–3** that follow this chapter.

Generally speaking, the role(s) of an engineer in their own discipline is to:

1. Analyze problems to see how existing or potential devices or processes might help solve them.
2. Design or redesign devices (and/or subsystems), using techniques including engineering analysis, computer-aided design, and prototype production.
3. Investigate component failures to diagnose faulty operation, then recommend remedies.
4. Collect and analyze the results of tests related to new designs or problem diagnoses.
5. Oversee the manufacturing processes related to new or existing devices and materials.
6. Describe, document, and discuss experimental results, design/materials-selection efforts, and outcomes related to fault diagnoses.

What kinds of systems and devices are involved varies between the engineering disciplines. Let's direct our attention to the materials-related needs of various engineering disciplines.

6.2 ROLES IN ENGINEERING

In your role as an engineer, your work will likely be limited to a small subset of all materials. Various engineering areas and their typical materials needs were outlined in **Table I.1**. Consider these more detailed descriptions of materials requirements for each area.

Mechanical Engineering. Mechanical engineers design, build, and test devices that operate as per mechanical and thermal principles. Such devices include (but are not limited to):

- combustion engines and electric motors,
- transmissions and drive systems,
- load-bearing components that support other devices,
- pumps and fans,
- control systems and sensors, and
- biomedical devices (including implants).

Manufacturing Engineering. Manufacturing or industrial engineers design and produce systems that integrate workers, machines, materials, information, and energy in the production process with an eye toward efficiency, safety, and product quality. Application areas include:

- characterization and preparation of raw material inputs,
- casting of metals,
- machining,
- mechanical forming of ductile materials,
- welding/joining,
- injection molding, and
- other materials treatments (heat treatments, coatings, etc.)

Aerospace Engineering. Aerospace engineering involves the design and testing of aircraft, launch vehicles/rockets, and spacecraft/satellites. Common designs incorporate

- power plants, turbines, and other propulsion systems,
- control systems/sensors/instrumentation,
- communication or telemetry systems,
- life-support systems, and
- test equipment.

Biomedical Engineering. Biomedical engineers (sometimes called "bioengineers") employ engineering principles and basic science in the design and manufacture of devices, systems, and software to improve human health. Common application areas involve

- artificial internal organs and prosthetics,
- imaging and diagnosis tools,
- quality control and safety validation of medical equipment, and
- data analysis and health analytics tools.

Electrical/Electronic Engineering. Electrical engineering is about the design, testing, and manufacture of electrical and electronic equipment, including

- electric motors
- radar/GPS navigation systems,
- communications systems,
- power-generation equipment,
- electrical systems for transportation systems, consumer electronics,
- sensors and control systems, and
- computer hardware.

It is important to recognize that the different engineering areas overlap across the various applications listed above; there is no reason that an electrical engineer cannot contribute to the instrumentation package in an aerospace-related design.

Example 6.1:

What type(s) of engineers might work on the following endeavors?

a. Control-panel readout system for a vehicle.
b. Fuselage section of an aircraft.
c. Blow-molding machine/tooling for polymer or glass bottles.
d. Lithium-ion battery for an insulin pump.

SOLUTION

Consider:

a. **Control panel.** The primary field associated with this application might be **electrical engineering**, but if the panel is integrated with other cockpit features/controls, then **mechanical** or **aerospace engineers** (if the vehicle is an aircraft) might be involved.
b. **Fuselage section.** An **aerospace engineer** would be primarily responsible for such work, but the mechanical design skills of **mechanical engineers** and the fabrication knowledge of **manufacturing engineers** are also relevant to this application.
c. **Blow molder.** A **manufacturing engineer** is the obvious choice, but many aspects of fluid and heat transport within the machine would involve the know-how of **mechanical engineers**. Also, the control systems for the machine might involve the design skills of an **electrical engineer**.
d. **Implant battery.** This application is primarily the domain of **biomedical engineers**, but an **electrical engineer** might consult on the associated circuitry and a **manufacturing engineer** might have input on the safety and reliability aspects of the device.

6.3 ENGINEERING ETHICS

The occupations of law, medicine, teaching, architecture, and engineering are all traditionally recognized as **professions**. These professions require significant advanced training and involve numerous intellectual skills, and their practitioners are expected to exercise the most rigorous judgment in their work. For these reasons, practitioners of these professions are subject to additional training and certification requirements and have special, expanded responsibilities to their communities. Typically, government statutes apply to the way such professionals approach their work. The work of engineers is crucial to the public welfare, and engineering designs, materials, and methods have a significant impact on public safety. For this reason, engineers who wish to operate at the highest levels of their profession must obtain licensure so that their technical competence and understanding of their responsibilities to the public can be verified.

PROFESSIONAL-DEVELOPMENT NOTE – NCEES MODEL RULES OF PROFESSIONAL CONDUCT

The National Council of Examiners for Engineering and Surveying (NCEES) is a nonprofit organization that offers resources for, administers examinations for, and maintains records of professional licensure for engineers in the United States. The stated mission of the NCEES is to "advance licensure for engineers and surveyors in order to safeguard the health, safety, and welfare of the public". An engineer licensed by an organization like the NCEES is a **professional engineer** (PE).

To this end, the NCEES maintains model rules for professional conduct.[1] The Model Rules embody a **code of ethics** that all engineers must adhere to in their professional capacities. They consist of a preamble/introduction and a list of obligations. The preamble describes how the rules are binding on all registered/licensed engineers, who must adhere to the highest standards of ethical and moral conduct. Furthermore, because the practice of professional engineering is a privilege and not a right, the privilege is earned by training and registration. These introductory materials mostly describe a procedural framework for licensure and the board required for doing so. This framework represents the "gold standard" for licensure requirements.

According to the code embodied by the Model Rules, PEs have obligations to three constituencies. These are, in order of importance

a. the licensee's obligation to the *public*,
b. the licensee's obligation to the *employer* and *clients*, and
c. the licensee's obligation to *other licensees.*

The obligations are summarized below. They mostly make a distinction between what is *right* and what is *expedient* in the practice of engineering.

a. **To the Public.** Public safety and public welfare are more important than moneymaking or meeting deadlines, whether for a PE's self or a PE's employer. PEs must represent themselves forthrightly in public affairs and must only practice within their respective disciplines. The following obligations apply when the PE is performing services for a client or employer:
   1. PEs' first and foremost responsibility is to safeguard the health, safety, and welfare of the public.
   2. PEs shall sign and seal only those plans and other documents that conform to accepted engineering standards and that safeguard the health, safety, and welfare of the public.
   3. PEs shall notify their employer, client, or other authority when their professional judgment is overruled regarding the health, safety, or welfare of the public.
   4. PEs shall include all relevant and pertinent information in an objective and truthful manner within all professional documents, statements, and testimony.
   5. PEs shall only express a professional opinion publicly that is based on facts and a competent evaluation of the subject matter.
   6. PEs shall issue no statements, criticisms, or arguments on engineering matters that are inspired or paid for by interested parties (unless they are explicitly declared).

7. PEs shall not practice with any person or firm that they know is engaged in fraudulent or dishonest business or professional practices.
8. PEs who think that any person or firm has violated any rules or laws applying to the practice of engineering shall report it to the authorities.
9. PEs shall not knowingly provide false or incomplete information regarding an applicant in obtaining licensure.
10. PEs shall comply with the licensing laws and rules governing their professional practice in each of the jurisdictions in which they practice.

b. **To the Employer and Clients.** PEs should only provide services in areas where they have competence. They should also strive to maintain confidentiality and avoid conflicts of interest.

1. PEs shall undertake assignments only when qualified by education or experience in the specific technical fields of engineering involved.
2. PEs shall not sign any plans or documents dealing with subject matter in which they lack competence, nor any such plan or document not prepared under their responsible charge.
3. PEs may accept assignments and assume responsibility for the coordination of an entire project if each technical segment is signed and sealed by the responsible licensee.
4. PEs shall not reveal facts, data, or information obtained in a professional capacity without the prior consent of the client, employer, or public body they work for (unless authorized).
5. PEs shall not solicit or accept gratuities from contractors, etc., in connection with work for employers or clients.
6. PEs shall disclose to their employers or clients all known or potential conflicts of interest that could influence their judgment on a project.
7. PEs shall not accept compensation, financial or otherwise, from more than one party for services pertaining to the same project without disclosure.
8. PEs shall not solicit or accept a professional contract from a governmental body on which a principal or officer of their organization serves as a member.
9. PEs shall not use confidential information received in the course of their assignments as a means of making personal profit without consent.

c. **To Other Licensees.** False advertising, bribery, or spurious attacks on the reputation of others are prohibited for PEs.

1. PEs shall not falsify or permit misrepresentation of their, or their associates', qualifications or experience.
2. PEs shall not offer, give, solicit, or receive, either directly or indirectly, any gift in order to secure work.
3. PEs shall not injure or attempt to injure, maliciously or falsely, the professional reputation of other PEs.

4. PEs shall make a reasonable effort to inform another licensee whose work is believed to contain a material discrepancy, error, or omission that may impact the health, safety, or welfare of the public, unless such reporting is legally prohibited.

Further information and resources can be found online via organizations like the NCEES, the National Society of Professional Engineers (NSPE), or other professional engineering societies (ASME, ASM, IEEE, etc.).

There are numerous reasons why an engineer might stray from their ethical obligations in the course of their work. They might be overconfident in their work and therefore not assess it objectively. This may cause small mistakes to go overlooked, even though these small mistakes might be the very thing that causes a disaster. Engineers may also become impatient in their work, judging a design to be ready for public use even though it might not be fully validated by data. Sometimes, engineers might accede to authority figures in their workplace (motivated by deadlines or finances) regarding the readiness of a design or project. Thus, it is recommended that engineers check their work at least twice, including assessment by their peers, regardless of the cost in wealth and time. Engineers should be open to other ideas and be able to admit that they could be wrong.

Example 6.2:

Suppose that you are the engineer responsible for the materials selection in the design of a medical implant. The production of the design is behind schedule due to delays in the delivery of one of the material constituents. There are other materials, available immediately, that lack the same degree of performance in terms of reliability and lifespan but are still mostly within the margin of safety of the design. How might you assess the following options?

a. Order the new material and deploy it in production, but refuse to certify it as you did with the original material.
b. Order the new material and deploy it in production, but issue a warning to the end user that they use the device at their own risk.
c. Order the new material and deploy it in production; since it is really just as good as the original material to within some tolerance, there is no need to inform anyone else.
d. Inform your supervisors that production of the device will be delayed until the original material arrives.

SOLUTION

An assessment of the options might look like:

a. This might seem like a good option, but since the final authorization of the material is part of your responsibility, you should not try to "hand it off" in this way. The health and safety of the end user are more important than internal deadlines in your operation.

b. This is another "copout" in the performance of your duties. The concerns are the same as in **A** above; simply disclosing the shortcomings of the product does *not* fulfill your ethical obligations.
c. This is obviously unethical. The margin of safety is determined by the design, and "stretching" it by any amount risks negative consequences for health and safety.
d. **This is the preferred ethical course.** There may be negative consequences in your working environment imposed by your supervisors, but it is important to recognize that these are a small consideration in comparison to the public good.

6.4 VALIDATING MATERIALS AND DECISION-MAKING

"Decision-making" is commonly thought of as a cognitive process that underlies the determination of a particular course of action. Engineers are taught to approach problems from a pragmatic standpoint and proceed via logical reasoning. Engineering decision-making emphasizes efficiency, utility, cost, and aesthetics over other considerations. The decisions made involve the allocation of resources and often have some impact on the well-being of the public. For these reasons, decisions should be based on the best information possible and use the most rigorous possible reasoning.

The primary source of data as inputs to the decision-making process comes from experiments. In materials science and engineering, these experiments are typically aimed at determining the properties of a material. Some descriptions of characterization experiments are given in **Chapter 3**. The reliability and usefulness of any experiment depend on the quality of the sampling practices used. Some important aspects of the sampling process include

1. careful consideration of the population from which the sample is taken,
2. the selection and collection of a sample of appropriate size, condition, etc. from this population,
3. attention to the speed and/or economy of obtaining and evaluating the sample, and
4. transformation of this sample into a condition suitable for the laboratory analysis techniques required.

Sampling practices come in many forms and degrees of sophistication. Consider the following basic strategies:

- **Random Samples.** If information on the distribution of some property is desired, a sample can be selected randomly. Sometimes, the entire distribution is not necessary, such as in the case where only the highest and lowest property values are of interest. Typically, the population of materials to be sampled is divided into segments, one of the segments is selected at random to pull a component from, and the selection proceeds by pulling another component from every n^{th} segment according to a predetermined criterion.
- **Systematic Samples.** The samples are selected to test some hypotheses related to anticipated systematic variation in their properties. These include

expected differences related to the location, temperature, or time elapsed between samples. Careful consideration of the variables involved is required to establish the sampling plan.

- **Representative Samples.** A representative sample is one that is intended to represent the typical condition of a population. There is no guarantee that a sample selected at random will be "representative", so the judgment of the experimenter or reference to some preexisting criteria is required for selection. Representative sampling can be revealing in its own way but does not provide the same information as a random sample, such as distribution parameters or uncertainties (see below).

The ultimate goal of testing across a population is obtaining information on the condition of your material with minimal uncertainty. When we measure experimentally a certain property of a material x, we obtain a result that has two components:

$$x = \overline{x} \pm \Delta x \tag{6.1}$$

The first component $\overline{x}$ is the **best estimate** of the property x produced by your experiment. The second component Δx is the **uncertainty** (sometimes called the "error") in your experiment. This absolute uncertainty value is always positive: $\Delta x > 0$ and has the same units as the best estimate. The "±" symbol that joins the two components reflects the fact that your best estimate of the property can either overestimate or underestimate x to a degree given by Δx. Uncertainties may also be represented as relative uncertainties, like

$$x = \overline{x} \pm \frac{\Delta x}{\overline{x}} \tag{6.1$'$}$$

In this case, the relative uncertainty is unitless, and it is frequently given as a percentage or some equivalent ("parts-per-million", "parts-per-billion", etc.). When the value of Δx is small, we say that the result of the experiment is *precise*. When reporting the result of any experiment aimed at measuring a materials property, *always* provide both $\overline{x}$ and Δx.[2]

Example 6.3:

You have obtained experimental results on the tensile strength σ_u of a chromium-molybdenum steel alloy that indicates a best estimate of 4.29×10^8 Pa with an uncertainty of 3×10^6 Pa. What are some ways of representing this result?

SOLUTION

Consider the following representations for reporting the result:

1. Absolute error, standard form:

$$\sigma_u = 4.29 \times 10^8 \text{Pa} \pm 3 \times 10^6 \text{Pa} \quad \text{or} \quad \sigma_u = 429 \text{ MPa} \pm 3 \text{ MPa}$$

2. Absolute error, compact form:

$$\sigma_u = (429 \pm 3)\ \text{MPa}$$

3. Absolute error, "last-digit" form:

$$\sigma_u = 429(3)\ \text{MPa}$$

4. Relative (percentage) error:

$$\sigma_u = 429\ \text{MPa} \pm 0.7\%$$

All of these representations convey the same information (best estimate, uncertainty, units) in an unambiguous fashion. Note that in the third example the "(3)" refers to the uncertainty in the final digit of "429". This format is useful when you have many digits of precision, e.g.

$$e = (1.602176634 \pm 0.000000001) \times 10^{-19}\ \text{C} = 1.602176634(1) \times 10^{-19}\ \text{C}$$

The latter representation is more compact.

A certain proportion Δx_s of the uncertainty is associated with variation in the sample, and a certain proportion Δx_e is associated with variation that arises during the experimental process:

$$(\Delta x)^2 = (\Delta x_s)^2 + (\Delta x_e)^2$$

The uncertainty in the measurement process can be controlled by careful practices: use of representative standards, maintenance of equipment, control of intervening variables, etc. When Δx_e is small compared to Δx_s (say, $\Delta x_e < \Delta x_s/3$), we obtain $\Delta x \approx \Delta x_s$, of course. In our discussion, we will assume that this is the case and take $\Delta x = \Delta x_s$. Performing such careful experiments requires time and resources, so it is important to strike a good balance between the number of samples analyzed and the effort dedicated to the individual tests themselves.

How are the parameters $\bar{x}$ and Δx determined, given a particular experiment? The question is rife with statistical subtleties, which we do not have the space to address here. We introduce a scheme for finding the components of an experimental quantity that is typical of the physical sciences. Suppose that we have performed n measurements in an attempt to determine the property x. Our sampling practices are sufficient so that these N measurements are fully independent of one another and N has a sufficient size ($N \geq 5$). The N measurements are $\{x_1, x_2, \ldots, x_n\}$. The best estimate has the form of the **mean** or the average of the N measurements:

$$\bar{x} = \frac{1}{N} \sum_{i=0}^{N} x_i \tag{6.2}$$

This should make intuitive sense when we imagine that the true value of the property x lies near the "middle" of all the x_i we determined. Our uncertainty, then, reflects how well the best estimate matches the actual value based on the "scatter" in our sample. The parameter that best captures how accurate that match is called the **standard error of the mean** (sometimes just "standard error"):

$$\Delta x = \frac{\sigma}{\sqrt{N}} \tag{6.3}$$

In **Equation 6.3**, the parameter σ is the familiar **standard deviation**, computed as[3]

$$\sigma = \sqrt{\frac{1}{N}\sum_{i=0}^{N}(\bar{x} - x_i)^2} \tag{6.4}$$

When N is small, the quantities $\bar{x}$ and Δx are easily computed, but when the number of measurements is large, software-based approaches are most effective (see MATLAB® **Exercise M6.2**).

We know that when we report measured property values, we must use an expression of the form $x = \bar{x} \pm \Delta x$, so that we can communicate both the essential result and its precision. The precision of our reported value must also be captured in the number of digits that we use in the representation. Meaning: the result of a calculation of $\bar{x}$ and Δx will yield strings of digits whose length depends mostly on the display capacity of our calculator. Though the numerical value may be computed correctly, not all of these digits will be meaningful in our description of our experimental result, i.e., we must be able to determine how many digits are significant. To this end, we employ a number of rules.

1. The numerical value of Δx is reported with one significant digit; this is the most significant (leftmost) digit. *Exception*: when this leftmost digit is "1", Δx is reported using only the two most significant digits.
2. In the numerical value of $\bar{x}$, all digits that are less significant than the digit(s) of Δx are dropped, i.e., the rightmost digit in $\bar{x}$, should have the same order of magnitude as Δx.

By following these rules, the results reported will always reflect the best of our knowledge and not overstate the precision of our experiment.

Example 6.4:

A random sample of an aluminum alloy produced under the same conditions was sent to the lab, where Rockwell hardness measurements (on the B scale) were performed. The results were

$$\{35.6, 32.3, 39.0, 39.4, 36.5, 37.5, 32.6, 33.3, 35.0\}$$

What would you report as the hardness of this material?

SOLUTION

The best estimate for the hardness of this lot of $N = 9$ samples is

$$\overline{\text{HRB}} = \frac{35.6 + 32.3 + \cdots + 35.0}{9} = 35.6899$$

(Recall that HR is unitless.) For now, we retain many more digits than we expect to need in the end. The standard deviation of this data set is

$$\sigma = \sqrt{\frac{1}{9}\left[(35.6889 - 35.6)^2 + \cdots + (35.6889 - 35.0)^2\right]} = 2.64313$$

and the standard error/uncertainty is

$$\Delta HRB = \frac{\sigma}{\sqrt{N}} = \frac{2.64313}{\sqrt{9}} = 0.881042$$

Next, consider the uncertainty value of 0.881042.... Following rule #1, we round this value to one significant digit, i.e.

$$\Delta HRB = 0.8\cancel{81042} \rightarrow \underline{\mathbf{\Delta HRB = 0.9}}$$

This single significant digit of ΔHRB is in the *tenth place*. Therefore, our best estimate can report no digits in the hundredths, thousandths, etc. place:

$$\overline{HRB} = 35.6\cancel{889} \rightarrow \underline{\mathbf{\overline{HRB} = 35.7}}$$

Our final result is then

$$HRB = \overline{HRB} \pm \Delta HRB = \underline{\mathbf{35.7 \pm 0.9}}$$

Values with uncertainties can be compared to accepted values, required values (minimum or maximum), or each other. These scenarios are illustrated in **Figure 6.1**. **Figure 6.1(a)** illustrates a comparison between the measured density of some samples of aluminum alloys. Results for samples "A", "B", and "C" are reported as best estimates $\bar{\rho}$ (the ●s) and uncertainties $\Delta\rho$ (as "error bars" bracketing the ●s). The **discrepancy** between any two values ρ_1 and ρ_2 is the difference in their best estimates:

$$\text{discrepancy} = |\bar{\rho}_1 - \bar{\rho}_2| \tag{6.5}$$

If this discrepancy of two values is *greater* than their combined uncertainty (= $\Delta\rho_1 + \Delta\rho_2$), then we say that the discrepancy is *significant*. If $|\bar{\rho}_1 - \bar{\rho}_2| < \Delta\rho_1 + \Delta\rho_2$, then the discrepancy is *not significant*. In terms of the data from **Figure 6.1(a)**, the discrepancy between sample A and sample B is significant (since their error bars do not overlap), and the discrepancy between sample B and sample C is not significant. Roughly, the values ρ_A and ρ_B are "statistically different", and the values ρ_B and ρ_C are statistically the same. The explanation for these differences may lie in the properties of the material or the conduct of the experiment, but a comparison of the discrepancies to the uncertainties gives an indication that a closer look is appropriate.

The discrepancy concept can be extended to a comparison of a measured value to some accepted or standard value, as shown in **Figure 6.1(b)**. Suppose that you have three materials subjected to different hardening treatments and wish to determine

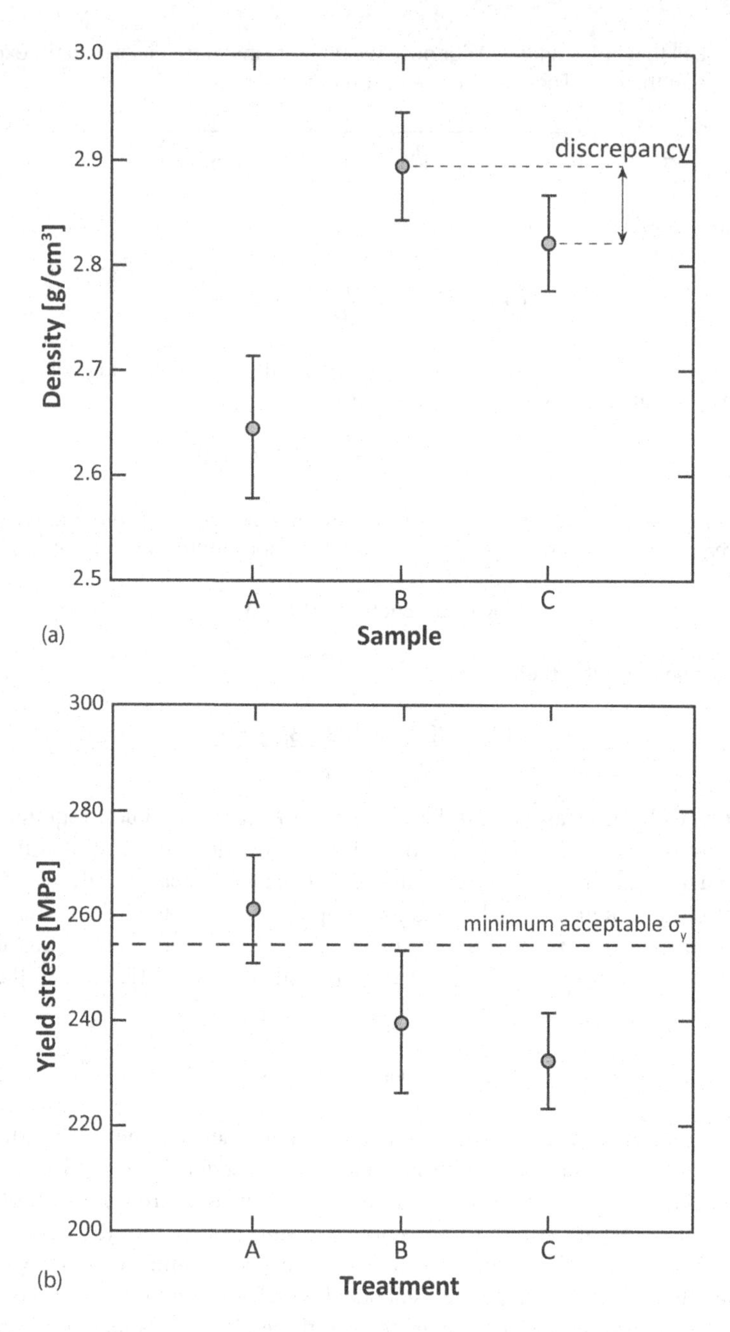

FIGURE 6.1 The use of the discrepancy in decision-making. In (a), the ability to resolve the differences in density ρ between the three samples A, B, and C depends on the size of the discrepancy relative to the uncertainties. Samples B and C have experimentally indistinguishable values of ρ, but that of sample A is almost certainly different. In (b), the comparison is not between the values of σ_y of two or more samples but between the sample values and an exact "threshold" value.

whether the treatments produced a yield strength σ_y that equals or exceeds the required value σ_{y0}. We can calculate a discrepancy like that of **Equation 6.5**, but relative to the required or accepted value:

$$\text{discrepancy} = \left|\bar{\sigma}_y - \bar{\sigma}_{y0}\right| \tag{6.5'}$$

As you can probably infer, when this discrepancy is larger than the uncertainty $\Delta\sigma_y$, it is significant. If the discrepancy $|\bar{\sigma}_y - \bar{\sigma}_{y0}| < \Delta\sigma_y$, we say that σ_y is "consistent" with σ_{y0}. For treatment A in **Figure 6.1(b)**, we observe that, to a good degree of confidence, the measured σ_y equals or exceeds the requirement. Treatment B is a case where the discrepancy is larger than the uncertainty, but the uncertainty bounds are so close to the requirement that we might provisionally accept this result as "borderline". Obviously, treatment C is unsatisfactory in producing the required level of strength.

The ability to resolve the differences between measured property values plays an important role in quality control and quality improvement. **Figure 6.2** is an example of a **control chart** for a measured process variable (dimension, weight/mass, materials property, etc.). The "control limits" indicate the maximum and minimum allowable values of the property for the material to meet specifications. Whether or not the material is adjudged to fall within these limits depends on the statistics associated with a sample. Generally, if the scatter in a sample property x is small compared to

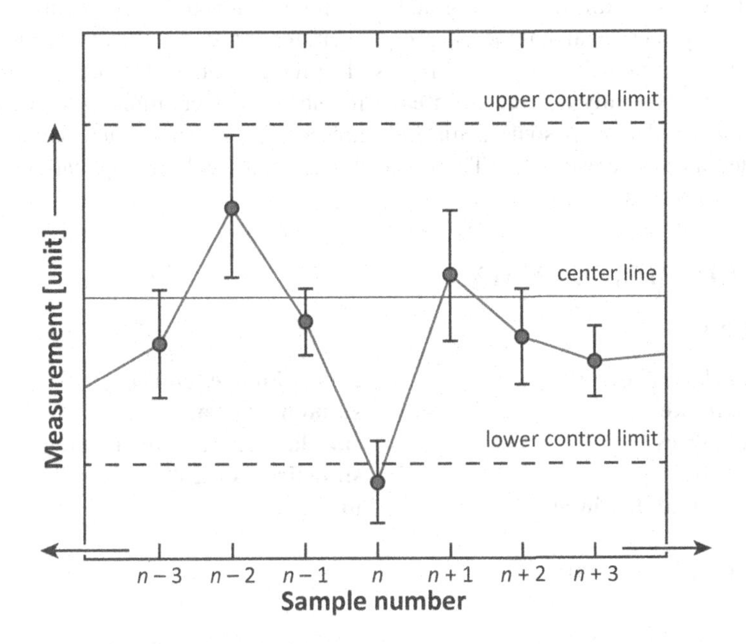

FIGURE 6.2 An example of a schematic control chart for some measured variables. The measurements are obtained from samples, and these values exist within the specification limits or outside of them. A value that lands outside of the limits indicates trouble, as well as a sequence of values within the limits that indicate an unwanted trend.

the span of the acceptable limits $x_{min} - x_{max}$, then the process is deemed capable. To this end, the **process capability index** C_p is defined:

$$C_x = \frac{x_{max} - x_{min}}{6\sigma} \tag{6.6}$$

(The "6σ" parameter represents the "natural" variation possible in a normally distributed variable with standard deviation σ.) When C_x is large, the process is capable; when it is small, difficulties may arise when it comes to producing materials with properties within the specification limits. When the samples yield values of the parameter $\bar{x}$ that fall outside of the limits, the process is "out of control" and requires adjustment. When the variability of a process is entirely random, the process is said to be in a state of **statistical control**. This state is typically achieved by eliminating sources of variation related to human operators, worn equipment, inadequate raw materials, etc. These are called **assignable variations**, and their elimination is the object of quality control and improvement.

6.5 CLOSING

Engineers, working within their respective domains, analyze and solve problems and make decisions based on those solutions. The engineer's designs are required to satisfy constraints, and their decision-making process is expected to be robust, data-driven, and ethical. Ethical guidance can be obtained through many avenues, including professional engineering organizations, trusted colleagues, and legal frameworks established by governments. The data inputs to the decision-making process are typically derived from experiments and observations, and these data must be handled with some statistical care so that associated uncertainties are reported alongside estimates. These statistical variations have applications in (for instance) process control.

6.6 CHAPTER SUMMARY

Key Terms

assignable variations
best estimate
code of ethics
control chart
process capability index
professional engineer
standard deviation
standard error of the mean
statistical control
uncertainty

Important Relationships

$$x = \bar{x} \pm \Delta x$$ (experimentally determined value)

$$\bar{x} = \frac{1}{N}\sum_{i=0}^{N} x_i$$ (sample mean)

$$\Delta x = \frac{\sigma}{\sqrt{N}} \qquad \text{(standard error of the mean)}$$

$$\sigma = \sqrt{\frac{1}{N}\sum_{i=0}^{N}(\bar{x} - x_i)^2} \qquad \text{(standard deviation)}$$

$$\text{discrepancy} = |\bar{x}_1 - \bar{x}_2| \qquad \text{(discrepancy of two values)}$$

$$\text{discrepancy} = |\bar{x} - \bar{x}_0| \qquad \text{(discrepancy for a fixed limit)}$$

$$C_x = \frac{x_{max} - x_{min}}{6\sigma} \qquad \text{(process capability index)}$$

6.7 QUESTIONS AND EXERCISES

Concept Review

C6.1 What type(s) of engineers might work on the following projects?

1. Frame for a mobility device for disabled people.
2. Controller for a piece of industrial equipment.
3. Mechatronic sorting system for automatic inventory control.
4. Optical sensor for measuring the color of applied paint.

C6.2 An engineer at a large manufacturing firm has been overseeing the development of a production technique for a particular material for some years. Desiring a change of career, the engineer leaves the manufacturing firm and establishes themselves as an independent consultant. Another firm approaches the consultant and requests their assistance in producing the same material that they worked with at their previous employer. It is apparent that the best possible way to produce the material is the one that the engineer developed at the first firm. Should they accept the new assignment, and why?

a. **Yes**, as long as there was no non-disclosure agreement in place with the previous firm.
b. **No**, since information originating with the first employer would be required for developing the process at the new firm.
c. **Yes**, it is only knowledge the consulting engineer possesses, and such a transfer of knowledge via individuals is normal and natural.
d. **No**, accepting money to do the same thing twice is not good practice.

C6.3 An engineer working for a design and manufacturing firm that makes vehicles for consumers discovers that the low-cost material selected for the design of a particular component may have different properties than advertised by the supplier because of differences in test methods. These differences are significant and can influence the performance, lifespan, and/or safety of the vehicle. What should the engineer do?

a. Inform the firm of the discrepancy and then inform the authorities that regulate personal transportation.
b. Send anonymous warnings online and in the media warning the public about the potential for component failure.

c. Continue to use the material since it satisfies the low-cost aspect of the design in accordance with company needs.
d. Inform the lead engineer on the vehicle project and prepare a resignation in the event that the issue is not resolved in a consumer-oriented manner.

C6.4 In hypothesis testing, the "z statistic" is given by

$$z = \frac{\bar{x} - x}{\sigma / \sqrt{N}}$$

The parameter z indicates whether we should accept or reject the hypothesis $\bar{x} = x$ at a given level of confidence. Explain the relationship between our representation of the experimentally measured value $x = \bar{x} \pm \Delta x$ and the z statistic.

Discussion-forum Prompt

D6.1 A new engineering firm has been formed with a roster of engineers with experience from other firms in the sector. A partner at the new firm has developed promotional materials for the firm's website that include a list of clients of the firm. The "client" list consists of the previous employers that the firm's engineers were recruited from. Do you think this is ethical? Explain your reasoning in a forum post.

PROBLEMS

P6.1 As a quality control engineer, you collect data on the yield strength of the material produced by the operation you supervise. A sample of material from a given interval of time yields the following:

$$\sigma_y = \{125, 134, 162, 122, 136, 139, 150., 144\} \text{ MPa}$$

What would you report as the yield strength of this material sample? If the minimum required yield strength for the process is 140. MPa, would you say that the process is under control?

MATLAB Exercises

M6.1 The spring constant k of a spring may be measured by timing the oscillations of a mass m attached to it. Since the period T of the oscillations is $T = 2\pi\sqrt{m/k}$, we can determine

$$k = \frac{4\pi^2 m}{T^2}$$

Suppose that you have measured the period of oscillation of a spring for various masses, obtaining the results below. Make a plot of T vs. $\sqrt{m}$. Given this result, what are the implications for the behavior of the spring at different levels of load?

Mass, m [kg]	Period, T [s]
0.501	1.22
0.596	1.32
0.648	1.41
0.710	1.46
0.799	1.54
1.002	1.73

M6.2 Using the data of **Exercise M6.1**, calculate the spring constant and the associated uncertainty.

NOTES

1. National Council of Examiners for Engineering and Surveying, "Model Rules" (National Council of Examiners for Engineering and Surveying, Greenville, SC, USA), 2023.
2. This requirement can apply to representative values, whose uncertainties are determined by the testing process itself.
3. Sometimes, when N is small, the "sample" standard deviation is used. This calculation replaces the denominator N in **Equation 6.4** with $N - 1$. The justification for this can be found elsewhere, but note that when $N = 5$, the difference between the two calculations is almost insignificant. If you are unsure if the difference is important or not, it is best to state which form of the standard deviation you are employing.

Topic 1: Materials for Structures

Engineering Components and Systems

LEARNING OBJECTIVES

After completing this chapter, you should be able to:

1. Define an engineering structure and provide some example application areas that involve structural design.
2. Describe the buckling of a structural element and solve for the critical buckling stress in a given element.
3. Describe the principles underlying fracture and the materials properties that govern a material's resistance to fracture. Discuss the principles of toughening.
4. Solve critical fracture stresses or flaw sizes in loaded structural members that contain cracks.
5. Define fatigue and describe the expected features of a σ–N curve.
6. Estimate the fatigue lifetime of a component under given conditions.
7. Calculate the stress inside the wall of a pressure vessel and compare it to the materials limitations.

T1.1 DESIGN REQUIREMENTS FOR STRUCTURES

When we talk about **structures**, we are talking about components whose primary function is to transmit force or support weight. Typical applications are given in **Table T1.1**. The exact nature of the requirements of the application determines what materials are necessary and (of course) how much is required. Depending on the exact nature of the application, the materials involved might be metallic, ceramic, polymeric, or composite. The materials might be required to perform in dimensions involving rigidity (or flexibility), weight, environmental tolerance, safety, etc. We introduce a few more important behaviors of structures that involve materials below.

DOI: 10.1201/9781003214403-8

TABLE T1.1
A Short Catalog of Structural Applications by Sector

Application Area	Examples	
Transportation	• Aircraft • Automobiles • Trains • Naval vessels • Human-powered vehicles	
Construction	• Buildings • Stadiums/amphitheaters • Roads • Bridges	
Human safety	• Helmets • Eye protection • Body armor • Impact protection for other body parts	
Industrial and energy production	• Turbine blades • Power-transmission shafts • Equipment housings	
Other small devices	• Tools • Appliance cases • Furniture • Recreational equipment	

T1.2 RELIABILITY OF STRUCTURES – BUCKLING, FRACTURE, AND FATIGUE

We know from the discussion in **Chapter 3** that materials have intrinsic limitations on the amount of load that they can sustain. These limits are referred to as strengths, and we discussed in some detail the signatures and origins of the yield strength and the ultimate tensile strength of materials. Exceeding these limits often leads to a degradation or elimination of the material's ability to perform its engineering function, so understanding strengths is crucial to engineering design. When it comes to the design of structures, there are some additional strength-related considerations that require our attention.

Buckling is the *lateral* deflection of a structure loaded along its axis. The buckling deformation mode is contrasted with uniform compression in **Figure T1.1**. Under the influence of a compressive force F, the column may deform elastically according to the principles outlined in **Chapter 2**; the column shortens along the loading axis (and widens along the transverse axes). Alternately, the column may buckle, bowing outward away from its axis. Buckling is an example of a kind of "elastic instability". The abrupt change in beam geometry at the point of instability leads to an impairment of function. There is a critical force F_c at which buckling occurs, given by Euler's buckling formula

$$F_c = \frac{\pi^2 EI}{(KL_0)^2} \tag{T1.1}$$

where E is the Young's modulus of the material, I is the second moment of area of the column, and L_0 is its undeformed length. The factor $0.5 \lesssim K \lesssim 2$ accounts for the way the endpoints of the column are attached. Example values of K for several

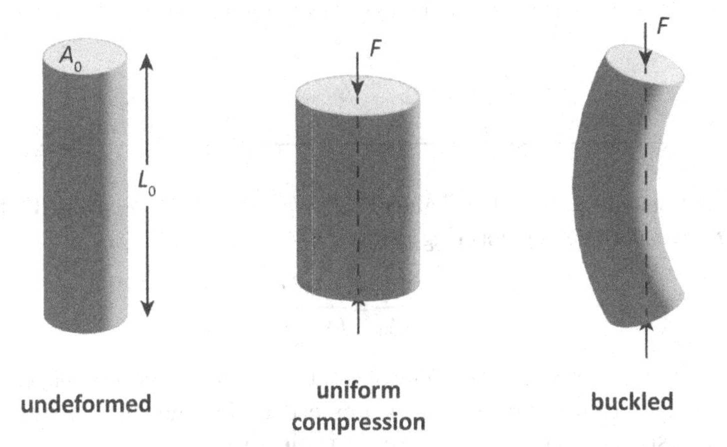

FIGURE T1.1 Comparison of uniform compressive and buckling deformations. The deformations are exaggerated for the sake of clarity. Columns that are buckled cannot support their full, expected load and can lead to design failure.

TABLE T1.2
Values of *K* for the Euler Buckling Formula by Case

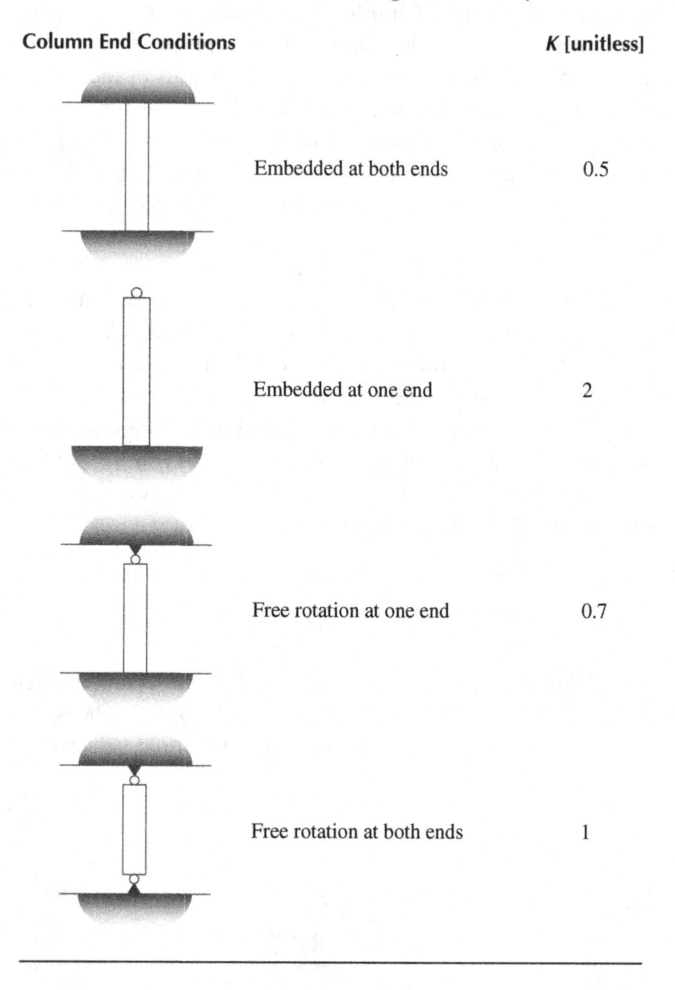

Column End Conditions	*K* [unitless]
Embedded at both ends	0.5
Embedded at one end	2
Free rotation at one end	0.7
Free rotation at both ends	1

configurations are provided in **Table T1.2**. We can sensibly convert the buckling force F_c into a critical (engineering) stress:

$$\sigma_c = \frac{F_c}{A_0} = \frac{\pi^2 EI}{(KL_0)^2 A_0} \tag{T1.1'}$$

When doing mechanical design involving beams under compression, care should be taken that buckling strengths are not exceeded. (For guidance on designing for buckling resistance, refer to those cases in **Table 5.1**.)

In **Chapter 3**, we introduced the idea of fracture strength σ_f as the limiting parameter in materials that lack ductility/are brittle. Fracture is a mechanical process that can affect the performance of *all* types of materials if the correct conditions are

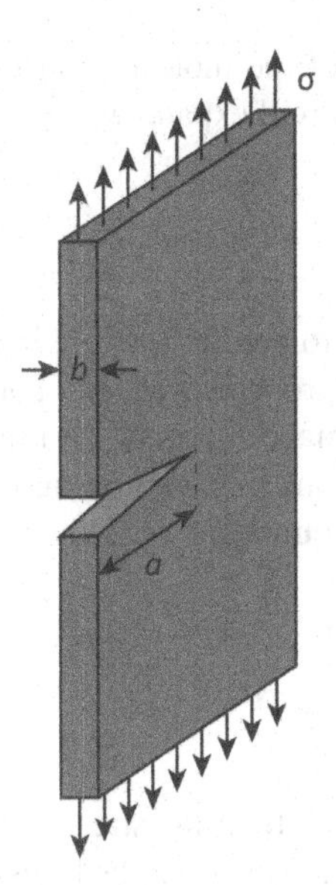

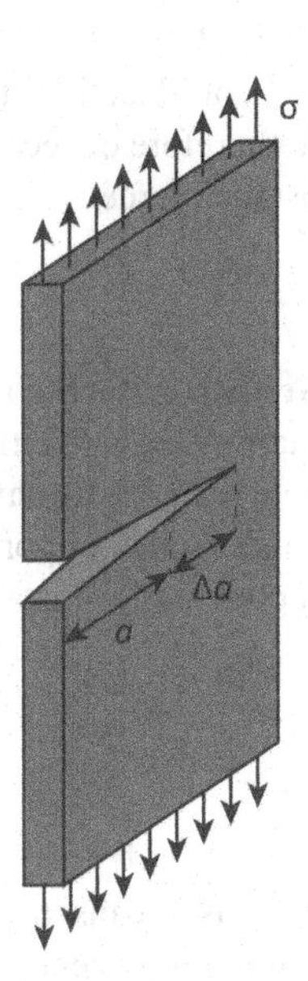

FIGURE T1.2 The geometry of cracking in a material. A flat, sharp crack of length a is present in a plate loaded by a stress σ. Under the influence of this stress, the crack tip can open further, expanding the cracked region by some increment Δa. The mechanical energy required to do this is $b\mathcal{G}_c\Delta a$.

present. When considering the behavior of materials, we must consider the combined influence of the material's intrinsic properties as well as the nature of the extrinsic flaws that it contains. The most frequently considered type of flaw is a **crack**. A crack has the form of a (typically narrow) slit in the material's structure. As long as a crack remains stable, the material can continue to support the loads applied to it. If, on the other hand, the crack becomes unstable and expands without stopping, the material will fail and suffer (like irreversible) loss of function. The expansion of a crack is what we call the **fracture process**, and the exact nature of this process depends to a large degree on the structure of the material.

Consider the situation illustrated in **Figure T1.2**. A (very large) plate that contains a crack through its thickness is subjected to a tensile stress σ. Initially, the crack has a length a. Under the influence of the applied stress, the crack may expand, i.e., the stress tends to pull the crack open. According to the Griffith theory of fracture [formulated by A. A. Griffith (1893–1963) and G. R. Irwin (1907–1998)], mechanical deformation energy produced by the applied stress becomes available for opening the crack. The "crack extension force" $\mathcal{G}$ that results (per unit crack length) is given by

$$\mathcal{G} = \frac{\pi\sigma^2 a}{E} \tag{T1.2}$$

where E is the young's modulus of the material and $[\mathcal{G}]$ = [force/length] = [energy/length^2] = J/m^2. The crack extension force $\mathcal{G}$ is counterposed against a crack resistance force $\mathcal{G}_c$ produced by the cohesion of the molecules at the leading edge of the crack. We therefore expect that the crack will expand when the opening force exceeds the resistance force:

$$\mathcal{G} > \mathcal{G}_c \tag{T1.3}$$

This is the **Griffith criterion** that allows us to predict fracture in this component based on the state of stress (σ), the material's properties (E, $\mathcal{G}_c$), and the material's condition (a). Note that, for **Equation T1.3** to make sense, $[\mathcal{G}_c] = \mathrm{J/m^2}$.

The Griffith theory has important applications in determining the reliability of a particular component. For instance, taking **Equations T1.2** and **T1.3** and solving for σ gives

$$\sigma > \sqrt{\frac{E\mathcal{G}_c}{\pi a}} \tag{T1.4}$$

during fracture. This means there is a critical fracture stress σ_f given by $\sigma_f = (E\mathcal{G}_c/\pi a)^{1/2}$ that cannot be exceeded lest the crack expand. In the limit of a very brittle material, the cohesive force is constant and given by the interfacial energy: $\mathcal{G}_c = 2\gamma$. (γ is related to the strength/configuration of the bonds at the crack tip.) This gives

$$\sigma_f = \sqrt{\frac{2E\gamma}{\pi a}} \tag{T1.4'}$$

at fracture. In such fragile materials, exceeding σ_f produces *unlimited* rupture since γ (and hence $\mathcal{G}_c$ as well) is a constant. We call such a fracture process *unstable*. In many "tough" materials, $\mathcal{G}_c$ is not constant but may increase as fracture proceeds: $\mathcal{G}_c(a + \Delta a) > \mathcal{G}_c(a)$. Under conditions of fixed stress, a crack may expand up to a point and then stop; this is *stable* fracture. Note the crucial difference between these two cases. Unstable fracture is typically instantaneous and catastrophic, while stable fracture might not occur all at once. In the latter case, the negative effect of cracking on reliability might be mitigated by an inspection process that can reveal fracture damage in a component and hasten its replacement. Some example fracture toughness and fracture strength data are provided in **Table T1.3**.

When designing using materials that are prone to fracture, the parameter $\mathcal{G}_c$ becomes an important constraint. This makes the determination of $\mathcal{G}_c$ all the more critical for safe design. To this end, we evaluate the fracture properties of materials using a variety of tests.[1] Consider the schematic "double-cantilever–beam" and "three-point–flexural" tests shown in **Figure T1.3**. In the double-cantilever–beam test of **Figure T1.3(a)**, a premade notch is cut into the specimen so that the parameter a is controlled at the start of the experiment. The tip of the notch is placed under conditions similar to that of the crack in **Figure T1.2** so that the fracture behavior can be assessed. From an analysis of the double-cantilever configuration, we can determine

$$\mathcal{G}(F) = \frac{12F^2a^2}{Eb^2d^3} \tag{T1.5}$$

TABLE T1.3
Values of $\mathcal{G}_c$ and σ_f for Selected Materials

Material	$\mathcal{G}_c$ [J/m²]	σ_f [MPa]
Silica glass	5–10	200
Concrete, brick, stone	3–20	3–5
Silicon	4–8	300
Silicon carbide	5–8	186
Alumina	30–80	344
Wood (with grain)	150	–
High-alloy steel	300–10,000	–
Wood (across grain)	12,000	–
Mild steel	10,000–20,000	–
Stainless steel	2000–20,000	–
Al alloys	7000–20,000	–
Brasses	10,000–50,000	–
Ni alloys	70,000–100,000	–
Ti alloys	3000–60,000	–
Rubber	1000–200,000	–
Carbon-fiber reinforced polymer	8000–20,000	11,000–20,000
Glass-fiber–reinforced polymer	10,000–80,000	7000–10,000

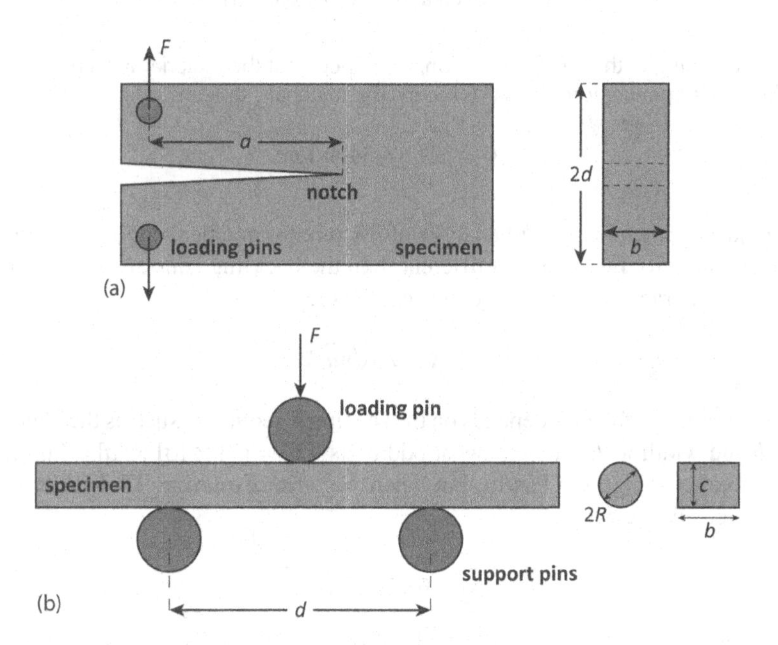

FIGURE T1.3 Test geometries for fracture-toughness/fracture-strength testing. The double-cantilever setup in (a) permits the determination of $\mathcal{G}$ directly; $\mathcal{G}_c$ is the value of $\mathcal{G}$ at fracture/crack expansion. A three-point–bending test is shown in (b). The fracture strength σ_f can be identified as the stress σ at rupture.

Determining $\mathcal{G}_c$ for this material is then as simple as determining what forces produce expansion of the crack. For the three-point–flexural configuration of **Figure T1.3(b)**, we compute the stress in the material as

$$\sigma = \frac{FL}{\pi R^3} \quad \text{or} \quad \sigma = \frac{3FL}{2bc^2} \tag{T1.6}$$

depending on the shape of the specimen cross section. The stress at failure in this case is called the **modulus of rupture** and can be inferred from the force at breaking.

Example T1.1:

Consider a double-cantilever test coupon like that of **Figure T1.3(a)** with a = 10. cm, b = 2.0 cm, d = 2.5 cm. If the coupon is made of a mild steel (with Young's modulus E = 30. GPa) and the force when cracking is first observed is F_c = 2900 N, estimate $\mathcal{G}_c$ for this material.

SOLUTION

From **Equation T1.5**, we can compute the crack-opening force $\mathcal{G}$ at fracture as

$$\mathcal{G}(F_c) = \frac{12F_c^2a^2}{Eb^2d^3} = \frac{12(2900\text{ N})^2(0.10\text{ m})^2}{(30.\text{GPa})(0.020\text{ m})^2(0.025)^3} = 5400\ \frac{\text{J}}{\text{m}^2}$$

According to the Griffith criterion, we expect that the material will crack when this crack-opening force equals/exceeds the material's toughness limit $\mathcal{G}_c$, i.e.

$$\mathcal{G}_c = \mathcal{G}(F_c) = \mathbf{5400\ J/m^2}$$

Frequently, the most useful analysis of fracture for practical engineering involves a **stress intensity factor** K. K (different than the buckling constant!) is the product of a stress, a crack size, and a geometrical factor:

$$K = Y\sigma\sqrt{\pi a} \tag{T1.7}$$

The correction factor Y depends on the specimen geometry, such as the plate thickness b and width w. K has somewhat odd units: $[K] = [Y] \times [\sigma] \times [a]$ = [unitless] × [force/area] × [length] = $\text{Pa}\sqrt{\text{m}}$, but when we write **Equation T1.4** in terms of a stress intensity, we get

$$K_\text{I} = \sigma\sqrt{\pi a} > \sqrt{E\mathcal{G}_c} \tag{T1.8}$$

That is, we have isolated the materials-related quantities $\{E, \mathcal{G}_c\}$ on one side and have taken $Y = 1$. The rewritten criterion for fracture then becomes

$$K_\text{I} > K_\text{Ic} \tag{T1.9}$$

where K_c is called the "critical stress intensity factor" and is another representation of a material's fracture toughness.[2] In **Equations T1.8** and **T1.9**, we indicate with the subscript "I" that this particular loading configuration "opens" the crack. There are two other modes ("II" and "III") besides, and the three modes are:

Mode I ("opening mode") – a tensile stress pulls the faces of the crack away from each other.
Mode II ("sliding mode") – a shear stress pushes one crack face across the other *in the direction* of the crack.
Mode III ("tearing mode") – a shear stress pushes one crack face across the other *transverse to the direction* of the crack.

These modes are illustrated in **Figure T1.4**. Fracture can occur when the stress intensity in any particular mode exceeds the toughness value:

$$K_{\mathrm{I}} > K_{\mathrm{Ic}} \quad \text{or} \quad K_{\mathrm{II}} > K_{\mathrm{IIc}} \quad \text{or} \quad K_{\mathrm{III}} > K_{\mathrm{IIIc}} \tag{T1.9'}$$

In cases where multiple modes are present simultaneously, a "combined" criterion may be written:

$$K_{\mathrm{I}}^2 + K_{\mathrm{II}}^2 + \frac{E}{2G} K_{\mathrm{III}}^2 > K_c^2 \tag{T1.10}$$

The stress intensity analysis of **Equations T1.7–T1.10** is typically more amenable to the analysis of material reliability than the Griffith theory since the Griffith theory only captures the behavior of an idealized system. A great number of stress-intensity factors are cataloged in engineering handbooks and other resources.

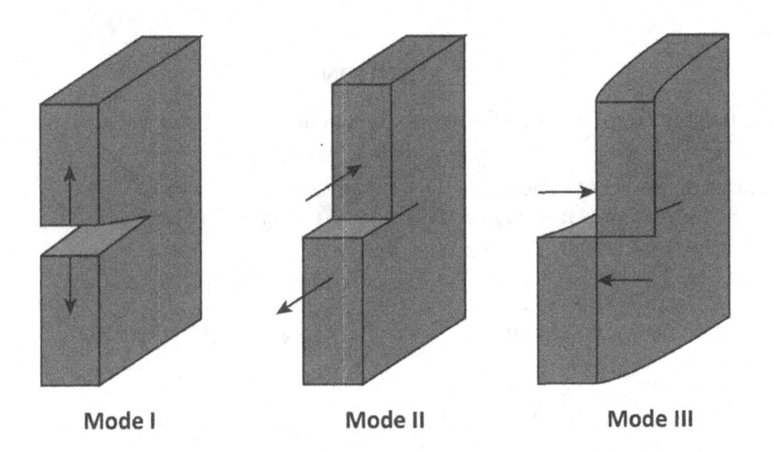

FIGURE T1.4 The three distinct fracture modes. Mode I is a crack-opening configuration; Mode II is a shear configuration with the crack faces, and Mode III is a shear configuration transverse to the crack faces.

Example T1.2:

A large plate is loaded along two axes, as shown below, by different stresses. The plate contains a crack that is oriented at an angle θ to the horizontal, and the stress intensity factors are

$$K_I = \left(\cos^2\theta + \frac{1}{2}\sin^2\theta\right)\sigma\sqrt{\pi a} \quad \text{and} \quad K_{II} = \frac{1}{2}(\cos\theta\sin\theta)\sigma\sqrt{\pi a} \quad \text{and} \quad K_{III} = 0$$

Suppose that $\sigma = 50.$ MPa, $a = 2.0$ mm. Make a plot of $K_I^2 + K_{II}^2$ for $0° \leq \theta \leq 180°$. If the material the plate is made of has a K_c value of 3.0 MPa$\sqrt{m}$, what values of θ indicate failure?

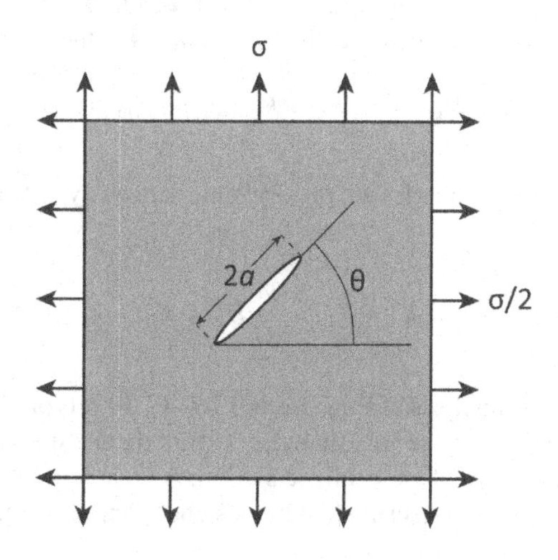

SOLUTION

It is straightforward to set up the necessary functions for `KI` and `KII` in MATLAB®:

```
function K = KI(theta)
      % sigma in MPa; a in m; K in MPa*m^(1/2)
      sigma = 50;
      a = 0.002;
      (cosd(theta).^2 + ...
             0.5*sind(theta).^2)*sigma*sqrt(pi*a);
end
function K = KII(theta)
      % sigma in MPa; a in m; K in MPa*m^(1/2)
      sigma = 50;
      a = 0.002;
      0.5*cosd(theta).*sind(theta)*sigma*sqrt(pi*a);
end
```

We can now plot over the specified range:

```
>> x = 0:1:180;
>> y = KI(x).^2 + KII(x).^2;
>> plot(x, y);
>>
```

To obtain

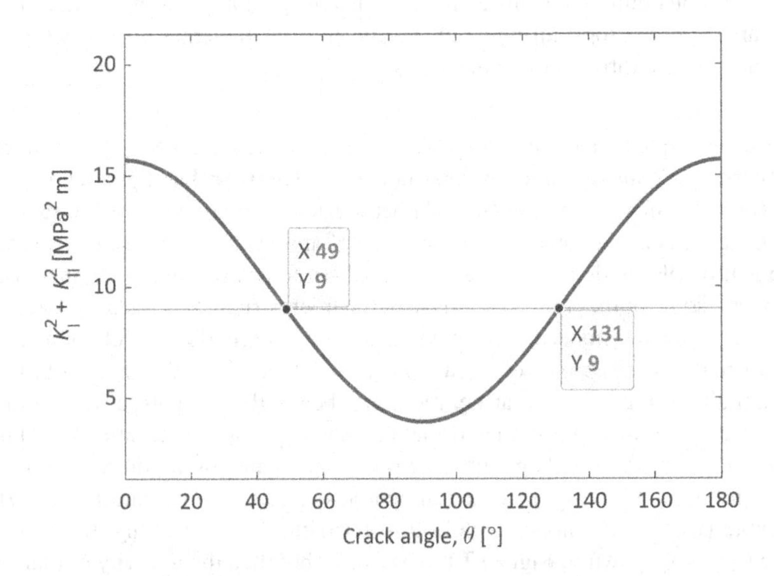

We have added a horizontal line that indicates the value of K_c^2. From this line, we can evaluate the criterion of **Equation T1.10**:

$$K_I^2 + K_{II}^2 > K_c^2$$

where $K_c^2 = 3.0^2 = 9.0$ MPa²m. We identify the conditions for failure as when $\boldsymbol{\theta < 49°}$ or $\boldsymbol{\theta > 131°}$.

Note that the parameters $\{\mathcal{G}_c, \sigma_f, K_c\}$ are all related in some way and typically involve many different materials phenomena. However, the physical representation of the toughness parameter $\mathcal{G}_c$ gives us some clues as to how to modify a material to, say, increase its toughness. The fracture toughness is given as an *energy* (per unit crack area). Increasing the energy required for crack propagation can be effected using a number of strategies. For example:

1. **Dissipation of energy near the crack tip.** Deformation processes in materials that are *not* elastic (i.e., during which energy is not conserved) can reduce the mechanical energy delivered to the crack that would otherwise cause its expansion. For example, any plastic deformation that occurs during the fracture process absorbs energy. In polymeric viscoelastic materials,

relaxation mechanisms in the macromolecules near the crack tip also reduce the energy available for crack expansion. (See **Topic 2.**)

2. **Redirecting cracks as they develop.** The energy input required to expand the crack is given as an energy per unit area. Fracture processes in which the crack does not propagate straight through the material but takes an indirect path correlate to a higher toughness value. Materials that contain a reinforcing phase, like many composites, are toughened in this way. The particulate or fibrous reinforcements artificially increase the amount of crack surface area that develops during crack advancement as the crack passes between, across, and through these obstacles.

As an example of a toughening process, consider the treatment called tempering. The martensite transformation was introduced in **Chapter 4** and produces the very hard (high H) and very brittle (low $\mathcal{G}_c$) bct martensite phase from fcc austenite (γ). This material is not suitable for a wide range of applications because it is so fragile; there is little plastic deformation in the bct phase to absorb energy. To increase the toughness, some of the fragile bct phase in the microstructure is replaced by tougher constituents via the tempering treatment, as shown in **Figure T1.5**. Following **Figure T1.5(a)**, martensite is produced via a rapid quench that misses the "nose" of the TTT diagram and that carries the austenitic metal below the martensite transformation temperature. The martensite transformation, having produced the lath-like bct phase in the microstructure, is then halted by raising the temperature during the temper. The metastable martensite and the remnant austenite transform into ferrite (α) and cementite (Fe_3C) via a non-eutectoid transformation. The resulting distribution of the new phases, shown in **Figure T1.5(b)**, is tougher than the majority of martensite structures because the phases are less brittle and have a shape that is less prone to crack development.

Another structural failure mode worth discussing alongside buckling and fracture is fatigue. Fatigue is a process that is similar to the fracture process defined above, except that fatigue is driven by repetitive or cyclic loading rather than monotonic (i.e., "strictly increasing") loading. In fatigue, failure is the result of deformation that is applied, removed, and applied again for many cycles at levels of stress that are *lower* than the Griffith fracture stress at which we would normally expect fracture. The fatigue process is driven by the gradual accumulation of microstructural damage rather than bulk tearing of the material.

Essential fatigue principles are illustrated in **Figure T1.6**. Fatigue failure initiates with a preexisting crack that is too small to produce overall rupture in the material, as shown in **Figure T1.6(a)**. When the crack is subject to a repetitive, time-dependent stress $\sigma(t)$, it expands gradually over time: $a = a(t)$. Eventually, the crack reaches a size ($= a_c$) where it is subject to unstable expansion according to the fracture mechanics principles above. The model for the nature of the repetitive stress is a sinusoid like

$$\sigma(t) = \sigma_0 \sin\left(\frac{t}{\tau}\right) \tag{T1.11}$$

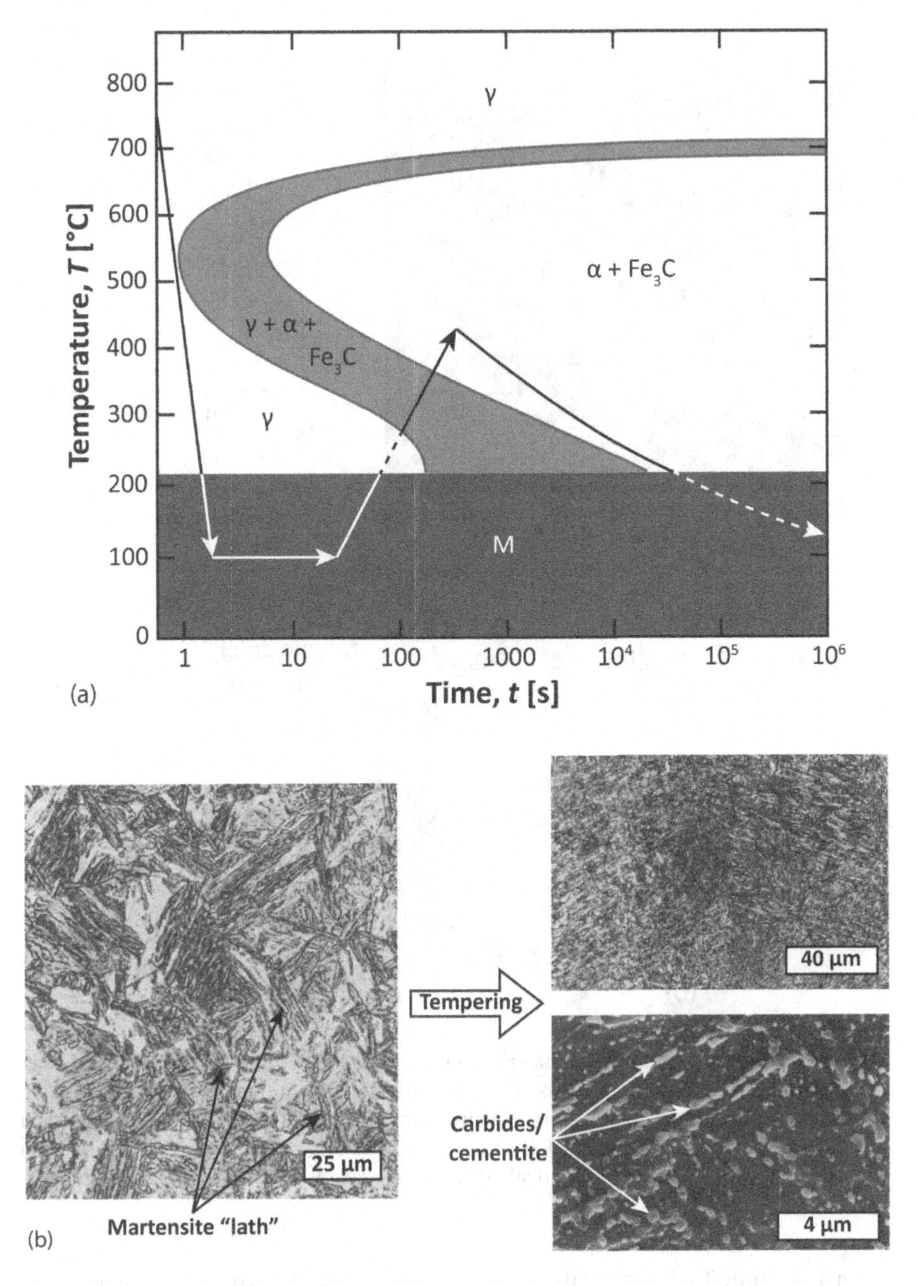

FIGURE T1.5 The tempering process and its effects. In (a), a tempering scheme is superimposed on a plain-carbon steel TTT diagram. A rapid quench generates martensite when the temperature drops below ca. 200°C. By subsequently raising the temperature, the eutectoid transformation can proceed, converting retained austenite and metastable martensite into ferrite and clusters of cementite particles. The microstructures that result are shown in (b). The structure is a "blend" of hard phases (martensite and cementite) and softer material (ferrite) in a distribution that better resists fracture.

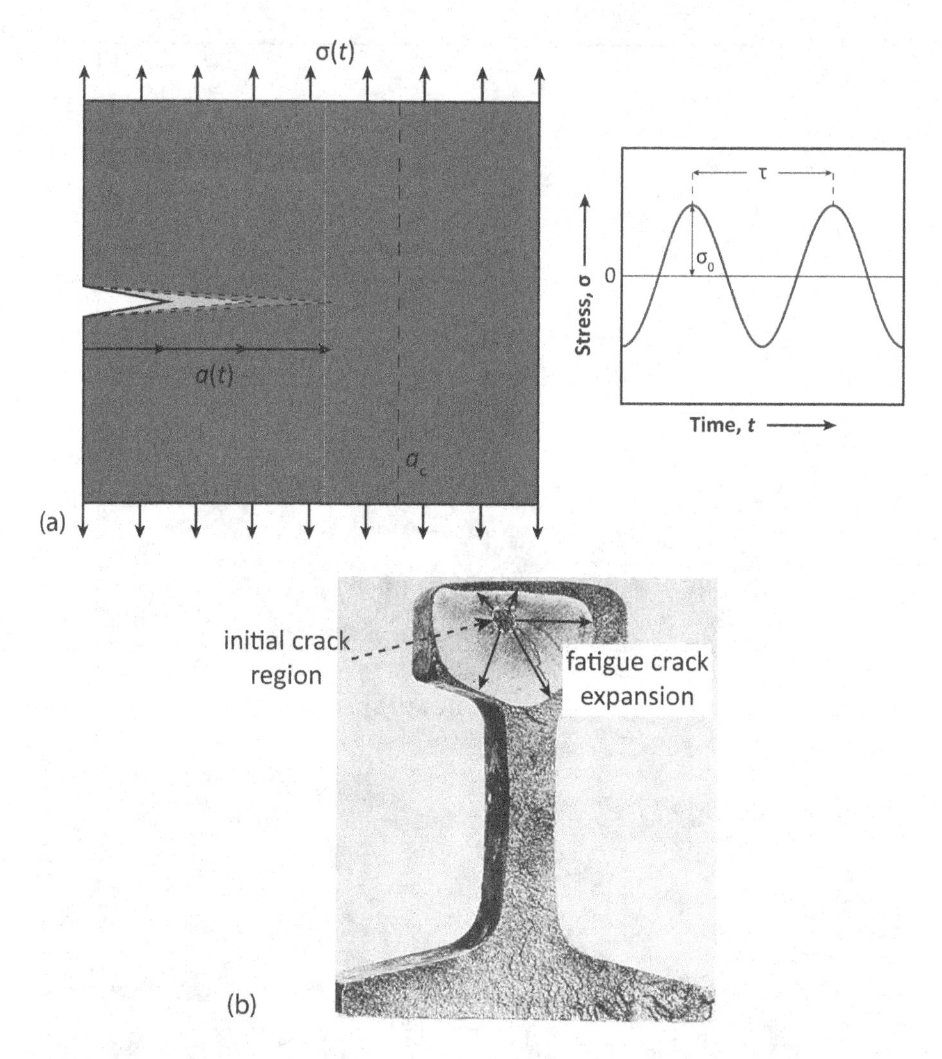

FIGURE T1.6 Fatigue failure of materials. (a) is a schematic of the fatigue process. Under the influence of an oscillating stress $\sigma(t)$, a preexisting crack in a component can expand, even if the applied stress amplitude σ_0 remains below the fracture strength σ_f. A photograph of a fatigued section of railroad track is shown in (b). The original "incipient" cracked region expanded outward slowly until a critical crack size a_c was reached. Unstable crack propagation followed.

In **Equation T1.11**, σ_0 is the stress amplitude, and τ is the period. The stress amplitude is an important determining factor in the development of the fatigue crack; as σ_0 increases, the crack expansion rate da/dt increases as well, and the component lifespan therefore decreases. **Figure T1.6(b)** shows a component that failed during service – a section of railroad track – that exhibits the signs of fatigue. An internal defect/crack in the "head" of the track expanded under the stress of repeated train passings and eventually split the entire cross section of the rail.

The lifetime of a component subject to fatigue is typically tracked using the number of cycles to failure N. N is related to the period of the loading as

$$N(t) \approx \frac{t}{\tau} \qquad \text{(T1.12)}$$

During **fatigue testing**, a component or material coupon is subjected to repetitive stress cycling at a fixed amplitude, and the number of cycles to failure is recorded. A materials **fatigue strength** $\sigma_N = \sigma_0(N)$ is the stress amplitude at which it fails after N cycles. The relationship between σ_N and N is displayed on a **σ–*N* plot** (sometimes called an "S–N plot"). Note that the appropriate axes for a σ–N plot are logarithmic (for N) and linear (for σ) and that sometimes the number of stress *reversals* $2N$, rather than the number of complete stress cycles N, is used in the presentation of fatigue data. Some σ–N plots are shown in **Figure T1.7**. Schematic plots contrasting two different behaviors are shown in **Figure T1.7(a)**. Materials typically follow the lower trajectory; lifespan decreases as σ_0 increases. Other types of materials, mostly ferrous alloys, exhibit a **fatigue limit** (or "endurance limit") σ_L, below which the component's lifespan is effectively unlimited. Materials with a fatigue limit are most frequently Fe or Ti alloys. Some real σ–N data for various materials are provided in **Figure T1.7**.

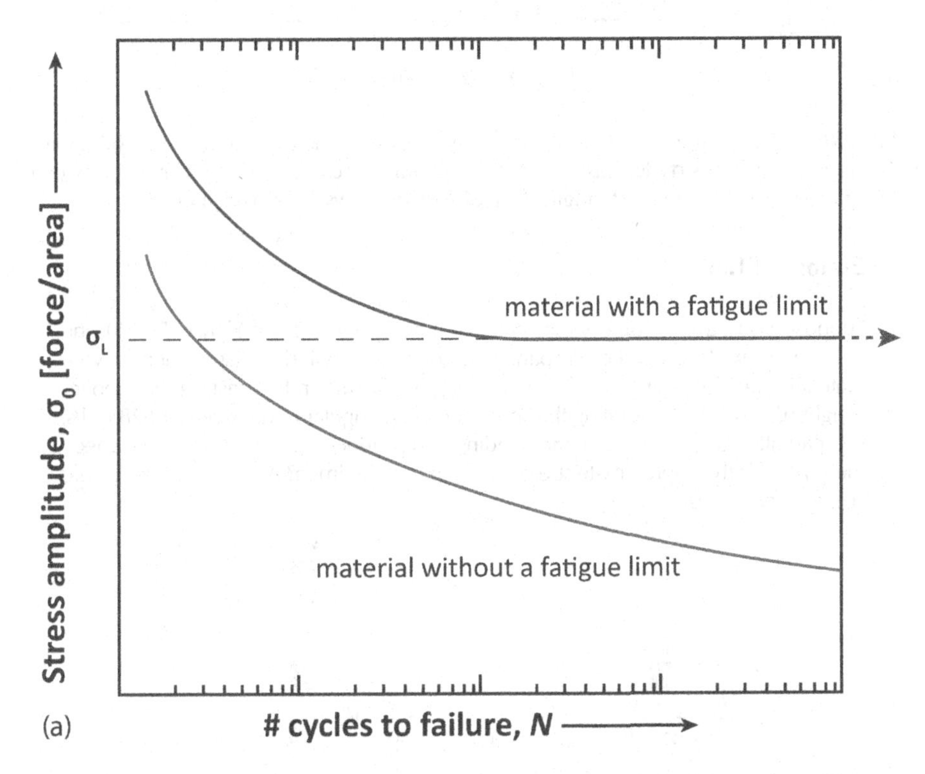

FIGURE T1.7 Idealized fatigue behavior and actual fatigue data. Schematic σ–N curves are shown in (a). *(Continued)*

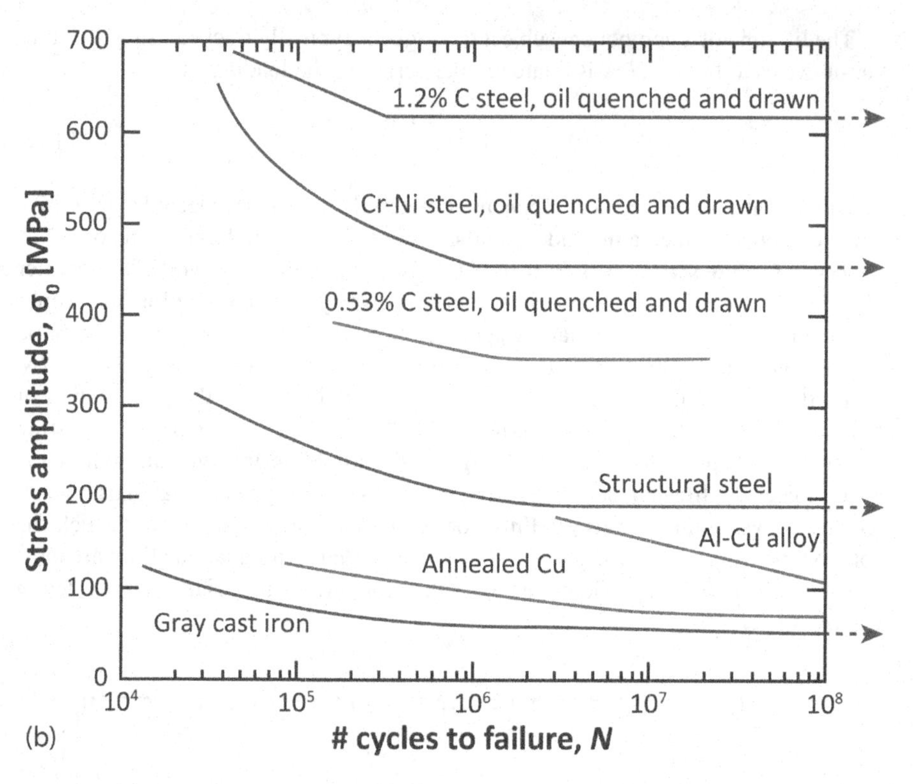

FIGURE T1.7 *(Continued)* The fatigue lifespan, given as the number of stress cycles until failure N, increases as σ_0 decreases. Note the logarithmic scale for N and the possibility of a "fatigue limit" σ_L. Example of fatigue data for various alloys are provided in (b).

Example T1.3:

Suppose you have a component made of the copper alloy of **Figure T1.7(b)** and wish to verify its expected lifespan. You connect a cylindrical specimen of your material with diameter d = 1.8 cm and length L = 10. cm to a test rig that applies weight w = 5$\underline{0}$0 N. By rotating the specimen at an angular frequency ω = 628 rad/s, you can alternate the sense of the bending and produce a cyclic stress in the specimen. What is the approximate stress profile in the specimen? About how long would you expect the test to take?

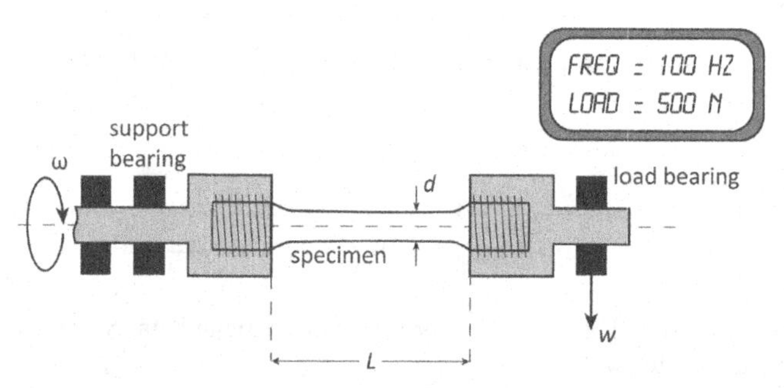

SOLUTION

From **Table 2.5**, the maximum stress σ_{max} that develops in the rod subject to end loading is given by

$$\sigma_{max} = \frac{FLd}{2I}$$

where

$$I = \pi\frac{(d/2)^4}{4} = \pi\frac{(0.0090\text{ m})^4}{4} = 5.2\times10^{-9}\text{ m}^4$$

from Table 2.4. We therefore now have

$$\sigma_0 \approx \sigma_{max} = \frac{FLd}{2I} = \frac{(500\text{ N})(0.10\text{ m})(0.018\text{ m})}{2(5.2\times10^{-9}\text{ m}^4)} = 87\text{ MPa}$$

Note that, as an amplitude, σ_0 is always positive, though the instantaneous stress in the specimen varies between $\pm\sigma_0$. This variation is captured by the complete stress profile

$$\sigma(t) = \sigma_0\sin\left(\frac{t}{\tau}\right) = \sigma_0\sin\omega t = \underline{\mathbf{87\sin628}\boldsymbol{t}}$$

in MPa. From the data in **Figure T1.7(b)**, we determine that the expected number of cycles to failure is $N = 2.1\times10^6$. Given that the frequency $f = \omega/2\pi = 100.$ cycles/s during the test, we obtain a test lifetime t_{test} of

$$t_{test} = \frac{N}{f} = \frac{2.1\times10^6}{100.} = 21{,}000\text{ s} = \underline{\mathbf{5.8\text{ h}}}$$

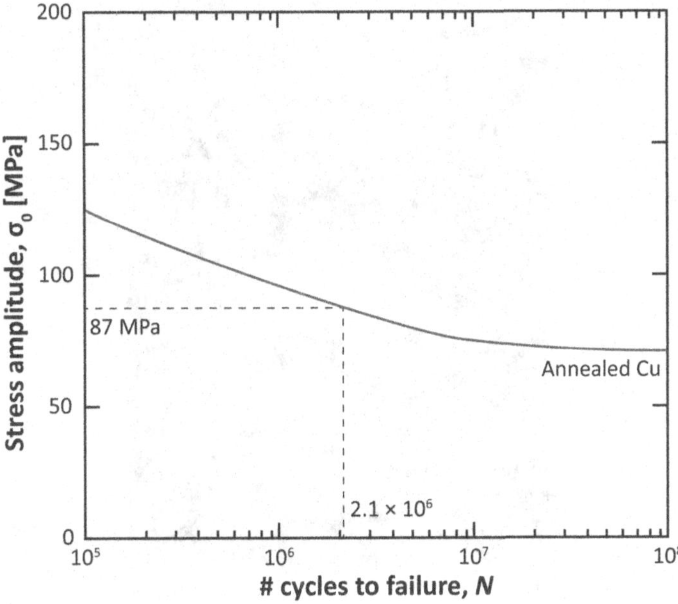

Stress amplitude, σ_0 [MPa]
200
150
100
50
0
87 MPa
Annealed Cu
2.1×10^6
10^5 10^6 10^7 10^8
 # cycles to failure, *N*

The reasons why different materials have different fatigue strengths (for the same value of N) are not straightforward. As a fracture process, fatigue life is improved by introducing microstructural features that make crack propagation more difficult or convoluted. Additionally, extrinsic factors like the size and orientation of preexisting cracks are important for fatigue life. Quality-control processes that can detect flawed components or treatments that can decrease incipient crack size at the surface (such as shot peening) will toughen the component. Stress concentrators, such as sharp corners, can induce or accelerate fatigue, so good component design in this regard is also important.

T1.3 STRUCTURAL DESIGN

Various engineering areas and their materials needs were outlined in **Table I.1**. **Structural design** is the application of the engineering design process to produce stability, resistance to failure, and optimized load-carrying capacity in structures. As an example of the application of structural design principles, consider the following case study on materials design in pressure vessel.

CASE STUDY – MATERIAL SELECTION FOR A FAILURE-RESISTANT PRESSURE VESSEL[3]

Pressure vessels contain gases or liquids at a pressure higher or lower than the environment outside of the vessel. They can be used for static, long-term storage or as a local reservoir to feed some active process or device with pressurized fluid. They also have applications in some human-centric systems, like the pressurized cabin of an aircraft or submersible. They span a range of sizes, from large boilers to hand-operated containers of propellant and other chemicals.

Much of the design of the pressure vessel depends on the substances to be contained, the pressure the contents are to be maintained at, the size of the vessel, and the cost. One overriding factor in pressure-vessel design that establishes the primary design criterion is safety. If the pressurized contents are released in an unregulated manner (or the vessel is breached and the passengers are exposed to the exterior environment), harm to humans and nearby facilities can result. Failure of a design can happen in two ways:

1. *Yielding* of the vessel wall impairs its function or produces observable distortion.
2. *Fracture* of the vessel wall produces catastrophic loss of function and likely instigates a hazard.

There are possible conditions that produce the behaviors above but in a minimal way. For example, the material could yield and distort, but the effect might only remain a cosmetic concern until the vessel can be replaced. Microcracks could develop in a manner that is stable and that does not produce fracture, only leakage of the contents of the vessel.

Based on the above discussion, we identify the requirements for this design as

- **Function.** The vessel must contain fluid at a (significant) internal pressure p.
- **Constraints.** The vessel must deform plastically before fracturing so that over-pressurization may be diagnosed as distortion of the vessel. Additionally, the vessel must leak before fracturing, so that crack development may be detected as a loss of internal pressure. Additionally, the wall thickness t should be small to reduce material cost and vessel weight.
- **Objectives.** Maximize safety at pressure p.

The essential mechanical features of a pressure-vessel design are shown below. The circular cylindrical vessel has a radius R and wall thickness t, where $t << R$. The vessel is pressurized to a gauge pressure p. Under the influence of this pressure, a tensile stress σ develops in the vessel wall:

$$\sigma = \frac{pR}{t}$$

This "hoop stress" can drive plastic deformation of the vessel wall (if $\sigma > \sigma_y$) or fracture (if $\sigma > \sigma_f$) if the wall contains a flaw (with length $2a$).

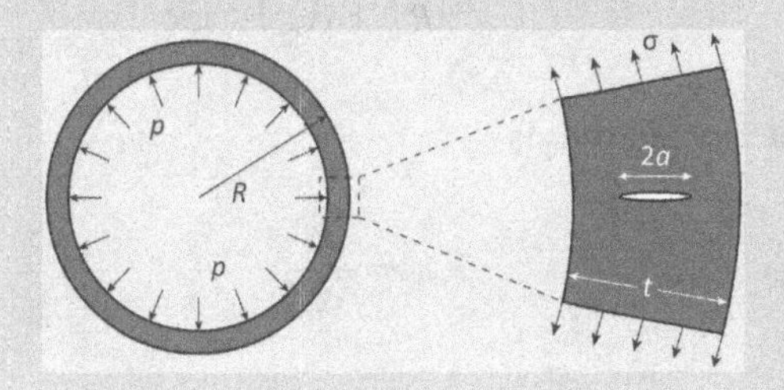

Consider first the conditions under which the vessel will rupture. If the vessel wall has built-in flaws of length $2a$, the fracture stress would be

$$\sigma_f = Y\sqrt{\frac{E\mathcal{G}_c}{\pi a}} \approx \frac{K_{Ic}}{\sqrt{\pi a}}$$

where $Y \approx 1$ depends on the crack orientation. If the stress σ exceeds this value, catastrophic failure is expected. Since we also require that $\sigma < \sigma_y$ for "stable" deformation without rupture, we can reorganize the above to reflect the modified criterion

$$a < a_c = \frac{1}{\pi}\left(\frac{K_{Ic}}{\sigma_y}\right)^2$$

We can now identify a material index h_1 for the vessel design as

$$h_1 = \left(\frac{K_c}{\sigma_y}\right)^2 \quad \text{or} \quad h_1 = \frac{K_c}{\sigma_y}$$

The criterion above expresses the maximum or critical crack size a_c in terms of the materials properties K_c and σ_y. Consider now that determination of a for a particular manufactured vessel may be difficult or impractical. Furthermore, during service, emptying and refilling a tank will produce cyclic stresses in the vessel wall. These cyclic stresses will fatigue existing cracks, expanding a. Corrosion can also erode existing cracks, causing further expansion. The goal, then, is a design in which a crack can expand through the entire wall thickness t but still remain stable. This means that the failure criterion for $2a = t$ is then

$$\sigma_f \approx \frac{K_c}{\sqrt{\pi t/2}}$$

Since $\sigma = pR/2t = \sigma_y$ at yield, we obtain the maximum pressure/performance parameter P as

$$P = p = \frac{4}{\pi R} \times \left(\frac{K_c^2}{\sigma_y}\right) = \frac{4}{\pi R} h_2$$

where the material index h_2 is

$$h_2 = \frac{K_c^2}{\sigma_y}$$

Finally, since smaller values of t correspond to larger values of σ, the thinnest possible wall is obtained when the material has the highest possible σ_y:

$$h_3 = \sigma_y$$

The materials indices $\{h_1, h_2, h_3\}$ together govern materials selection for our design. These indices may be further refined by the introduction of any necessary safety factors.

The materials property chart that provides data on σ_y alongside K_c is below. Several guidelines, indicating contours of constant h_1 and h_2, are indicated. (Lines of constant h_3 are vertical.) Consider the meaning of these guidelines; they mark off contours of constant h, and the further above a given contour a material is, the better suited it is in terms of that index. Various materials that are most suitable for the pressure-vessel application are provided alongside their indices in the data table. Consider the fixed, explicit constraints of $h_1 \geq 0.1\sqrt{m}$ and $h_3 \geq 100$ MPa. (As a constraint, h_2 doesn't change much about the selection, as we shall see.) The materials in the upper-right quadrant in between the h_1 and h_3 guidelines are best suited to the application. Some data are collected and presented in the subsequent selection table.

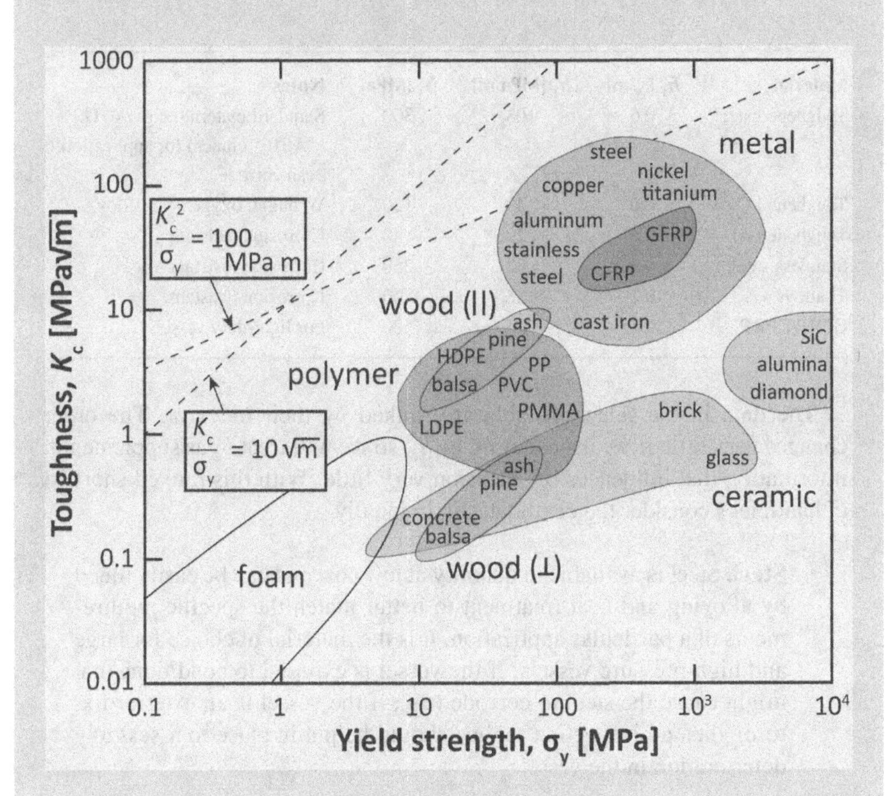

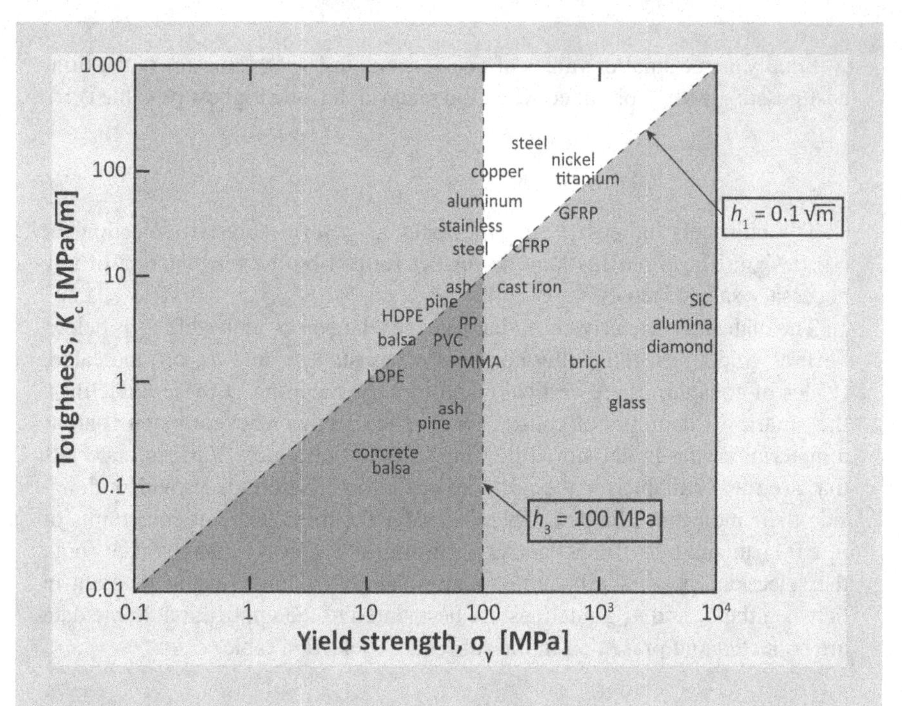

Material	h_1 [$\sqrt{m}$]	h_2 [MPa m]	h_3 [MPa]	Notes
Toughened steel	0.6	108	300	Standard material (e.g., ASTM "A516" grades) for application; can corrode
Toughened Cu	0.6	43	120	Wrought, oxygen-free alloy
Toughened Al	0.6	29	80	Corrosion resistant
Stainless steel	0.4	32	250	Corrosion resistant
Ti alloys	0.2	28	700	Corrosion resistant
GFRP/CFRP	0.1	5	500	For light-duty vessels

The data in the selection table are ranked by their index h_1. The order changes very little if we instead rank by h_2, so we will treat h_2 as supplemental information that influences the decision very little. With this ranked shortlist in hand, let's consider the candidates individually.

- **Steel.** Steel is available in quantity at low cost and can be easily tuned by alloying and heat treatment to better match the specific requirements of a particular application. It is the material of choice for large and high-pressure vessels. If the vessel is exposed to conditions that might cause the steel to corrode (e.g., if the vessel is an exterior fixture), then an inspection regime should be put in place to assess any deterioration in the vessel.

- **Copper.** Copper compares favorably with steel except in economic dimensions. Its lower yield strength necessitates a thicker (and therefore heavier, costlier) vessel. Copper does have some environmental stability, so it is expected to provide a longer service life in some cases.
- **Aluminum.** Aluminum that has been toughened has a suitable safety margin, but a larger quantity of Al will be required since the low h_3 value indicates greater t is required. Al is also more costly to weld together, in the case of vessels that are assembled from sections. Al has good environmental stability.
- **Stainless steel.** Stainless steel has corrosion resistance that other alloy steels lack. However, its relative fragility makes for a less safe design.
- **Titanium.** Ti alloys are a bit too fragile and expensive application in commercial vessels, but they do find use in some specialized aerospace applications because of their lightweight and high toughness at elevated temperatures.
- **CFRP/GFRP.** These materials are relatively cheap and easy to fabricate, but are lacking in the safety index h_1. They can be useful for smaller, lower-pressure vessels.

The engineering science of pressure vessels is much more sophisticated than we have presented here. There are numerous regulations in their design and manufacture that introduce additional constraints. This example does have value in that it introduces the most significant materials properties that govern the design of such vessels.

T1.4 CLOSING

The properties of materials that influence their performance in structures are important because they appear in the majority of materials design and selection problems and practices. The principles underlying buckling behavior, fracture behavior, and fatigue behavior have been introduced in a way that indicates the role of the material in the behavior. This affords us the opportunity either to establish explicit constraints for materials selection or to design processes and treatments that improve a material's performance in a given application.

T1.5 CHAPTER SUMMARY

Key Terms

buckling
crack
fatigue
fatigue limit
fatigue strength
fatigue testing
fracture process
Griffith criterion
modulus of rupture
stress intensity factor
structural design
structures (engineering)
σ–N plot

IMPORTANT RELATIONSHIPS

$$F_c = \frac{\pi^2 EI}{(KL_0)^2} \qquad \text{(buckling force)}$$

$$\sigma_c = \frac{\pi^2 EI}{(KL_0)^2 A_0} \qquad \text{(buckling stress)}$$

$$\mathcal{G} = \frac{\pi\sigma^2 a}{E} \qquad \text{(crack-extension force)}$$

$$\mathcal{G} > \mathcal{G}_c \qquad \text{(Griffith criterion)}$$

$$\sigma_f = \sqrt{\frac{2E\gamma}{\pi a}} \qquad \text{(Griffith fracture strength)}$$

$$K = Y\sigma\sqrt{\pi a} \qquad \text{(stress intensity factor)}$$

$$K_I > K_{Ic} \text{ or } K_{II} > K_{IIc} \text{ or } K_{III} > K_{IIIc} \qquad \text{(stress-intensity criterion)}$$

$$K_I^2 + K_{II}^2 + \frac{E}{2G}K_{III}^2 > K_c^2 \qquad \text{(multi-mode criterion)}$$

$$\sigma(t) = \sigma_0 \sin\left(\frac{t}{\tau}\right) \qquad \text{(cyclic stress)}$$

$$\sigma = \frac{pR}{t} \qquad \text{(hoop stress in pressure vessel)}$$

T1.6 QUESTIONS AND EXERCISES

Concept Review

CT1.1 Describe how the fatigue properties of a material influence its lifespan in the sense of Chapter 5.

CT1.2 Some (typically fragile) ceramic materials are reinforced with "whiskers", which are very short and very tough fibers that are added to the ceramic. How do you expect that these whiskers serve to toughen the ceramic matrix against fracture?

PROBLEMS

PT1.1 According to **Table 5.1**, the material index that indicates maximum buckling load with minimum cost is $h = (\sqrt{E}/\rho c)^{-1}$. Show why this is the case.

PT1.2 A steel plate has a through-thickness crack on its edge, as shown in the diagram below. The plate width w = 80. mm, and its thickness t = 15 mm. The fracture toughness for this material is K_{Ic} = 80 MPa$\sqrt{\text{m}}$. If the plate is expected to experience a stress of $\sigma = 3\underline{0}0$ MPa along the axis perpendicular to the crack, how large a crack (a) can the plate sustain? For this geometry, $K_I = 2.5\sigma\sqrt{a}$.

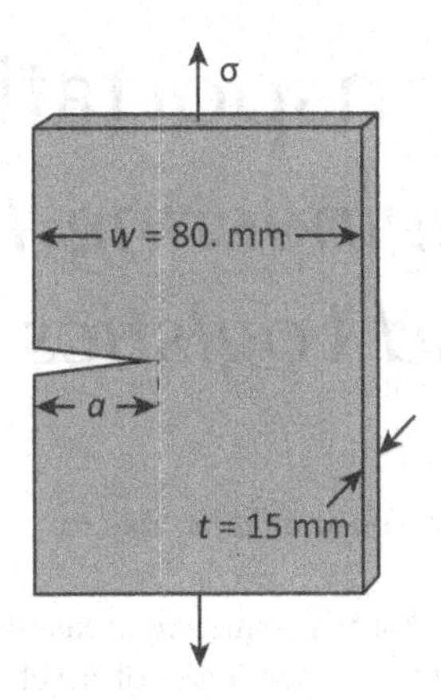

PT1.3 Suppose the S–N behavior of a piece of material can be represented by

$$\log_{10} N = 10 \times \left(1 - \frac{\sigma_a}{\sigma_f}\right)$$

where σ_f is the fracture strength in monotonic loading (i.e., $N = 1$ at $\sigma_a = \sigma_f$). If the material is subjected to continuous loading cycles at $\sigma_a = ½\sigma_f$, how many cycles do you predict it will last?

MATLAB Exercises

MT1.1 Write a MATLAB function `fatiguelife(sigmaa, sigmaf)` that predicts the fatigue lifespan of a component according to the model of `problem T1.3`, above.

NOTES

1. For instance, see ASTM Standard E1820–23b "Standard Test Method for Measurement of Fracture Toughness."
2. Note that, from this criterion, we can make the association $\mathcal{G}_c = K_{Ic}^2/E$.
3. After Michael F. Ashby. *Materials Selection in Mechanical Design*. Fourthth edition. Oxford, England, UK: Butterworth-Heinemann, 2011.

Topic 2: Materials for Transportation

Engineering Transit and Logistics Systems

LEARNING OBJECTIVES

After completing this chapter, you should be able to:

1. List and describe the major performance areas in transportation.
2. Describe quantitatively the dependence of a vehicle's power requirements on its characteristics and interaction with the environment.
3. Differentiate the specific energy and energy density of a power source and obtain one from the other.
4. Describe the essential thermodynamic principles underlying the operation of a heat engine and compute its ideal efficiency.
5. List the important materials properties that influence the operating parameters of heat engines.
6. Quantitatively describe heat transport in materials and the application of materials principles in thermal insulation.
7. List the stages of high-temperature creep and classify the different creep mechanisms according to the operating conditions.
8. Describe the creep and load relaxation of viscoelastic materials using time-dependent moduli/compliances.

T2.1 DESIGN REQUIREMENTS FOR TRANSPORTATION

It should come as no surprise to you that the performance of vehicles depends to a large extent on the materials they are constructed of. Example application areas are given in **Table T2.1**. The requirements of the main body (or "chassis" or "frame" or "fuselage") are similar to those of typical structures: the body must accommodate or transmit loads without failing, or (sometimes!) must fail in as graceful manner as possible (see **Topic 1**). The materials that make up the power plants and some portions of the drivetrains of vehicles must also maintain their performance at elevated temperatures. Other vehicular components must provide some mechanical **damping** of jolts transmitted by uneven surfaces or provide traction at the interface between the vehicle and the ground. Finally, materials principles play an important role in the shipping of commodities such as packaging and insulation.

DOI: 10.1201/9781003214403-9

TABLE T2.1
A Short Catalog of Transportation Applications by Sector

Application Area		Examples
Main bodies/ Frames	• Car chassis • Aircraft fuselages • Ship hulls • Bicycle frames	
Power plants	• Internal-combustion engines • Electric drives • Turbines	
Shipping and logistics	• Packaging • Insulation	
Ancillaries	• Vibration damping • Traction control • Occupant amenities	

T2.2 VEHICLE DYNAMICS

The basic mechanics underlying a vehicle's motion and/or internal operation can be illustrated using simple, discrete models with a handful of parameters or degrees of freedom. Some models are shown in **Figure T2.1**. The wheeled vehicle model (passenger bus) in **Figure T2.1(a)** has a few degrees of freedom: the x position of the center of mass, the rotation of the vehicle about the center of mass (φ), and the internal

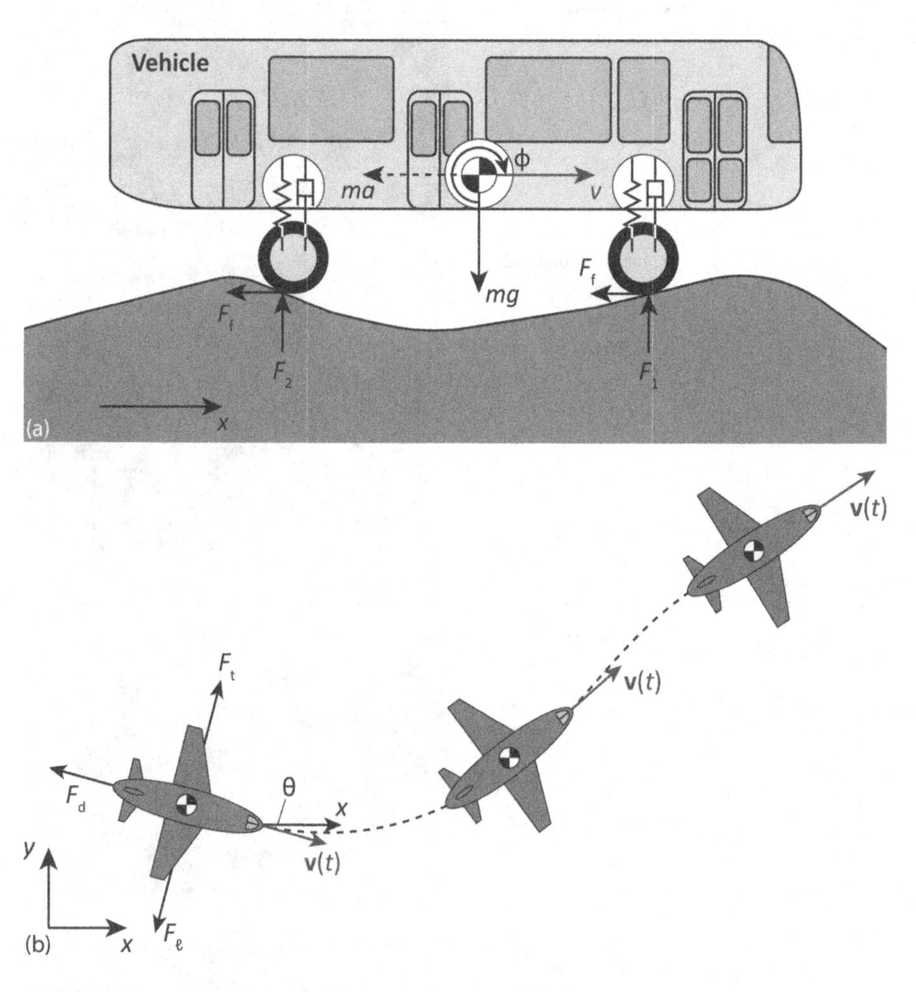

FIGURE T2.1 Physical representation of the dynamics of some vehicles. In (a), the vehicle (a passenger bus) is represented as a collection of parameters like the center-of-mass position with a concentrated weight mg, a (constant, forward) velocity V, angular orientation φ, etc. Changes to the bus's velocity are resisted by the bus's inertia, and the bus interacts with the ground via reaction forces F_1, F_2, and F_f. In the depiction of flight in (b), the aircraft has distinct degrees of freedom, such as the net velocity $V(t)$ and the yaw angle θ. The aerodynamic forces F_d and F_ℓ must be corrected by the craft's controls, as well as setting the trajectory with a turning force F_t. [The turning force is typically related to the lift force F_{lift} (not shown) of the craft: $F_t \propto F_{lift}$.]

states of the suspension springs and dampers. The bus interacts mechanically with the environment through the forces present where it contacts the road: normal forces F_1 and F_2 and lateral forces F_f. **Figure T2.1(b)** shows a simple model of an aircraft in transit. At constant altitude, the aircraft has degrees of freedom x position, y position, and "yaw" angle θ. The plane is driven forward via a thrust force F_t and encounters an opposing drag force F_d during motion. Lateral forces F_ℓ are produced by cross-winds and control surfaces on the aircraft.

Consider first the necessity of applying power to sustain the motion of the vehicle. The total power required P_{tot} depends on the power necessary to accelerate the vehicle (= P_{acc}), to overcome ground or air resistance forces (= P_r), and to resist forces during turning (= P_t):

$$P_{tot} = P_{acc} + P_r + P_t \tag{T2.1}$$

For the bus depicted in **Figure T2.1(a)**, we have

$$P_{acc} = F_{acc} \times v$$

where $F_{acc} = ma + mg \sin \varphi$ is the force required for acceleration a when the vehicle is on an incline (giving $\varphi > 0$ or $\varphi < 0$). (On relatively flat terrain, $\varphi \approx 0$ and $F_{acc} \approx ma$.) The rolling resistance

$$P_r = F_r \times v$$

is related to the friction force at an inclination of the road/vehicle like $F_r = mg\mu_f \cos \varphi$. Here, μ_f is a friction factor related to the interface between the vehicle and the road surface. When driving straight, $P_t = 0$ and

$$P_{tot} = m\left(a + g\sin\varphi + g\mu_f \cos\varphi\right) \times v = ma_{eff} \times v \tag{T2.1'}$$

where a_{eff} is an effective acceleration. For the airplane of **Figure T2.1(b)**, we obtain

$$P_{acc} = F_{acc} \times v = ma \times v$$

as before, but the drag/air-resistance force F_d is more important in this case. In air with density ρ_{air} and an aircraft with drag factor μ_d with velocity Δv relative to the air, we have

$$F_d = \frac{1}{2}\rho_{air}\mu_d\left(\Delta v\right)^2$$

When turning, an additional force $F_t \approx bF_{lift}$ (b constant) is applied, and so

$$P_{tot} = \left[ma + \frac{1}{2}\rho_{air}\mu_d\left(\Delta v\right)^2 + bF_{lift} + F_\ell\right] \times v \tag{T2.1''}$$

In both cases, we notice that the mass of the vehicle is an important contributing factor to the power required to sustain its motion:

$$P_{tot} \propto m$$

This is an unsurprising result, but it serves to underscore the importance of using lightweight materials (i.e., materials with low mass density ρ). Densities of example materials are given in **Table T2.2**.

Additionally, the power required is supplied by the power plant, which expends energy stored in some type of fuel or power cell. The mass of this fuel or the charged cell is, of course, part of the mass m of the vehicle, so it is imperative that the power source provide as much energy as possible for its weight. The capacity of the source to provide a certain quantity of energy is its **specific energy** U. Specific energy has units $[U]$ = [energy/mass] = J/kg. Equivalently, we might find the **energy density** $u = U\rho$ with $[u]$ = [specific energy] × [mass density] = J/m³ or J/L. Gasoline and aviation fuels can have specific energies around U = 47 MJ/kg (or u = 34,000 MJ/m³) and low-density diesel fuels (for larger vehicles and watercraft) can have U = 45 MJ/kg (u = 37,000 MJ/m³). Lithium-ion battery packs provide around U = 0.3 MJ/kg or u = 400 MJ/m³. Gasoline thus has about 100 times the energy density of battery storage. Hydrocarbon fuels appear to have much better performance in terms of energy storage, but this performance is offset by the relative inefficiency of the combustion process compared to electric plants (see **Example T2.2**). The propulsion efficiency (i.e., the fraction of energy provided by the energy-storage system that can be delivered

TABLE T2.2
Densities (or "Specific Masses") of Selected Materials (Values May Vary with Alloying/Treatment)

Material	ρ [g/cm³]	Material	ρ [g/cm³]
C	2.27	Al_2O_3	3.98
Si	2.33	TiO_2 (anatase)	3.84
Fe	7.87	SiC	3.21
Steel	7.84	Fe_3C	4.93
Stainless steel	7.9	SiO_2 (quartz)	2.85
Cu	8.93	SiO_2 (glass)	2.2
Al	2.7	Borosilicate glass	2.02
Ag	10.50	Acrylic	1.19
Mg	1.74	Low-density poly(ethylene)	0.93
Ti	4.51	High-density poly(ethylene)	0.96
Au	19.32	poly(vinyl chloride)	1.3
Zn	7.133	poly(propylene)	0.91
Brass	8.47	poly(tetrafluoroethylene)	2.2
Ni superalloy	8.90	polystyrene	1.05
Pb	11.34	Butyl rubber	1.20
W	19.25	Latex rubber	0.92

as propulsive power) of combustion-based vehicles is 20–30%; this can be as high as 70% for electrical drives. Furthermore, battery-based plants produce no atmospheric pollutants at the point of energy consumption.

Example T2.1:

Suppose you have a composite of 30 wt% SiO_2 glass fibers in an epoxy matrix. What are the volume fractions of each component? What is the overall density of the composite? The mass density of epoxy is 1.15 g/cm³.

SOLUTION

From **Chapter 4**, the rule of mixtures for the density of the blended material is

$$\rho = \rho_{epoxy}\phi_{epoxy} + \rho_{glass}\phi_{glass}$$

where the ρ_i are the densities of the two constituents and the ϕ_i are their volume fractions. You can find the volume fraction ϕ_1 of a component from its weight fraction W_1 using

$$\phi_1 = \frac{W_1/\rho_1}{W_1/\rho_1 + W_2/\rho_2}$$

Thus, we can find

$$\phi_{epoxy} = \frac{W_{epoxy}/\rho_{epoxy}}{W_{epoxy}/\rho_{epoxy} + W_{glass}/\rho_{glass}} = \frac{0.70/\left(1.15\ \text{g/cm}^3\right)}{0.70/\left(1.15\ \text{g/cm}^3\right) + 0.30/\left(2.2\ \text{g/cm}^3\right)} = \underline{\mathbf{0.82}}$$

and

$$\phi_{glass} = 1 - \phi_{epoxy} = 1 - 0.82 = \underline{\mathbf{0.18}}$$

We can now compute

$$\rho = \rho_{epoxy}\phi_{epoxy} + \rho_{glass}\phi_{glass} = 1.15(0.82) + 2.2(0.18) = \underline{\mathbf{1.3}\frac{\mathbf{g}}{\mathbf{cm}^3}}$$

T2.3 ENGINES, EFFICIENCY, AND THERMAL PROPERTIES

The power plants that vehicles depend on for propulsion typically consist of

1. A *chemical fuel* or *electrochemical cell* that stores the energy required to drive motion in the plant.
2. A *drive mechanism* for converting the stored energy of the fuel or cell into mechanical energy/work.
3. *Linkages* that mechanically connect the power output of the plant with the ancillary components necessary for propulsion (e.g., the remainder of an automotive drivetrain).

The details vary widely between designs, but overall, the plant must provide sufficient mechanical power P to sustain motion of the (sometimes quite large) vehicle.

One of the most common sources of power is a **heat engine**. A heat engine converts heat energy into work via some intermediary substance, typically a fluid. The schematic operating principles of a heat engine are shown in **Figure T2.2**. A cyclical

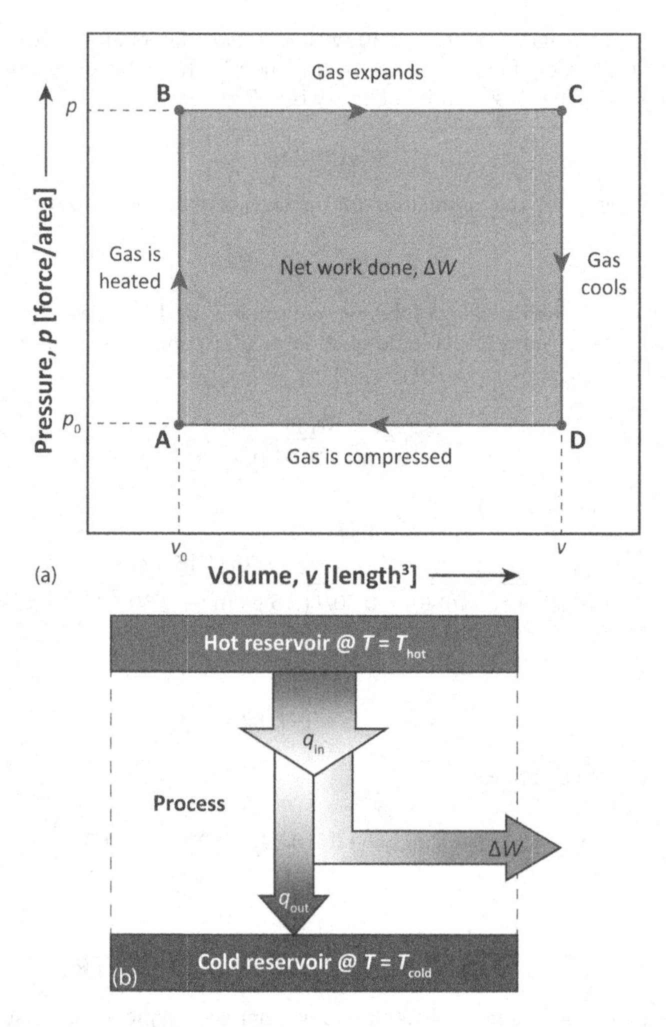

FIGURE T2.2 Schematic representations of the heat-engine process. (a) illustrates the *p-v* cycle for a heat engine. Going clockwise from **A**: the gas in the reservoir is heated, increasing the pressure (**AB**). Next, the expansion of the gas drives the mechanism producing mechanical work (**BC**; this work is done *by* the gas). Heat is released, decreasing the pressure (**CD**). Finally, the mechanism performs work *on* the gas, returning the gas to its beginning state. The network produced by the process cycle is the work done by the gas, less the work done on the gas. In (b), the "energy balance" of a process is depicted. Of all the heat that is supplied (= q_{in}) by the source reservoir, some is exhausted as "waste" into the cold reservoir, and some is transformed into the mechanical work output.

heat-engine process is depicted in **Figure T2.2(a)**. During this process, heat is added to the intermediary substance (here, n moles of a gas), and this addition produces an increase in the temperature $\Delta T = T - T_0$. According to the ideal-gas law, this increase in temperature is connected to an increase in pressure $\Delta p = p - p_0$ (at a constant volume V_0), according to

$$\Delta p V_0 = nR\Delta T \tag{T2.2}$$

where $R \approx 8.314$ J/mol K is the ideal-gas constant. This elevated pressure drives expansion of the gas against the mechanism (e.g., a piston) to a new volume v; during this stage, the gas does some amount of "pressure-volume" work $+W$ *on* the mechanism. The subsequent cooling of the gas (at constant volume v) lowers the temperature from T back to T_0, again following the ideal gas law:

$$\Delta p V = nR\Delta T \tag{T2.3}$$

Finally, the cycle is "reset" when the mechanism compresses the gas back to volume v_0. This final stage requires work W_0 to be done on the gas *by* the mechanism. All told, the net mechanical work ΔW provided is

$$\Delta W = W - W_0 \tag{T2.4}$$

(More correctly, the total work extracted is the area inside the cycle region.) If the **A** → **B** → **C** → **D** → **A** cycle takes an interval of time Δt to complete, then the overall process provides power $P = \Delta W/\Delta t$.

The primary input to the cyclic heat-engine process is heat energy. An important question is: how much work output do we get for how much heat input? This question is connected to the **efficiency** of the heat-engine process. According to the first law of thermodynamics, the energy input of the process U_{in} must be equal to the energy output U_{out}:

$$\Delta U = U_{in} - U_{out} = 0 \tag{T2.5}$$

In the context of our heat engine, this means that

$$(q_{in} + W_0) - (q_{out} + W) = 0$$

or

$$q_{in} - q_{out} = W - W_0 = \Delta W \tag{T2.6}$$

From these quantities, we define the efficiency η of the process cycle as

$$\eta = \frac{\Delta W}{q_{in}} = \frac{q_{in} - q_{out}}{q_{in}} \tag{T2.7}$$

In the ideal case (or "Carnot limit"), this is equivalent to

$$\eta = \frac{T_{\text{hot}} - T_{\text{cold}}}{T_{\text{hot}}} = 1 - \frac{T_{\text{cold}}}{T_{\text{hot}}} \tag{T2.7'}$$

where the temperatures T_{hot} and T_{cold} are represented in Kelvin. Since the temperature of the cold reservoir is typically that of the environment it is operating in, T_{cold} is a fixed value in most cases. Also, by the second law of thermodynamics, no heat-engine process can be loss-free, and so

$$\eta < 1$$

always.

Example T2.2:

Suppose the effective temperature of burning fuel can reach 375°C inside the cylinder of a small generator's engine. What is the ideal efficiency of a heat engine that uses this fuel? Suppose that the fuel has a specific energy content u = 35 MJ/L and that all of this energy can be realized as input heat q_{in} in the engine. If the fuel feed rate during operation is V' = 0.2 mL/cycle and a cycle lasts 0.25 s, what is the maximum power P produced by the engine?

SOLUTION

Since T_{hot} = 375°C = 648 K and $T_{\text{cold}} \approx T_r$ = 300. K, the upper, ideal limit to the efficiency η of the engine is

$$\eta = 1 - \frac{T_{\text{cold}}}{T_{\text{hot}}} = 1 - \frac{300}{648} = 0.537 = \mathbf{53.7\%}$$

The fuel energy input per cycle is $q_{\text{in}} = u \times V' = (35 \times 10^6) \times 0.2 = 7000$ J, so we can compute the mechanical output ΔW using **Equation T2.4** as

$$\Delta W = \eta q_{\text{in}} = 0.537(7000) = 3800 \text{ J}$$

This gives a power production P of

$$P = \frac{\Delta W}{\Delta t} = \frac{3800}{0.25} = 15000 \frac{\text{J}}{\text{s}} = 15 \text{ kW} = \underline{\mathbf{20\ hp}}$$

An important implication of **Equation T2.7′** is that the higher the operating temperature of the heat engine, the more efficient it will be. There are a number of materials properties that are associated with higher operating temperatures. First among these is the melting temperature T_m of the material. The melting temperature is the temperature at which the material/s solid phases and liquid phases have equal stability and is where a phase transformation can take place (**see Chapter 4**). Recall from **Chapter 1** how the mutual interaction between two particles via intermolecular forces produces a bonding potential that holds the molecules together. A schematic

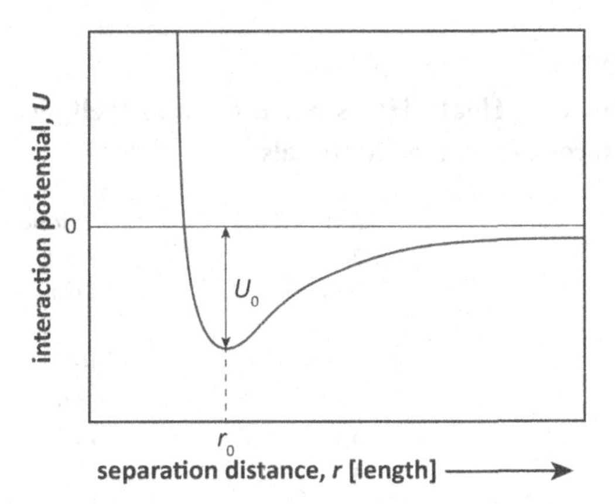

FIGURE T2.3 The bonding potential between molecules in a solid. The deeper the potential "well" U_0 the higher the melting temperature T_m. The bonds can be broken by the input of thermal energy; this is the heat of fusion ΔH_f.

bonding potential is shown in **Figure T2.3**. The depth of the potential U_0 indicates the work that must be done to separate the particles, so the deeper the potential, the more strongly bound the system is. Roughly, this is the energy that needs to be added to pull the molecules apart, and high U_0 is correlated with high T_m.

In order to melt a substance

1. the temperature must be raised to the melting temperature and
2. heat energy must be supplied to enact the phase transformation.

Raising the temperature of a substance requires an input of heat energy, but different amounts of heat correspond to different temperature increases in different materials. The materials property that governs this energy/temperature equivalence is the **specific heat** c_p:

$$c_p = \frac{dq}{dT} \tag{T2.8}$$

per mole of substance This quantity has units $[c_p]$ = [energy/(mass × temperature)] = J/kg K or $[C_p]$ = [energy/(number × temperature)] = J/mol K. (The subscript "p" indicates that the pressure of the environment (e.g., 1 atm) is fixed during heat uptake and the material is free to expand as it heats up.) The specific heat capacity, which typically depends on temperature $c_p = c_p(T)$, defines how much a particular amount of heat energy raises the temperature of the substance. The transformation from solid to liquid at the melting temperature also consumes energy equal to the **heat of fusion** ΔH_f. $[\Delta H_f]$ = [energy/number] = J/mol or $[\Delta H_f]$ = [energy/mass] = J/kg. This energy is required to "fill in" the potential well during a melting phase transformation and do pressure-volume work during melting. Some specific heats and heats of fusion for various materials are given in **Table T2.3**, alongside their melting temperatures.

TABLE T2.3
Melting Specific Heats, Heats of Fusion, and Melting Temperatures of Selected Materials

Material	C_p [J/mol °C][a]	ΔH_f [kJ/mol]	T_m [°C]
Hg	27.7	2.331	–38.9
H_2O	4.184[b]	6.008	0[b]
Sn	26.4	7.196	231.7
Al	24.4	10.67	660.4
TiO	55.3	58.6	991
Au	25.8	12.7	1064.4
Cu	24.5	13.0	1083.4
Fe_3C	104	51.59	1226.8
$CaCO_3$	83.5	53.1	1282
Si	20.2	39.6	1410
Fe	25.0 (α) 32.8 (γ, 727°C)	14.9	1535
Mn_3O_4	148 (α)	163[c]	1590
SiO_2 (cristobalite)	62.7	8.79	1723
TiO_2	55.3	48[c]	1825
Al_2O_3	81.0	110[c]	2045.0
MgO	37.2	77.4	2642
SiC	30.1	71.55	2830
W	25.0	2[c]	3410

[a] At room temperature T_R = 300 K and 1 atm pressure unless otherwise noted; [b]exact; [c]very approximate

Example T2.3:

Estimate the total heat energy input Q required to melt a cubic block of Au 3.0 cm on a side that is initially at room temperature.

SOLUTION

A block of Au with volume $v = 3.0 \times 3.0 \times 3.0 = 27$ cm³. Consulting **Table T2.2**, we find this material has a mass density $\rho = 19.32$ g/cm³, so the cubes mass m is

$$m = \rho v = 19.32(27) = 58 \text{ g}$$

From **Figure 1.7**, the molar mass of Au is $A = 196.91$ g/mol, and we find

$$n = \frac{m}{A} = \frac{58}{196.97} = 0.29 \text{ mol}$$

as the number of moles of Au. Since the melting temperature of Au is T_m = 1064.4°C and the specific heat is 25.8 J/mol/°C (**Table T2.3**), we can now determine the heat energy q required to raise the temperature of the block to T_m is

$$q = nC_p\Delta T = 0.29(25.8)(1064.4 - 27) = 7.9 \text{ kJ}$$

Added to the temperature required to raise the temperature of the material to its melting temperature, we add the energy required to melt the block at $T = T_m$: $\Delta H_f = 12.7$ kJ/mol. This gives a total energy

$$Q = q + n\Delta H_f = 7.9 + 0.29(12.7) = \underline{\mathbf{12\ kJ}}$$

The final important thermal property we wish to discuss is the **thermal conductivity** κ. The thermal conductivity determines how rapidly heat energy (= q) can be transported through a substance. The heat flux $\dot{\mathbf{q}}$ is a vector that describes the direction and overall rate of transport. It has magnitude

$$|\dot{q}| = \dot{q} = \frac{q}{At} \tag{T2.9}$$

where A is the area across which the heat flows and t is the interval of time. and is similar to the mass flux of **Chapter 4**. It has magnitude and has units $[\dot{q}]$ = [energy/length²/time] = J/m² s. The orientation of the heat flux is shown in **Figure T2.4**. Heat flows from regions of higher temperature to regions of lower temperature. The (1D) heat flux is proportional to the temperature gradient $\Delta T/\Delta x$ according to

$$\dot{q} = -\kappa\frac{\Delta T}{\Delta x} \tag{T2.10}$$

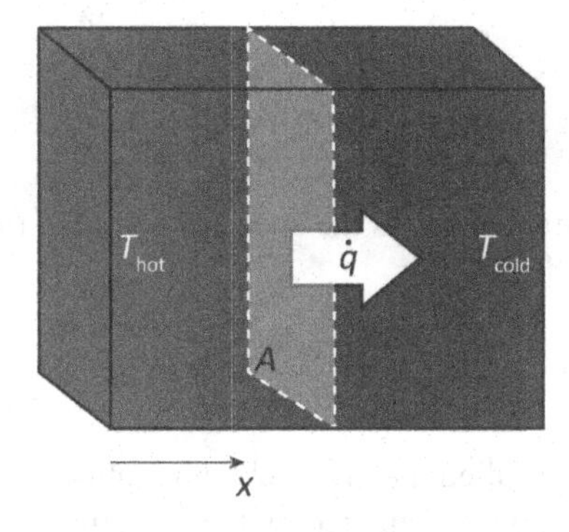

FIGURE T2.4 Heat flux in a material. Heat energy flows from the hotter region (with temperature T_h) to the cooler region (with temperature T_c). The overall heat flux $\dot{q}$ describes the energy exchanged across an area A in an interval of time t according to **Equation T2.9**. The heat flux is related to the materials property called the thermal conductivity κ.

This is **Fourier's law**. The thermal conductivity plays a similar role to the diffusivity in Fick's first law; it establishes the intrinsic mobility of heat in the material. It has units [κ] = [energy/length/temperature] = W/m K.

The origin of a material's thermal conductivity lies in the mechanisms by which kinetic energy can be transferred from one region of the material to another. The reapportionment of kinetic energy corresponds to changes in the temperature distribution within a material. The various mechanisms are connected to the types of *carriers* in the material that transport kinetic energy from one location to another. For a given carrier type i, the contribution of the carrier κ_i to the thermal conductivity is

$$\kappa_i = \frac{1}{3}\left(\frac{n_i c_i v_i L_i}{N_A}\right) \tag{T2.11}$$

In **Equation T2.11**, n_i is the carrier concentration, c_i is the carrier's specific heat, v_i is the average carrier velocity, and L_i is the carrier's "mean free path". (The mean free path is the typical distance that a carrier travels before being scattered by an obstacle in the structure.) The total thermal conductivity is then

$$\kappa = \sum_i \kappa_i = \kappa_{ph} + \kappa_e \tag{T2.12}$$

The two primary types of heat carriers in most materials are electrons and **phonons**.

Phonons are vibrational "modes" that arise in bonded atoms in a material. Phonon principles are illustrated in **Figure T2.5**. Examples of (1D) modes are shown in **Figure T2.5(a)**. The various N modes ($1 \le N \le N_{max}$) are nonexclusive, so there may be many overlapping modes at any given time, and they are distinguished by their wavelength $\lambda_N = 2\ell/N$. The wavelength can be small (on the order of λ_{min} ~ two atomic spacings for $N = N_{max} \sim N_A^{1/3}$) or relatively large (similar to the object size $\lambda_1 \sim 2\ell$ for $N = 1$). The modes are associated with a certain amount of quantized energy u_n given by

$$u_N = h\frac{v_s}{\lambda_N} \tag{T2.13}$$

where h is Plank's constant and v_s is the speed of sound in the material. The energy of the highest-order mode is

$$u_{N\max} = \frac{h v_s N_{\max}}{2\ell} = k_B T_D$$

where $T_D = h v_s N_A / 2\ell k_B$ is called the "Debye[1] temperature". The total phonon energy is then the sum over all modes, weighted by their likelihood of excitation P:

$$U = \sum_{i=0}^{N_{\max}} u_i P_i$$

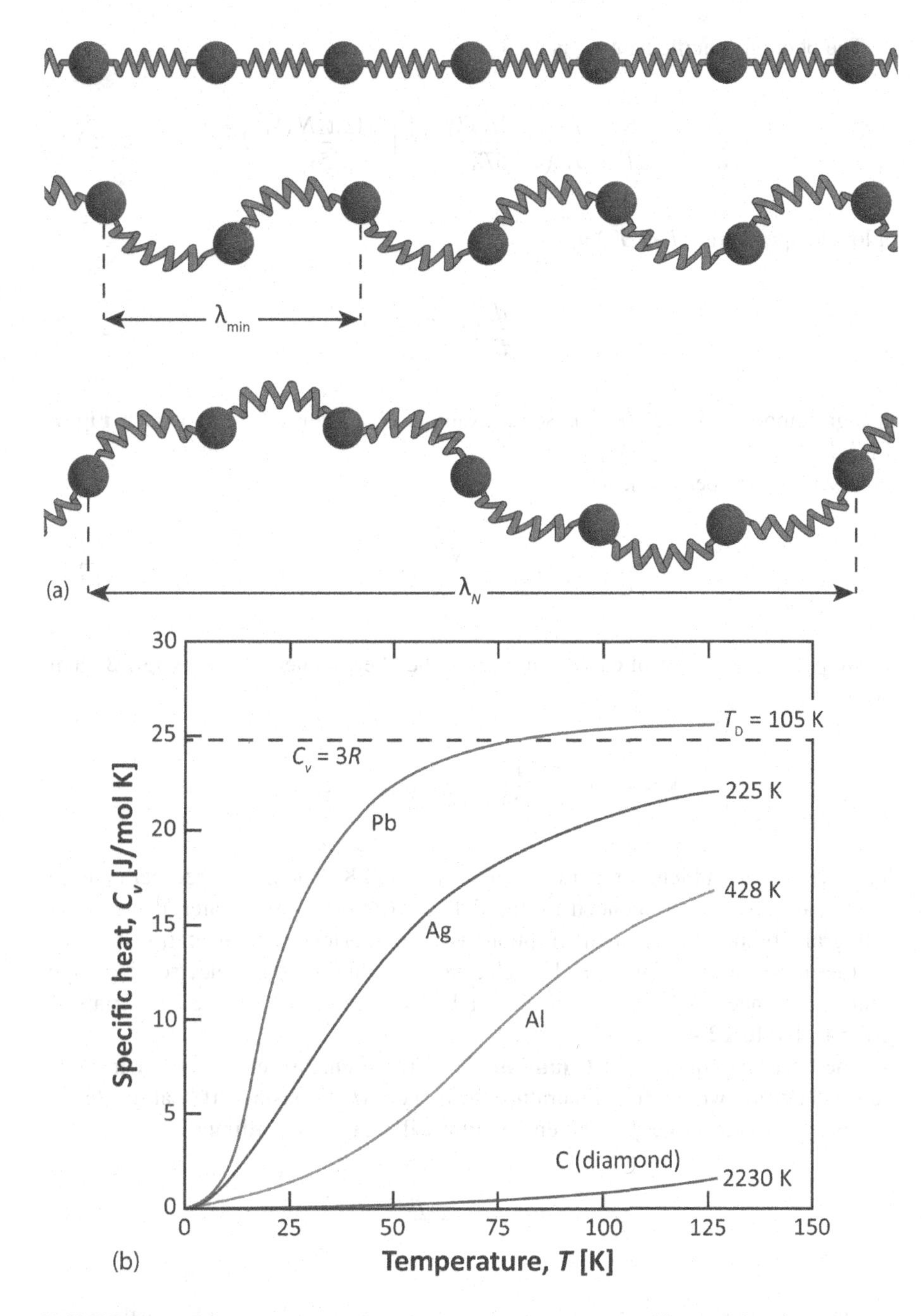

FIGURE T2.5 Phonons in materials and their influence on specific heat C_v. (a) shows how atoms in the lattice can vibrate in different modes with different wavelengths λ_N. These vibrations are similar to those of propagating sound waves. The various modes N are associated with discrete energies u_N, and these energies together determine the specific heat of the material, shown in (b). The values of C_v scale with the Debye temperature T_D and have a high-T limit of $\approx 3R$.

The phonon specific heat c_{ph} is then

$$c_{ph} = \frac{dU}{dT} = \frac{d}{dT}\left(\frac{3\pi^4 N_A k_B}{5T_D^3} T^4\right) = \frac{12\pi^4 N_A k_B}{5T_D^3} T^3 \tag{T2.14}$$

at low temperatures ($T \ll T_D$) and is

$$c_{ph} = \frac{d}{dT}(3RT) = 3R \tag{T2.14'}$$

at high temperatures ($T \gg T_D$). Some example $C_v(T)$ curves are shown in **Figure T2.5(b)**.

The electron-specific heat is

$$c_e \sim \frac{\pi^2 N_e k_B^2}{2E_F} T \tag{T2.15}$$

where N_e is the number of carriers and E_F is the "Fermi energy" (see **Topic 3**). This gives

$$\kappa = \kappa_{ph} + \kappa_e = \frac{1}{3}\left(n_{ph} c_{ph} v_{ph} L_{ph} + n_e c_e v_e L_e\right) \tag{T2.12'}$$

for materials at ambient temperatures $T = T_R = 300$ K. The mean-free-path parameters L are strongly influenced by the defect structure of the material: vacancies and grain boundaries and other phonons can interfere with heat transmission. Furthermore, in materials with low electrical conductivity, the electronic component $\kappa_e \approx 0$ since $n_e \approx 0$. Some thermal conductivity values for selected materials are given in **Table T2.4**.

The heat-flux equation of **Equation T2.10** represents a steady-state or instantaneous condition where the temperature difference ΔT is constant. If heat q enters or leaves a system at a rate $\dot{q}(t)$, its temperature will change according to

$$T(t) = T_0 + \frac{\dot{q}(t)}{c_p} \tag{T2.15}$$

per unit mass of substance with c_p the specific heat in J/kg °C or J/kg K. Furthermore, $[\rho c_p]$ = J/m^3°C or J/m^3 K. We may now find

$$\frac{T - T_0}{\Delta t} = \frac{1}{\rho c_p}\left(\frac{\Delta \dot{q}}{\Delta t}\right)$$

TABLE T2.4
Debye Temperatures and Thermal Conductivities of Selected Materials

Material	T_D [K][a]	κ [W/m °C][a]
Sn	254	66.6
Al	390.	237
Au	178	315
Cu	310	398
Fe_3C	604	8
Si	692	84
α-Fe	373	803
SiO_2	470	1.0
TiO_2	557	6.7
Al_2O_3	873	25
MgO	750	41
SiC	1150	260.
Ti	380.	91.6
W	312	178

[a] At room temperature $T_R = 300$ K and 1 atm pressure unless otherwise noted

The heat flux change per time $\Delta q/\Delta t$ must be the same as the change in heat flux over a layer of material of thickness Δx, and so

$$\frac{\Delta \dot{q}}{\Delta t} = -\frac{\Delta \dot{q}}{\Delta x}$$

This means

$$\frac{T - T_0}{\Delta t} = \frac{1}{\rho c_p}\left(\frac{\Delta \dot{q}}{\Delta t}\right) = -\frac{1}{\rho c_p}\left(\frac{\Delta \dot{q}}{\Delta x}\right) = \frac{1}{\rho c_p}\Delta\left[\left(-\kappa\frac{\Delta T}{\Delta x}\right)\Big/\Delta x\right]$$

or

$$\frac{\partial T}{\partial t} = \alpha\frac{\partial^2 T}{\partial t^2} \qquad \text{(T2.15)}$$

Equation T2.15 is called the "heat equation", and its solution gives the temperature as a function of time: $T(t)$. The factor $\alpha = \kappa/\rho c_p$ is called the **thermal diffusivity** with units $[\alpha]$ = [length²/time] = m²/s. The heat equation is similar to Fick's second law for mass diffusion and may be solved using similar techniques. Also, the distance d over which the temperature changes in a material exposed to an environment with a different temperature is (in one dimension)

$$d = \sqrt{2\alpha t} \qquad \text{(T2.16)}$$

CASE STUDY – MATERIAL SELECTION FOR AN INSULATED SHIPPING CONTAINER[2]

Protection from an unwanted thermal environment is an important concern in the shipping and storage of temperature-sensitive products such as pharmaceuticals, foodstuffs and other horticultural products, electronics, etc. Insulated packaging helps maintain temperatures appropriate for the cargo and thus inhibit deterioration until it reaches the end user. In addition to thermal-transport resistance, a good insulating material should have other characteristics, depending on the application. For instance, low moisture susceptibility, low cost, and (as we have discussed in the context of shipping) low mass. One-way or disposable materials have advantages (easy design and validation), as do reusable shipping containers (durability, better temperature control).

The choice of distribution-system material depends on the cargo, time in transit, and temperature sensitivity of the cargo. The longer that the container can maintain a low internal temperature, the better the resilience of the cargo to delays in shipping and the farther it can be shipped. The amount of material required for the container also influences its size and weight (both important considerations in logistics). Consider the following selection requirements:

- **Function.** The shipping container must provide short-term insulation.
- **Constraints.** For purposes of reducing container volume and weight, the maximum wall thickness w of the package must not exceed some limit.
- **Objectives.** A good design will maximize the time t that the interior of the container maintains a temperature below that of the environment (e.g., T_r), plus some margin T_0. That is, $T < T_r - T_0$ for as long as possible.

A schematic of a container is shown below. The circular cylindrical or spherical package has a wall thickness w, an internal temperature T_{cold}, and an external temperature T_{hot}. If the container material has a thermal conductivity of κ, then the heat flux $\dot{q}$ is

$$\dot{q} = \kappa \frac{\Delta T}{\Delta x} = \kappa \frac{T_{hot} - T_{cold}}{w}$$

according to **Equation T2.10**. It appears that, at first glance, minimizing κ will produce the best result for a given wall thickness w, but this relation does not give us much indication of how the internal temperature changes over time.

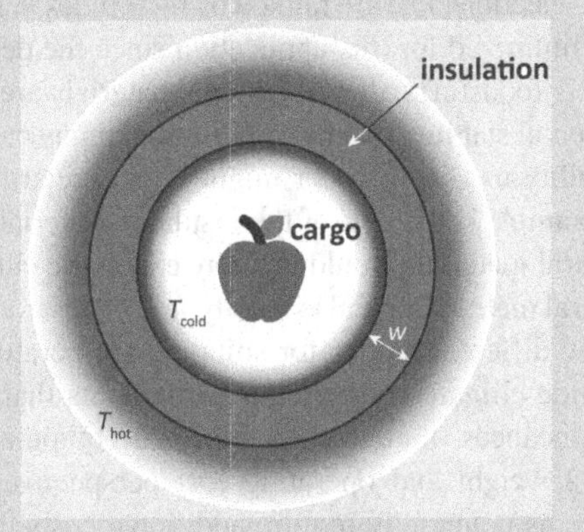

Equation T2.16 indicates that, for 1D heat transport across the container wall

$$d = \sqrt{2\alpha t}$$

This means that the interior of the container will maintain its temperature for a time t determined by how long it takes for the temperature-change distance d to span the wall thickness:

$$t = \frac{w^2}{2\alpha}$$

This gives a material index h for the container design as

$$h = \frac{1}{\alpha}$$

meaning that if α is minimized, the temperature-retention time will be *maximized* for a given w (assumed by constraint).

What materials might be best suited to the task? The following table gives the values of h for some materials of interest.

Material	$h \times 10^{-6}$ [s/m^2]	Notes
poly(vinyl chloride)	12.5	Food safe, low permeability to water
Cork (wood)	12.3	Natural material, more expensive
Pine (wood)	12.1	Natural material, absorbs water
Natural rubber	11.8	Natural material, low permeability
polystyrene	11.1	Food safe, moderate permeability to water
poly(propylene)	10.4	Food safe, low permeability to water
Glass-fiber–reinforced epoxy	~8	Low permeability to water
Glass	2.94	Frangible, high cost, impermeable to water

The data in the selection table are ranked by their index h. Note that the weight is automatically minimized for this application since the density ρ is part of h. Also, if the cargo is foodstuffs, then food-safe materials are typically required. Finally, environmental stability can be important in many cases, and the materials ability or inability to absorb water might be a crucial factor. Unaddressed in this selection example is the cost of the insulator material; a safe assumption would be that natural materials would be more expensive currency-wise in most cases, but the natural materials are also mostly free of environmental drawbacks.

There are a few different modes for shipping temperature-sensitive goods. Carriers can provide climate-controlled environments that permit transportation over longer distances, though the additional equipment required can be prohibitive from a weight and operating-cost perspective. Reusable insulating container systems potentially require additional costs in terms of cleaning and inventory maintenance. Nonreusable systems involve a significant amount of waste-disposal costs but can also incorporate additional materials effects besides thermal diffusivity, such as cooling via phase changes in the packaging.

T2.4 CREEP IN METALS AND POLYMERS

Another type of plastic deformation that becomes vitally important in materials subject to high temperatures that are loaded over a long period of time is plastic **creep**. Applications that subject materials to these conditions include turbine engines and steam-based reactors/generators. (Creep processes are also important in geology, where rocks deep in the earth are subject to high temperatures and pressures.) The mechanisms that underlie creep are partially stress-driven and partially heat-driven, so it is possible to have plastic creep deformation occur at stresses *below* the yield strength. But when we talk about temperature, how high is "high"? The **homologous temperature** is

$$T_h(T) = \frac{T}{T_m} \qquad \text{(T2.17)}$$

where T and T_m are in kelvin. Typically, creep deformation becomes possible when the material's homologous temperature $T_h > 1/2$.

The creep process is illustrated in **Figure T2.6**. During creep, plastic deformation accrues in the material as time passes; this is illustrated in **Figure T2.6(a)**. Suppose

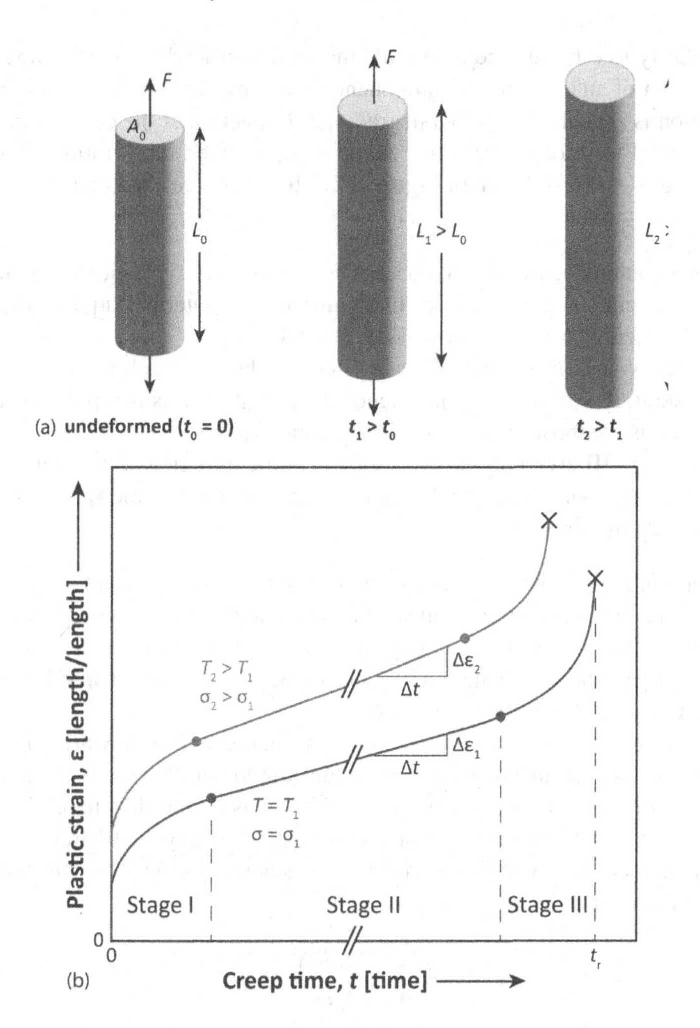

FIGURE T2.6 Creep behavior of metals and ceramics. (a) under the influence of an applied stress σ, a specimen of material will extend, growing progressively longer as time *t* passes. This extension can be represented as a time-dependent strain $\varepsilon(t) = \Delta L(t)/L_0$. Typical deformation behavior of metallic and ceramic materials ("creep curves") are in (b). Creep occurs in three stages: an initial transient stage (I), a "steady-state", constant-rate stage (II), and a final, tertiary stage (III). If either the stress or the temperature is increased, the curve is shifted to higher steady-state creep rates and shorter rupture times t_r.

you have a specimen that is subjected to a constant load F at some time $t = 0$. This establishes an engineering stress $\sigma = F/A_0$ that is also constant in time. What will the deformation look like over time? That is, what will the strain $\varepsilon = \varepsilon(t)$ be, and how will it depend on the applied stress σ, system temperature T, and materials characteristics? We expect that the strain rate $\dot{\varepsilon} = d\varepsilon/dt$ will have a dependence on the stress and the temperature and the materials structure:

$$\dot{\varepsilon} = \dot{\varepsilon}(\sigma, T, \text{structure}) \qquad \text{(T2.18)}$$

The ability to estimate the creep in a material is crucial in determining the ultimate lifespan of an engineered component since typically only a certain amount of deformation is acceptable without degrading the function of the component.

The typical form of the resulting creep strain at fixed temperature T_1 and fixed engineering stress σ_1 is shown in **Figure T2.6(b)**. There are a number of components to these creep curves:

1. An instantaneous strain that occurs at $t = 0$ related to elastic deformation.
2. A transient stage where work hardening in the material outpaces the thermal softening effect; this is called "stage-I creep".
3. A steady-state stage, called stage-II creep, during which the creep rate is constant: $\varepsilon^{\cdot} = c$; the material spends the majority of its lifespan in stage II, so this is the most important stage for analysis.
4. The stage-III creep region corresponds to the initiation of necking or fracture in the specimen; it ends when the specimen fails, and this establishes the "rupture time" t_r.

The elastic part of the strain is easy to understand and calculate (since the stress σ is fixed), and since the initial transient deformation of stage I is a small fraction of the materials lifespan, it is not as important as the behavior in stage II. Since stage III is typically connected to the fracture mechanics described in **Topic 1**, we turn our attention to the material's behavior in stage II.

There are many different kinds of creep mechanisms, but they are all dependent on the rate of mass transport/diffusion within the material. This is the reason why creep does not occur at low temperatures; diffusion is far too slow to produce significant deformation. The reorganization of atoms required is typically tracked using the diffusion of vacancies with diffusivity D_v. A "general" formula for the steady-state/constant creep rate is taken to be

$$\dot{\varepsilon} = A\left(\frac{\sigma}{G}\right)^m \left(\frac{b}{D}\right)^n D_v(T) \tag{T2.19}$$

where A is a constant, G is the shear modulus, b is the Burgers vector, and D is the grain size. The parameters G and b are introduced to "normalize" the stress and grain size, respectively. The exponents m and n describe how strongly the normalized stress and normalized grain size influence the creep rate, and their values depend on the creep mechanism involved. There are two primary creep mechanisms:

Diffusion creep. Diffusion creep involves the slow rearrangement of individual crystals/grains by the transport of atoms along the axis of the applied stress σ. Diffusion creep is most favorable when $\sigma/G \lesssim 10^{-4}$ and has $m = 1$ and $n = 2$.

Dislocation creep. During dislocation creep, dislocations absorb vacancies and climb into planes of easier glide. Plasticity can thereby occur at lower stresses, though it progresses very slowly. Dislocation creep is most favorable when $10^{-4} \lesssim \sigma/G \lesssim 10^{-2}$ and has $m = 5$ and $n \approx 0$. (This means that dislocation creep is more sensitive to the stress than diffusion creep is and is not affected by grain size.)

Example T2.4:

Suppose that the following strain rates $d\varepsilon/dt$ were recorded in identical specimens heated to the given temperatures and loaded at a constant stress level σ_0. What would you estimate is the activation energy Q for this creep mechanism?

Specimen	$d\varepsilon/dt$ [1/s]	T [°C]
1	2×10^{-6}	400
2	5×10^{-5}	500
3	6×10^{-3}	600
4	2×10^{-2}	700

SOLUTION

In MATLAB®

```
>> rates = [2*10^-6 5*10^-5 6*10^-3 2*10^-2];
>> temps = [400 500 600 700];
>>
```

The general creep formula (from **Equation T2.19**) is

$$\dot{\varepsilon} = A\left(\frac{\sigma}{G}\right)^m \left(\frac{b}{D}\right)^n D_v(T)$$

Since D_v is a diffusivity-like parameter $D_v(T) = D_{v0}\exp[-(Q/RT)]$ and the stress and structural parameters are fixed, the formula becomes

$$\dot{\varepsilon} = A\left(\frac{\sigma}{G}\right)^m \left(\frac{b}{D}\right)^n D_{v0}\exp\left(-\frac{Q}{RT}\right) = B\exp\left(-\frac{Q}{RT}\right)$$

where B is a constant. We can transform this relationship by taking the logarithm of both sides:

$$\ln\dot{\varepsilon} = \ln\left[B\exp\left(-\frac{Q}{RT}\right)\right] = \ln B - \frac{Q}{RT}$$

We can now transform the data, making a data series $y = \ln\dot{\varepsilon}$ vs. $x = 1/T$:

```
>> x = (temps + 273.15).^-1;
>> y = log(rates);
>>
```

(Note that we are using kelvin for the temperature, as is usual when working with this kind of data.) Since $y = \ln\dot{\varepsilon}$ and $x = 1/T$, we have the relation

$$y = mx + b$$

with $m = -Q/R$ and $b = \ln B$. These values look like

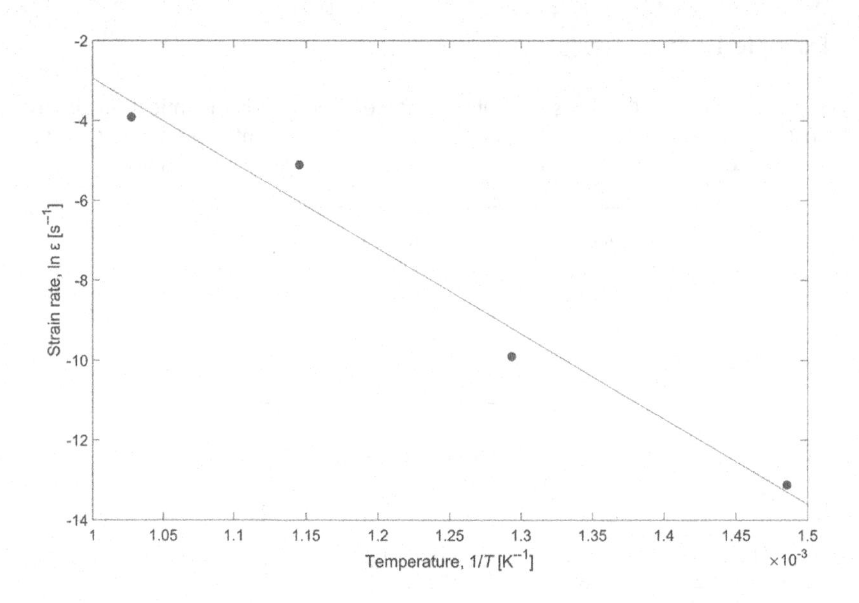

The superimposed trendline is a linear regression according to

```
>> p = polyfit(x, y, 1)
ans =
    1.0e+04 *
    -2.1336    0.0018
>>
```

From these coefficients, we obtain $m \approx -21000$, and so we can find (for $R = 8.314$ J/mol K)

```
>> Q = -8.314*p(1)
ans =
    -1.7739e+05
>>
```

or Q = 180,000 J/mol = **180 kJ/mol**.

In addition, the creep process makes a good description of the deformation behavior of polymers. Crystalline polymers tend to have low melting temperatures (200–300°C), and amorphous polymers have an even lower **transition temperature** T_g that separates the "liquid-like" amorphous state from the "solid-like" amorphous state. **Figure T2.7** shows typical transition behavior of an amorphous material. The transition temperature T_g indicates an abrupt shift in the **viscosity** $\eta = \eta(T)$ of the material. The viscosity property of a fluid relates its flow rate to the stress applied to it and has units $[\eta]$ = [mass/length/time] = [stress time] = Pa s.[3] Viscosity acts like an "internal friction" in a material: the higher the viscosity, the more force is required for flow. When the temperature is high, viscosity is uniformly low. As the temperature decreases below T_g, the viscosity rises rapidly. This rapid change reflects

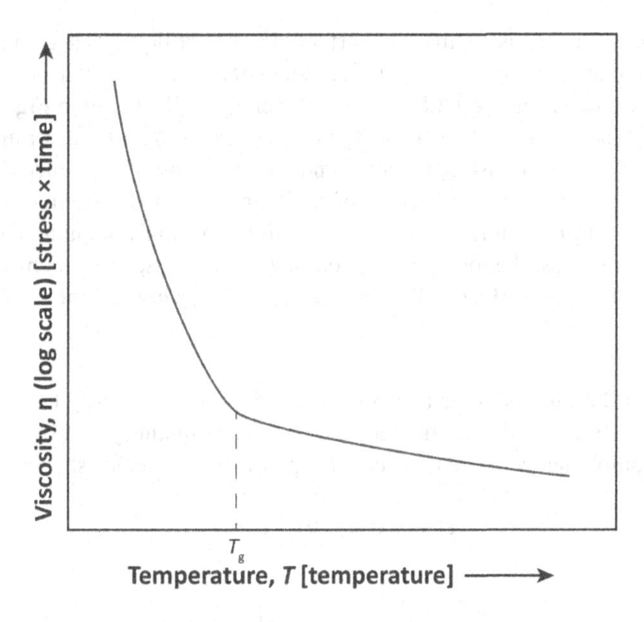

FIGURE T2.7 The transition behavior of amorphous polymers. The viscosity curve is very flat above T_g but rises rapidly when the temperature drops below T_g. The former low-viscosity condition is very liquid-like, and the latter high-viscosity presents as a solid.

the transition to a condition where internal friction is high and the material is akin to a solid in behavior. Some transition temperatures for amorphous polymer materials and glasses are given in **Table T2.5.**

For amorphous polymers, a redefinition of T_h would be

$$T_h(T) = \frac{T}{T_g} \tag{T2.17'}$$

Because most polymers see service at temperatures near room temperature, typically $T_h > 0.5$. What this means practically is that the deformation of amorphous

TABLE T2.5
Transition Temperatures of Selected Polymers

Material	T_g [°C]
Low-density poly(ethylene)	–110
High-density poly(ethylene)	–90
poly(vinyl chloride)	80.
poly(propylene)	–10
poly(tetrafluoroethylene)	110
poly(carbonate)	145
Polystyrene	100
poly(methyl methacrylate)	105

polymers has characteristics that are part-solid and part-liquid. This combination of deformation characteristics is called **viscoelasticity**. Some examples of viscoelastic curves are shown in **Figure T2.8**. Viscoelastic creep is illustrated in **Figure T2.8(a)**. Under the influence of a constant (engineering) stress σ_0, the material exhibits a deformation/strain response $\varepsilon(t)$ that depends on the elapsed time since the load was applied. Suppose that the stress is applied at time $t = 0$. The material undergoes instantaneous elastic deformation (according to the Young's modulus E) to a strain of ε_0. As time passes, the total strain ε continues to increase in a creep-like manner. The magnitude of this effect is determined by the **creep compliance** $J_E(t)$:

$$\varepsilon(t) = \sigma_0 J_E(t) \tag{T2.20}$$

Equation T2.20 is similar to Hooke's law; it is a relationship between a stress and a strain.[4] Because the strain is a time-dependent quantity and the stress is not, the creep compliance *must* be time-dependent. Furthermore, for short $t \approx 0$, we have

$$\varepsilon(0) = \sigma_0 J_E(0) = \varepsilon_0 = \frac{\sigma_0}{E}$$

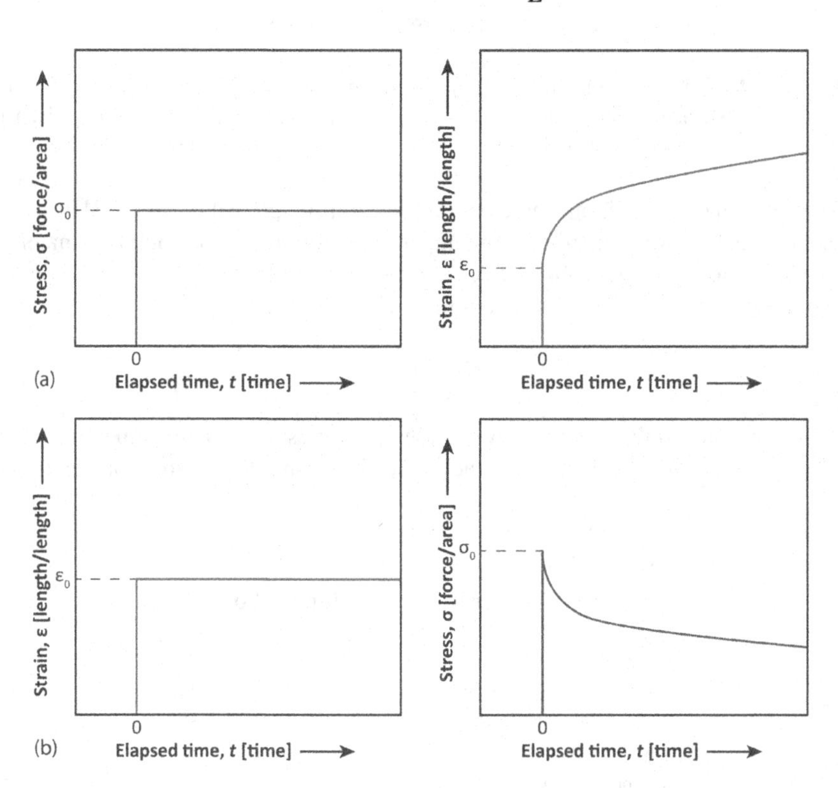

FIGURE T2.8 Creep behavior of polymers. (a) depicts constant-stress creep. A constant stress σ_0 is applied at time $t = 0$. During creep, the plastic strain in the material will accumulate for $t > 0$ according to $\varepsilon(t) = \sigma_0 J_E(t)$. The stress is the *cause*, and the time-dependent strain is the *effect*. In (b), the strain ε_0 is the cause, and the time-dependent stress is the effect. The stress decays according to $\sigma(t) = \varepsilon_0 E(t)$.

so $J_E(0) = 1/E$ and $[J_E(t)]$ = [force/area] = Pa. The creep effect in viscoelastic polymers is a result of the slow, permanent shift (or *flow*) in the positions of the macromolecules that make up the material's structure.

Figure T2.8(a) depicts the **load-relaxation** phenomenon. This behavior has a somewhat dual relationship with viscoelastic creep. In load relaxation, a material is subjected to a constant strain deformation ε_0. This deformation induces flow in the polymer's macromolecules, and this flow operates in a way that reduces (or "relaxes") the stress. The relationship that governs this behavior is

$$\sigma(t) = \varepsilon_0 E(t) \tag{T2.21}$$

Here, $E(t)$ is the time-dependent **relaxation modulus** with $E(0) = E$ (i.e., Young's modulus).[5] Additionally, consider the energy U (per unit volume) stored in a material:

$$U = \frac{1}{2}\sigma\varepsilon \tag{T2.22}$$

Since $\varepsilon = \varepsilon_0$ and $\sigma = \sigma(t)$, we have

$$U = \frac{1}{2}\sigma(t)\varepsilon_0$$

Since $\sigma(t)$ is *decreasing* over time, we observe that the system is losing stored energy. This loss is due to the internal friction in the amorphous polymer. It is frequently important to be able to assess this dissipative characteristic of these materials using a parameter for that purpose. The **loss tangent**, represented as $\tan\delta$, is the ratio of energy lost to energy stored over one cycle of deformation:

$$\tan\delta = \frac{1}{2\pi}\left(\frac{U_{\text{lost}}}{U_{\text{stored}}}\right) \tag{T2.23}$$

The mechanical damping capacity of a material indicated by $\tan\delta$ finds application in systems that are subject to vibrations.

Example T2.5:

A straight circular cylindrical rod of solid polymer has length L_0 = 1.2 m and diameter d = 15 mm. The polymer is viscoelastic with a creep compliance

$$J_E(t) = 3.0 - \exp\left[-\left(\frac{t}{10.}\right)\right] \text{ GPa}^{-1}$$

where t is in h. The rod is suspended vertically, and a mass of m = 15 kg is hung from it for 20. h. What is the change in length of the rod over this interval of time?

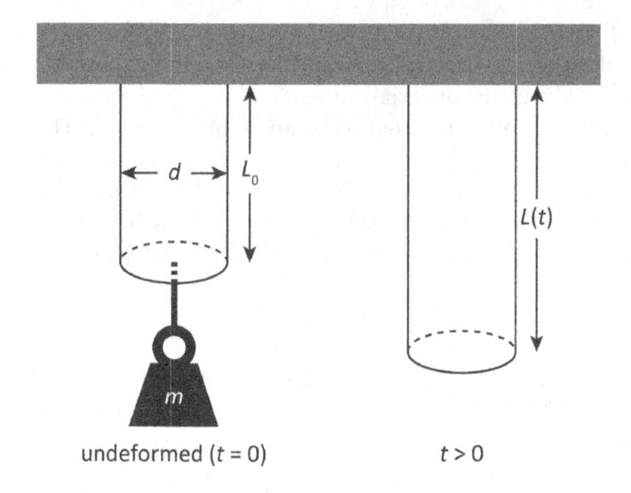

SOLUTION

This problem involves creep, which is the time-dependent viscoelastic deformation of a material under a constant applied load/stress. The time-independent stress σ_0 and time-dependent strain $\varepsilon(t)$ in the rod after a time t are connected by the relationship for viscoelastic creep in tension:

$$\varepsilon(t) = \sigma_0 J_E(t)$$

From the definitions of stress and strain, we have

$$\varepsilon(t) = \frac{L(t) - L_0}{L_0} = \frac{\Delta L(t)}{L_0}$$

and

$$\sigma_0 = \frac{W}{A_0} = \frac{mg}{\pi(d/2)^2} = (15\text{ kg})\left(9.82\ \frac{\text{m}}{\text{s}^2}\right)/\pi(0.0075\text{ mm})^2 = 0.83\text{ MPa}$$

We also find for t = 10 h that

$$J_E(10) = \left(3.0\text{ GPa}^{-1}\right) - \exp\left(-\frac{20.\text{ h}}{10.\text{ h}}\right) = 2.8\text{ GPa}^{-1} = 0.0029\text{ MPa}^{-1}$$

With these important pieces of information in hand, we can finally determine

$$\Delta L(20.\text{ h}) = \varepsilon(20.\text{ h})L_0 = \sigma_0 J_E(20.\text{ h})L_0 = (0.83\text{ MPa})\left(0.0029\text{ MPa}^{-1}\right)(1.2\text{ m})$$

which is

$$\sigma_0 J_E(20.\text{ h})L_0 = (0.83\text{ MPa})\left(0.0029\text{ MPa}^{-1}\right)(1.2\text{ m}) = 0.0029\text{ m} = \underline{\mathbf{2.9\text{ mm}}}$$

T2.5 CLOSING

The important materials characteristics that influence the performance of vehicles and logistical endeavors have been introduced and described. The important overall physical properties that determine much about the energy cost of moving and steering ground vehicles and aircraft are the mass density of the materials comprising the vehicle's frame and the energy density of its power source. Several important thermal properties of materials influence the performance of heat engines used in the power plants of vehicles, such as heat capacity, melting temperature, and thermal conductivity.

T2.6 CHAPTER SUMMARY

Key Terms

creep
creep compliance
damping
diffusion creep
dislocation creep
efficiency
energy density
heat engine
heat of fusion
homologous temperature
load-relaxation
phonon
relaxation modulus
specific energy
specific heat
thermal conductivity
thermal diffusivity
transition temperature
viscoelasticity
viscosity

Important Relationships

$$P_{tot} = P_{acc} + P_r + P_t$$ (power required for vehicle)

$$P_{tot} = m(a + g \sin \varphi + g\mu_f \cos \varphi) \times v$$ (P_{tot} for land vehicle)

$$P_{tot} = \left[ma + \tfrac{1}{2}\rho_{air}\mu_d (\Delta v)^2 + bF_{lift} + F_\ell\right] \times v$$ (P_{tot} for aircraft)

$$pV = nRT$$ (ideal-gas law)

$$\Delta U = U_{in} - U_{out} = 0$$ (1st law of thermodynamics)

$$q_{in} - q_{out} = W - W_0 = \Delta W$$ (1st law for heat engines)

$$\eta = \frac{q_{in} - q_{out}}{q_{in}}$$ (efficiency of heat engine)

$$\eta = 1 - \frac{T_{cold}}{T_{hot}}$$ (efficiency of heat engine)

$$c_p = \frac{dq}{dT}$$ (heat capacity)

$$\dot{q} = \frac{q}{At}$$ (heat flux)

$$\dot{q} = -\kappa \frac{\Delta T}{\Delta x}$$ (Fourier's law)

$$\kappa_i = \frac{1}{3}\left(\frac{n_i c_i v_i L_i}{N_A}\right)$$ (thermal conductivity by carrier)

$$\kappa = \kappa_{ph} + \kappa_e$$ (thermal conductivity of solids)

$$u_N = h\frac{v_s}{\lambda_N}$$ (phonon energy for mode N)

$$\kappa = \frac{1}{3}\left(n_{ph}c_{ph}v_{ph}L_{ph} + n_e c_e v_e L_e\right)$$ (thermal conductivity)

$$T(t) = T_0 + \frac{\dot{q}(t)}{c_p}$$ (temperature change with heat)

$$\frac{\partial T}{\partial t} = \alpha \frac{\partial^2 T}{\partial t^2}$$ (heat equation)

$$d = \sqrt{2\alpha t}$$ (thermal transport distance)

$$T_h(T) = \frac{T}{T_m}$$ (homologous temperature)

$$\dot{\varepsilon} = A\left(\frac{\sigma}{G}\right)^m \left(\frac{b}{D}\right)^n D_v(T)$$ (creep strain rate)

$$T_h(T) = \frac{T}{T_g}$$ (Th for polymers)

$$\varepsilon(t) = \sigma_0 J_E(t)$$ (viscoelastic creep)

$$\sigma(t) = \varepsilon_0 E(t)$$ (viscoelastic load relaxation)

$$\tan\delta = (1/2\pi)(U_{lost}/U_{stored})$$ (loss tangent)

T2.7 QUESTIONS AND EXERCISES

Concept Review

CT2.1 The "fuel efficiency" of a vehicle is the effective amount of mechanical energy that can be extracted from the fuel it utilizes. A "physical" measure of efficiency η could be the work W done on a vehicle (in J) and the energy content u' of the fuel (in J) that was consumed to produce W: $\eta = W/u'$. The units of η are therefore $[\eta]$ = [energy/energy] = J/J = unitless. However, the fuel efficiency of vehicles is sometimes quoted as $[\eta']$ = [distance/fuel volume] = km/L or "miles per gallon". How can you reconcile these two seemingly measures η and η'?

CT2.2 Given your answer to **Exercise CT2.1**, how would you describe the mass or density of their vehicles and its efficiency?

PROBLEMS

PT2.1 Suppose that the effective limiting temperature T_{hot} for a heat engine is around 80% of the melting temperature T_m of the materials the engine is made of. What is the approximate efficiency of a heat engine as a function of the melting temperature? What is the approximate efficiency of a heat engine made from Al?

PT2.2 Suppose that after 500 h, the extension in a creep specimen is $\Delta\ell$ = 0.5 cm. If the specimen's initial length was ℓ_0 = 10 cm, what was the creep rate (assumed constant)?

PT2.3 Suppose that you are making 1 mole of metallic U at room temperature using a reaction involving the reactive metal Mg: $UF_4 + 2Mg \rightarrow 2MgF_2 + U$. If the heat produced by this reaction is 330,000 J/mol of U produced, and if 1/10 of this heat is absorbed by your uranium product, how hot will the uranium be? For uranium, the molar heat capacity is $C_p \approx 25.10$ J/mol K.

PT2.4 During heat treatment, a flame with effective temperature $T = 1100°C$ is applied to an Al slab 1.2 cm thick initially at temperature T_R. Compute the thermal gradient $\Delta T/\Delta x$ and the heat flux at the beginning of the treatment. About how long would you expect it to take for the center of the slab to reach the applied temperature T?

PT2.5 The overall glass-transition temperature T_g (in K) of a random copolymer is given by

$$\frac{1}{T_g} = \frac{W_1}{T_{g1}} + \frac{W_2}{T_{g2}}$$

where W_1 and W_2 are the weight fractions of each of the comonomers and the T_gs are the respective transition temperatures of each. What is the transition temperature of an amorphous PE-PP copolymer that has $X_{PE} = 60\%$ and $X_{PP} = 40\%$?

PT2.6 For some viscoelastic polymers that are subjected to stress relaxation tests, the stress σ decays with time according to

$$\sigma(t) = \sigma(0)\exp\left(-\frac{t}{t_r}\right)$$

where σ(t) and σ(0) represent the time-dependent and initial stresses, respectively, and t_r is the "relaxation time" of the material (a constant). A specimen of this type of material was instantaneously pulled in tension to a strain of $\varepsilon_0 = 0.50$, and the stress was measured as a function of time. Determine $E(10)$ if the initial stress level was 3.5 MPa and dropped to 0.50 MPa after 30. s.

MATLAB Exercises

MT2.1 The bonding energies of some selected substances are provided below. Make a MATLAB plot of melting temperature T_m vs. bonding energy U_0.

Material	Bonding Energy, U_0 [kJ/mol]
H_2O	50.
Hg	63
Al	330
MgO	1000
SiC	1200
W	850

MT2.2 Recall that (phonon) specific heat of a material at low temperature is

$$c_{\text{ph}}(T) = \frac{12\pi^4 N_A k_B}{5T_D^3} T^3$$

Make a plot of this function for Fe over the domain $0 < T < 100$ K. Fe has Debye temperature $T_D = 470$ K.

MT2.3 Using your result from above, compute the phononic part of the thermal conductivity $k(T)$ of Fe from $T = 0$ to 100 K:

$$k(T) \approx \left(\frac{v_s T_m a}{B^2 V_m^2 \alpha^2}\right) \times \frac{\left[c_{\text{ph}}(T)\right]^3}{T}$$

Note that the average phonon velocity is approximately the speed of sound ($\approx v_s$) and that the speed of sound is related to the bulk modulus B and density ρ according to

$$v_s = \sqrt{\frac{B}{\rho}}$$

Property	Value [unit]
Coefficient of linear thermal expansion, α	12×10^{-6} m/m
Bulk modulus, B	170 GPa
Lattice parameter, a	2.87 Å
Debye temperature, T_D	470 K
Melting temperature, T_m	1538°C
Molar volume, V_m	7.09 cm^3/mol
Mass density, ρ	7.874 g/cm^3

MT2.4 A (circular) cylindrical specimen with initial diameter $d = 8.0$ mm and initial length $\ell_0 = 150$ mm creeps at an effectively constant rate $\dot{\varepsilon}_{ss} = 10^{-6}$ h^{-1}, when subjected to a constant load of $P = 1000$ N. Write a MATLAB function `creepdef(t)` that computes the deformation (in mm) of the specimen after an elapsed time t. What are the lengths of the specimen after 100, 1000, and 10,000 h? Plot the true stress in the specimen as a function of t.

NOTES

1. Peter Debye (1884–1966) was a Dutch-American physicist and physical chemist who was awarded a Nobel Prize for his work on dipole moments and X-ray diffraction.
2. After Michael F. Ashby. *Materials Selection in Mechanical Design*. Fourth edition. Oxford, England, UK: Butterworth-Heinemann, 2011.
3. An associated unit is the "poise" (P), with 10 P = 1 Pa s.
4. An equivalent relationship $\gamma(t) = \tau_0 J_G(t)$ holds in shear deformation.
5. An equivalent relationship $\tau(t) = \gamma_0 G(t)$ holds in shear deformation.

Topic 3: Materials for Electromagnetic and Optical Devices

Engineering with Charges and Fields

LEARNING OBJECTIVES

After completing this chapter, you should be able to:

1. Describe the essential operating principles of an electrical/electronic circuit.
2. List the most common passive and active components in electronic circuit design.
3. Describe how the materials properties like resistivity, dielectric constant/strength, and magnetization influence the behavior of passive components.
4. Solve problems related to the circuit characteristics of passive devices and the properties of materials they are made of utilizing real data.
5. Describe the band structure of materials and categorize materials according to their structure.
6. Utilize materials data in materials selection for electromagnetic components.
7. Describe the principles of *p*- and *n*-type doping and how they influence semiconductor characteristics, including the characteristics of a *pn* junction.
8. Describe the operation of a *pn*-junction diode and how to bias one.
9. Describe the biasing/operation of a bipolar junction transistor and its uses.

T3.1 DESIGN REQUIREMENTS FOR ELECTRICAL AND ELECTRONIC DEVICES

One of the defining features of the modern era is the ubiquity of **electromagnetic** devices and **electronics**. Electromagnetic devices convert electrical and/or magnetic energy into other useful forms, e.g., heat, light, mechanical, etc., during operation. Electronics utilize controlled fluxes of electrons to produce, collect, and interpret signals/information in both analog and digital contexts. Typical applications are provided in **Table T3.1**. The wide variety of physical phenomena involved in these applications means that a number of different classes of materials with tunable properties are required.

DOI: 10.1201/9781003214403-10

TABLE T3.1
A Short Catalog of Electromagnetic/Electronic Applications by Sector

Application Area	Examples	
Energy generation & transfer	• Power generation/ transmission equipment • Industrial power supplies • Domestic appliances/wiring	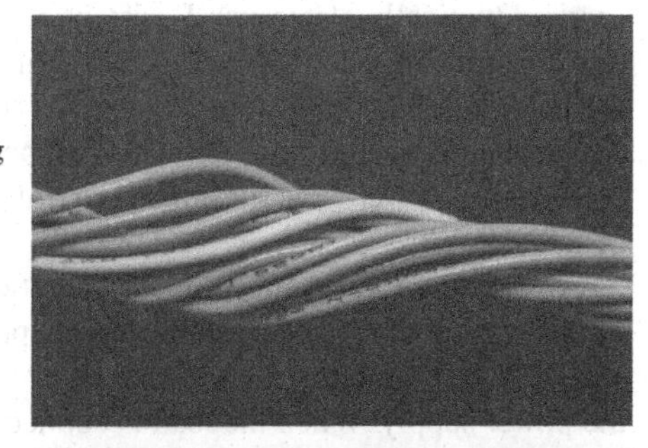
Computation & control	• Integrated circuits (ICs) • Circuit boards/IC packaging • User interfaces/displays	
Sensors	• Transducers for electrical measurement, physical measurement (distance, temperature, and weight), optical measurement, etc.	
Communication	• Antennas • Optical signal generation/ transmission	

T3.2 ESSENTIAL PROPERTIES – RESISTIVITY AND CONDUCTIVITY

You know from your introductory physics or electrical engineering course about the basic properties of "passive" circuit components: **resistors**, **capacitors**, and **inductors**. These devices exhibit a particular response when driven by an input of charge carriers (typically electrons). Resistors remove potential energy from a system and transform it into heat; capacitors store energy within the system in a static electrical field; and inductors store energy in a magnetic field. The combined behavior of circuits composed of these elements can be tuned to various tasks, such as providing conditioned power to a subsidiary circuit or transforming an electrical signal in a beneficial way.

A general circuit is illustrated schematically in **Figure T3.1**. The circuit consists of a power supply (e.g., a **voltage** source, potentially with its own internal resistance) and a "load", which is a collection of components and other devices energized by the power supply. When the load is connected and the power supply is activated, the resulting **current** through the load drives its effects. Recall the basic definitions of circuit quantities: a voltage v is a difference in electrical potential that exists between two points in a circuit or electrical system. Voltage v has units $[v]$ = [energy/charge] = J/C = V to reflect the fact that the net potential is apportioned among the charges in the electrical system. Differences in voltage/electrical potential produce an electromotive force that drives the motion of the charges. In particular, electromotive forces push charges through an object at a rate given by the current i, where $[i]$ = [charge/time] = C/s = A.

In typical resistors, the voltage and the current are related by **Ohm's law**:

$$v = iR \tag{T3.1}$$

where R is the resistor's **resistance** with $[R]$ = [voltage/current] = V/A = Ω. This important relationship is familiar, and it should not be surprising that resistors made of different materials will in general have different resistances, all other factors

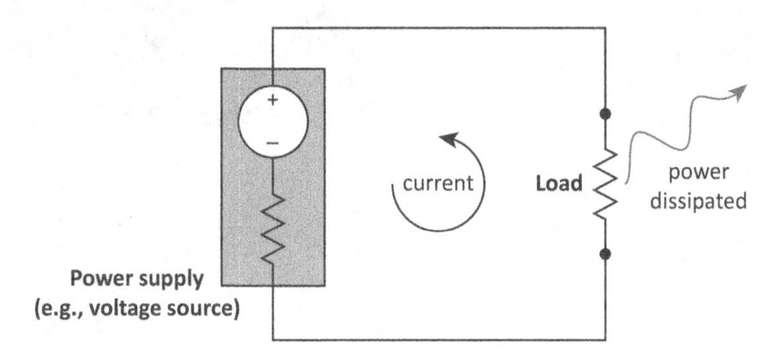

FIGURE T3.1 Schematic of a general circuit. A power supply drives current through a load. Electrons (at high potential energy) enter the load and emerge (at low energy) on the other side. The energy difference is dissipated by the load as motion, light, heat, etc., and is reflected in the power consumed by the circuit.

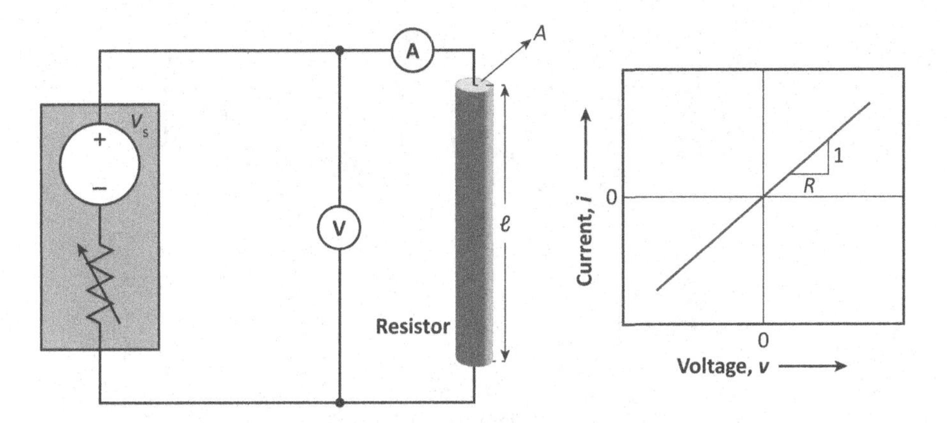

FIGURE T3.2 Schematic of a circuit with a cylindrical wire of material in place of the load. A reversible voltage source applies a potential v_s to the circuit, and an adjustable resistor produces a variable voltage across the specimen. Depending on the level of voltage v across the wire, a current i develops. The relationship between the current and the applied voltage is linear, and its slope depends on the resistance R of the wire.

being equal. Consider the diagram of **Figure T3.2**. The load of **Figure T3.1** has been replaced by a specimen of material with length ℓ and cross-sectional area A. Following Ohm's law, we expect a linear relationship between i and v in the specimen. This expectation is reflected in the associated i–v curve; the slope of the curve is $1/R$. We further expect that R will increase with the length of the specimen (i.e., it takes *more* electromotive force to push charges through a longer object) and that R will also decrease with area A. For these reasons, we surmise

$$R = \rho\frac{\ell}{A} \tag{T3.2}$$

In **Equation T3.1**, we have accounted for the effect of the geometry of the specimen on the measured value of R through the parameters ℓ and A. The remaining parameter ρ is called the **resistivity** and is a *material-dependent* property with [ρ] = [resistance × length] = Ω m. Resistivity values for some example materials are given in **Table T3.1**. Frequently, the equivalent property, the **conductivity** σ, is employed in calculations:

$$\sigma = \frac{1}{\rho} \tag{T3.3}$$

(σ then has units $[\sigma] = \Omega^{-1}\ \mathrm{m}^{-1}$.)

Example T3.1:

Suppose you have a sensitive ohmmeter ("Ω") that you are using to measure the resistance of a piece of wire. The wire has a diameter d = 1.5 mm and a length ℓ = 45 cm. If the ohmmeter reading is $R = 5.5 \times 10^{-5}\ \Omega$, what is the resistivity of the material? What electrical class of material is the specimen?

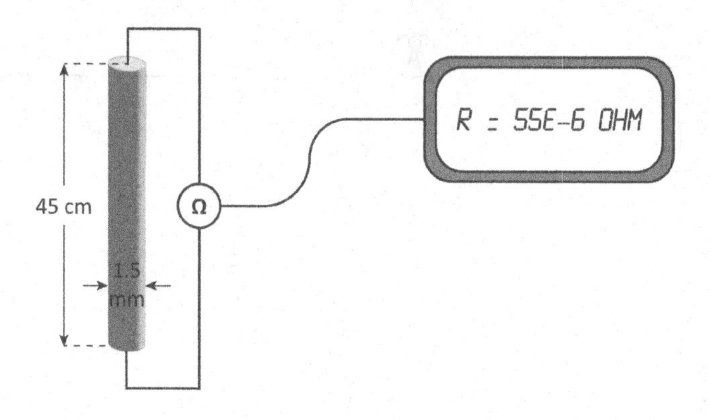

SOLUTION

Following **Equation T3.2**, we obtain

$$R = \rho\frac{\ell}{A} \rightarrow \rho = \frac{RA}{\ell}$$

where $A = \pi(d/2)^2 = \pi[(0.15\ \text{cm})/2]^2 = 0.0187\ \text{cm}^2$. We now calculate directly

$$\rho = \frac{RA}{\ell} = \frac{\left(5.5\times10^{-5}\ \Omega\right)\left(0.0187\ \text{cm}^2\right)}{45\ \text{cm}} = 2.2\times10^{-6}\ \Omega\ \text{cm} = \underline{\mathbf{2.2\times10^{-8}\ \Omega\ m}}$$

Comparison with the values in **Table T3.2** indicates that the material is a **conductor**.

Having by now developed a certain sensibility when it comes to materials properties like resistivity, you might ask, "Why does a particular material have the value of ρ that it does?" To understand the conductivity (or lack thereof) in bulk materials, we must inspect their electronic arrangement. Consider the case of a metallic material: Al. The lower-level electrons in an atom exist in filled 1*s*, 2*s*, and 2*p* orbitals (with 2, 2, and 6 electrons) and do not participate in bonding. The outer, unfilled-electron–shell, or "valence" electrons, however, are shared among the Al atoms in a metallic bond. The circumstances are illustrated in **Figure T3.3**. In **Figure T3.3(a)**, the ionic Al^{+3} "cores" are embedded in a "sea" of electrons that constitute the **valence band**; the individual Al atoms give up their valence electrons to the material as a whole. Outside of the cores, the rest of the electrons are free to move from location to location. (Within limits; see below.)

The complication here is that all of these electrons in the valence band cannot all share the same state, according to the Pauli Exclusion principle. The outer orbitals that were previously associated with a particular atom core ("localized") split into finely spaced energy levels to accommodate the larger number of electrons (say, N_A of them) in a bulk material. We refer to this aggregate of electrons in finely spaced states as an **electron band**, and understanding the properties of this band gives us insight into the electronic behavior of materials. We imagine that the delocalized electrons are confined in a *square-well potential*, or "box", that has a size ℓ and a

TABLE T3.2
Resistivity and Conductivity Values (at Room Temperature) of Some Common Materials

Class	Material	ρ [Ω m]	σ [Ω⁻¹ m⁻¹]
Conductor	Al	2.65×10^{-8}	3.77×10^{7}
	Cu	1.67×10^{-8}	5.99×10^{7}
	Ag	1.59×10^{-8}	6.29×10^{7}
	Pt	1.06×10^{-7}	9.43×10^{6}
	Au	2.35×10^{-8}	4.26×10^{7}
	C (graphite)	9.101×10^{-6}	1.099×10^{5}
Semiconductor	Ge	4.6×10^{-1}	2.2
	Si	2.3×10^{3}	4.3×10^{-4}
	SiC	10^{4}	10^{-4}
Insulator	ZrO_2	23	4.3×10^{-2}
	Al_2O_3	10^{13}	10^{-13}
	SiO_2	10^{16}	10^{-16}
	PET[a]	10^{21}	10^{-21}
	Rubber	10^{13}	10^{-13}
Superconductor[b]	Hg at $T < 4$ K	0	–
	$Bi_2Sr_2Ca_2Cu_3O_{10}$ (BSCCO) at $T < 110$ K	0	–
	$YBa_2Cu_3O_7$ (YBCO) at $T < 92$ K	0	–

[a] PET = poly(ethylene terephthalate). [b]Superconductors are *perfectly* conducting in their pure form. Since this effect is only observed at low temperatures and/or high pressures, these materials are uncommon.

potential-energy depth of V_0, as shown in **Figure T3.3(b)**. The electron's (1D) "stationary" wavefunction $\psi(x)$ inside this box obeys the Schrödinger equation. If the electron's energy is $E \ll V_0$, then

$$\frac{d^2\psi}{dx^2} + k^2\psi(x) = 0 \tag{T3.4}$$

where $k^2 = 2m_e(2\pi/h)^2E$ in the box. Since V_0 is very large, escape is impossible, and **Equation T3.4** has solution

$$\psi(x) = \sin(kx) \tag{T3.5}$$

Furthermore, we make the association $k = 2\pi/\lambda$ for a sine wave with wavelength λ.

Now, the electron must have a sine wavefunction that "fits" exactly in the box, i.e., *only* whole or half wavelengths will make $\psi = 0$ at the edges of the box, and so we have the condition

$$\lambda = 2\ell, \ell, \frac{2\ell}{3}, \ldots, \frac{2\ell}{n}$$

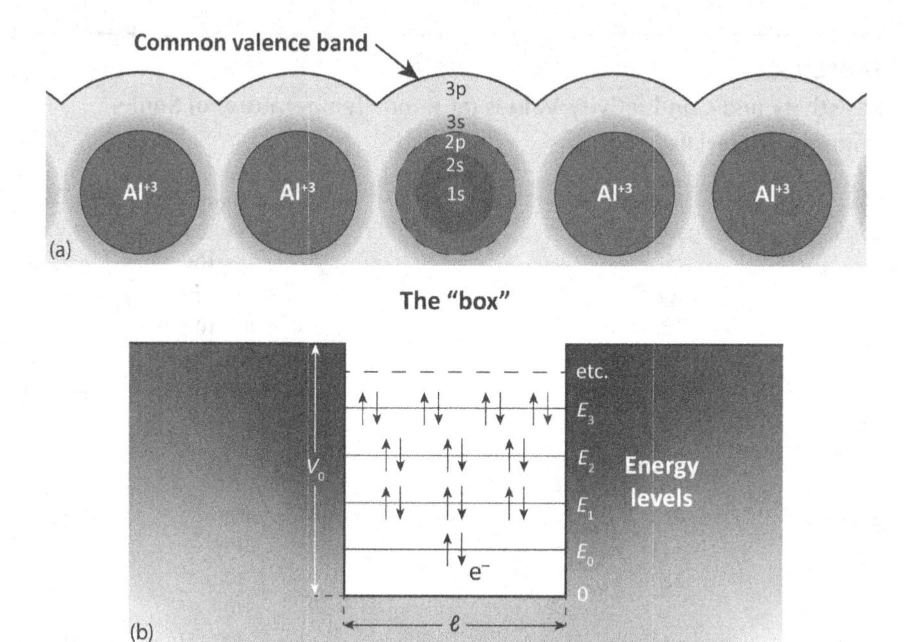

FIGURE T3.3 Material origin of electron-band structure. (a) The innermost orbitals of each Al atom form an Al^{+3} "ion core". The outermost valence electrons contribut to the metal generally becoming *delocalized.* (b) The delocalized electrons are still confined within the metal, and this confinement establishes energy levels $n = 1, 2, 3 \ldots$ with $E \propto n^2$. Electrons fill these levels, forming an electron band in the material.

where $n = 1, 2, 3, \ldots$ We can now establish

$$E = \frac{1}{2m_e}\left(\frac{h}{2\pi}\right)^2 k^2 = \frac{1}{2m_e}\left(\frac{h}{2\pi}\right)^2\left(\frac{2\pi}{\lambda}\right)^2$$

or

$$E(n) = \frac{1}{2m_e}\left(\frac{h}{2\ell}\right)^2 n^2 \tag{T3.6}$$

as the energy levels in the box. The parameter n therefore takes on the role of a quantum number that labels the possible states, like those that describe atomic structure. In a 3D box, we have the corresponding solution

$$E(n_x, n_y, n_z) = \frac{1}{2m_e}\left(\frac{h}{2\ell}\right)^2 \left(n_x^2 + n_y^2 + n_z^2\right) \tag{T3.6$'$}$$

where $\{n_x, n_y, n_z\}$ are the quantum numbers. The highest energy level E_F occupied by electrons in the material at a temperature $T = 0$ K is called the **Fermi level** [named after the physicist Enrico Fermi (1901–1954)].

The filling of energy bands and the position of the Fermi level determine the conductive properties of a material. The reason for this is that electrons are free to move when there are unoccupied states in a band to move between. If an electron is in a fully filled band, there is nowhere for it to go *unless* it can move to another band with unoccupied states. A complicating factor in moving between bands in a material is that the periodic structure of the lattice disrupts the idealized box-like behavior and introduces "forbidden zones", or **bandgaps,** that restrict electron motion. The situation is depicted in **Figure T3.4**. In a conducting material, like the Al discussed above, the valence band has unoccupied states that electrons are free to move into. There are even more unoccupied states in the next closest band, called the **conduction band**. However, the valence band and the conduction band are separated by the bandgap ΔE_g of forbidden electron energies. The nature of this bandgap is more important in semiconductor materials since they have completely filled valence bands. Without any additional energy input, electrons are trapped in the valence band and cannot move. A similar situation exists in insulating materials (like ceramics), whose valence bands are also filled and whose bandgap is even greater than that of the semiconductors.

It is important to recognize that the static picture presented in **Figure T3.4** is only an accurate depiction at a temperature $T = 0$ K. If we raise the temperature, the added thermal energy can give a "boost" to the charge carriers sufficient to enable conduction. The physics are similar to those of thermal activation we studied in **Chapter 4**. The odds F that a charge carrier has an energy E (relative to the Fermi level E_F) are given by the Fermi function:

$$F(E) = \frac{1}{1 + \exp\left(\frac{E - E_F}{k_B T}\right)} \tag{T3.7}$$

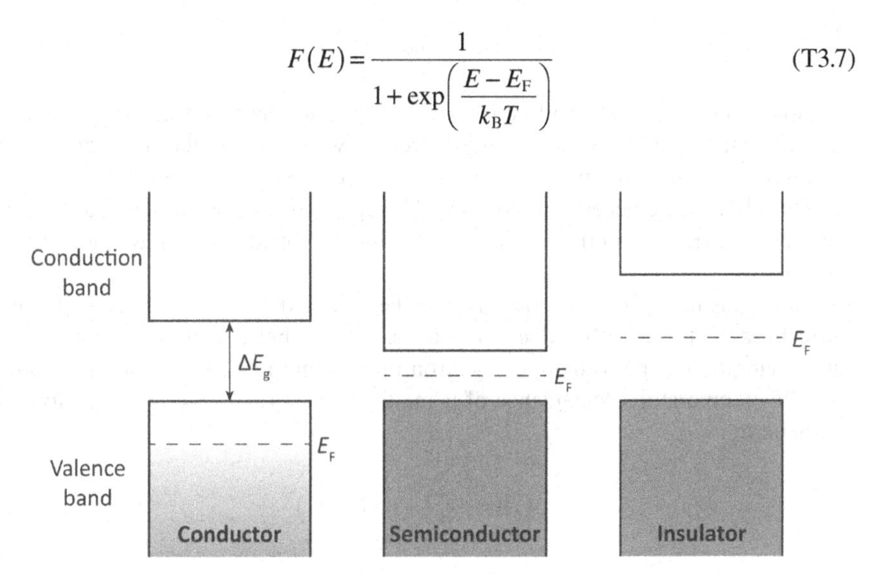

FIGURE T3.4 The band organization of various types of materials. Conductors have a partially filled valence band, and so charges can change states readily. Insulators have a filled valence band, so charge carriers are trapped unless they can cross a significant bandgap into the conduction band. Semiconductors are similar to insulators but have a smaller bandgap. In insulators and semiconductors, the Fermi levels are conventionally located halfway across the bandgap, though no charge carriers occupy those energies.

This means that, at elevated temperature, the occupancy of the bands is now a matter of *probability*. The influence of temperature and its consequences on F are shown in **Figure T3.5**. In conductors, as shown in **Figure T3.5(a)**, additional conductive states immediately above E_F become available when the temperature is increased above 0 K. This effect becomes more important in semiconductors. Since the bandgap in semiconductors is so small, moderate temperatures can elevate electrons into the conduction band from the full valence band, enabling conduction, as shown in **Figure T3.5(b)**.

Now consider the flux of electrons through a material. Following the discussion of **Chapter 4**, we can represent the flux J of electrons in a material as the product of a mobility and a driving force:

$$J = \text{mobility} \times \text{driving force}$$

The driving force is provided by the applied electric field $\mathbf{E}$ and has magnitude $Ne|\mathbf{E}|$, where N is the number of charge carriers (in coulomb per unit volume) and e is the electron charge (in coulomb). If the mobility is μ, then

$$J = \mu \times Ne|\mathbf{E}| \tag{T3.8}$$

Unit analysis gives $[|\mathbf{E}|]$ = [force/charge] = kg m/s^2 C. Therefore, we must have $[\mu Ne]$ = C^2 s/kg m^3 = $(\Omega\ \text{m})^{-1}$. That is,

$$J = \sigma|\mathbf{E}| \quad \text{and} \quad \sigma = \mu Ne \tag{T3.9}$$

This is an important result since it indicates the physical/material parameters that influence σ and ρ. Conductivity increases when the number of charge carriers increases and/or the mobility of the charge carriers increases. For instance, if a material has little access to unoccupied levels, as is the case in insulators whose electrons are trapped in the valence band below the bandgap, then μ, and hence σ, will be low.

An important associated effect is that the conductivity depends on the temperature because the mobility does: $\mu = \mu(T)$. This is because temperature-induced lattice vibrations tend to dampen electron motion and reduce mobility. In conductors, this temperature dependence of μ manifests as an increase in resistivity with temperature:

$$\rho(T) = \rho_0\left[1 + \alpha(T - T_0)\right] \tag{T3.10}$$

In **Equation T3.10**, ρ_0 is a reference resistivity at T_0, a reference temperature (e.g., room temperature $T_R \approx 300$ K), and α is a materials-dependent parameter (the "temperature coefficient of resistivity"). Additionally, the resistivity/conductivity of a material also depends on its defect content. Impurities and grain boundaries can increase the resistivity of materials by "scattering" electrons as they move through the material.

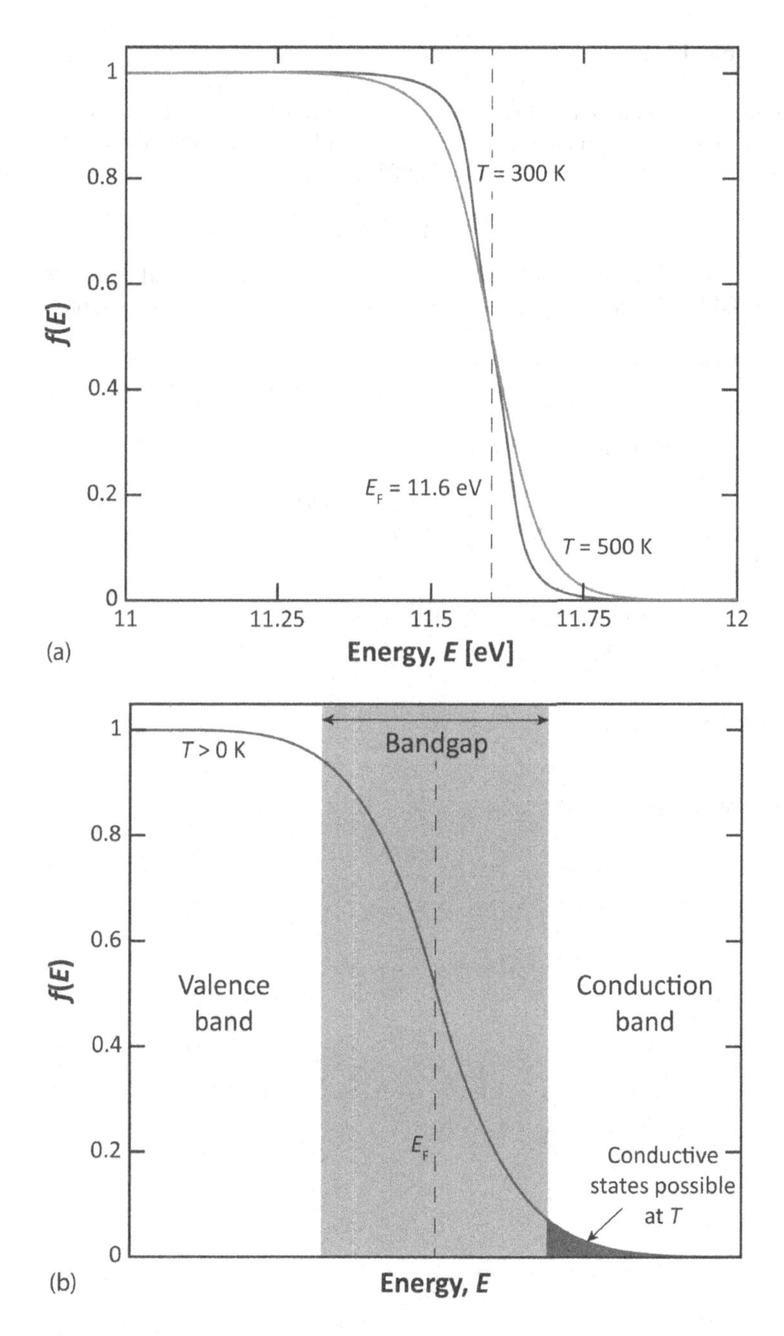

FIGURE T3.5 Fermi functions. (a) gives the function for the conductor Al. When $T = 0$ K, all electron energies fall below E_F. At temperatures $T > 0$ K, electrons become less likely to inhabit energy states below E_F and more likely to inhabit higher energy states. As the temperature increases, the effect becomes more dramatic, leading to an overall increase in conductivity. (b) is for a schematic semiconductor. The states within the bandgap are forbidden, but states above the bandgap in the conduction band become possible.

Example T3.2:

Cu has a temperature coefficient of resistivity of $\alpha = 3.9 \times 10^{-3}$°C^{-1}. Assuming that the temperature coefficient remains constant, plot the conductivity σ of Cu over the temperature range $-100°C < T < 500°C$.

SOLUTION

We can compute the conductivity using the given value of α and the value of ρ_0 from **Table T3.1**: $\rho_0 = 1.67 \times 10^{-8}\ \Omega$ m. In MATLAB®, this calculation looks like

```
function sigma = sigma(T)
      % rho0 in ohm m; T0 in deg. C; alpha in (deg. C)^-1
      rho0 = 1.67*10^-8;
      T0 = 27;
      alpha = 0.0039;
      rho = rho0*(1 + alpha*(T - T0));
      sigma = 1./rho; % sigma in (ohm m)^-1
end
```

We can now plot over the specified range:

```
>> x = -100:1:500;
>> y = sigma(x);
>> plot(x, y);
>>
```

to obtain something like

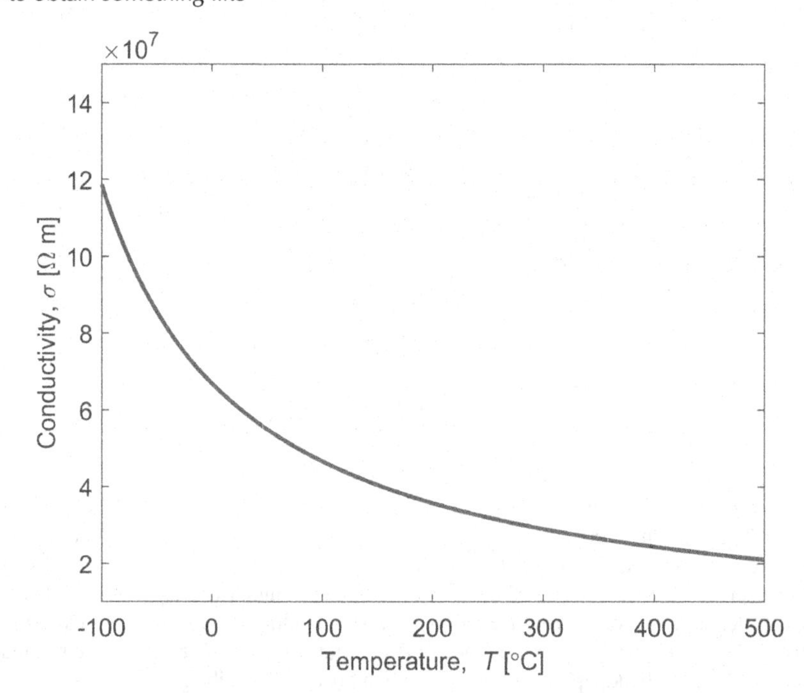

APPLICATION NOTE – THERMOCOUPLES

The concepts underlying electron mobility and conductivity above can be applied to temperature-sensing devices called *thermocouples*. A thermocouple is made from two wires of dissimilar metals ("A" and "B") that have been joined together, forming a junction. Since the Fermi levels of the two metals that meet at this junction will in general be different, the electrons on one side will be at higher potential than those on the other side. This potential will ideally drive the motion of charges. If the two wires are connected at both ends, and both ends are maintained at the same temperature, then the potential difference between the two materials will be effectively zero (equal and opposite at each junction) and no current will flow.

However, if the temperatures of the two junctions are different, as shown in the diagram below, then a net potential will develop in the wires. Furthermore, this potential will be *temperature-dependent*. This is called the **Seebeck effect** and is used in the design of temperature-sensitive thermocouples. The voltage V of the voltmeter **V** is given by

$$V(T) = -S(T_{\text{high}} - T_{\text{low}}) = -S\Delta T$$

Here, S is the Seebeck coefficient that determines the magnitude of the effect; S is essentially a materials property and typically has a magnitude of around 500 µV/°C. This material response of thermocouples enables their use as sensors, meaning they convert a physical parameter (the temperature) into another one (a voltage).

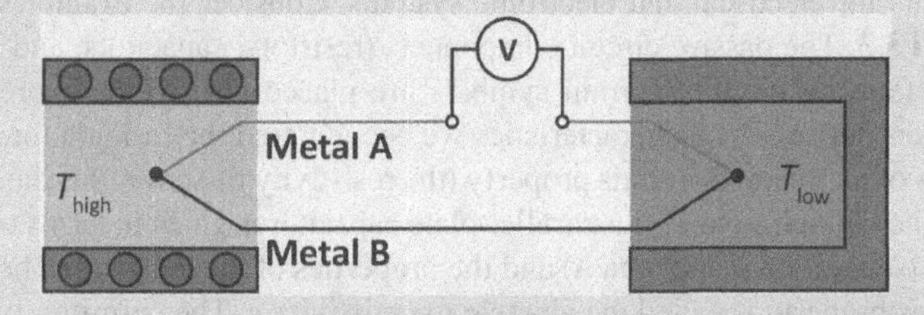

Different materials combinations will give thermocouple sensors with different limitations and sensitivities. Consider the types of thermocouples and their behaviors given below. Some example types are

Thermocouple Type	Materials	Uses
E	Ni-Cr + Ni-Cu	High sensitivity
J	Fe + Cu-Ni	
K	Ni-Cr + Ni-Al	Wide temperature range
C	W-Re(5%) + W-Re(26%)	High-T use; not useful below 750°C
T	Cu + CuNi	Low-T and cryogenic temperatures

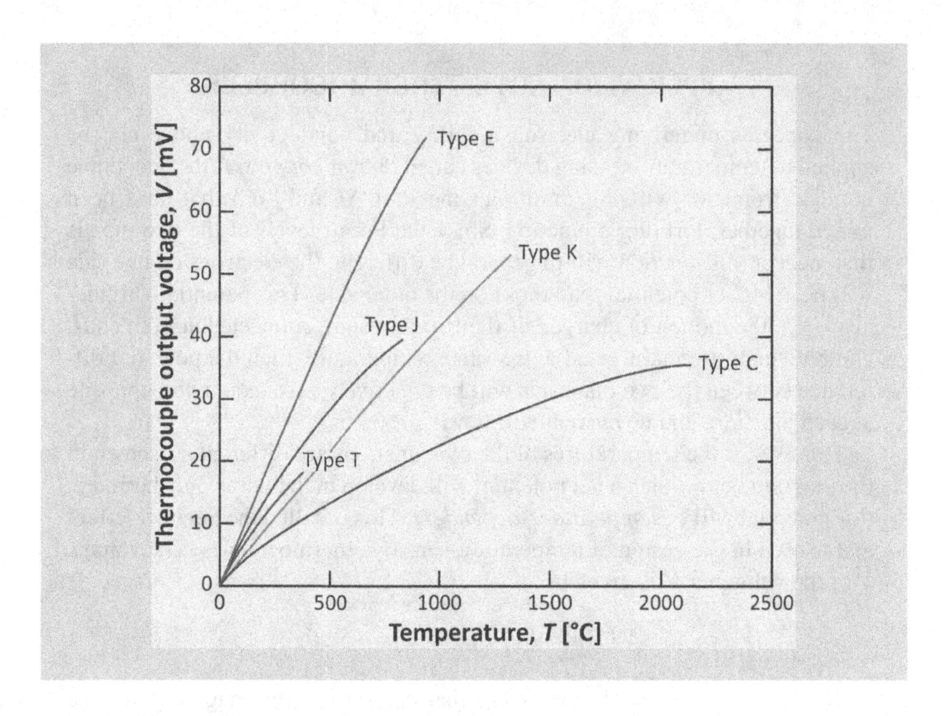

T3.3 ELECTROMAGNETIC COMPONENTS AND MATERIALS PROPERTIES

Materials understanding has important applications in the design and production of components for electrical and electronic systems. Consider the examples provided in **Table T3.3**. The passive circuit components (resistors, capacitors, and inductors) and their familiar circuit-diagram symbols are placed next to the expression that defines their performance characteristic. We encountered the formula for resistance *R* in terms of an innate materials property (the resistivity ρ) above. Similarly, the formula for the **capacitance** *C* of a parallel-plate capacitor is given in terms of the plate geometry (separation *d* and area *A*) and the properties of the material in between the plates. This property is called the **electric permittivity** ϵ. The permittivity indicates how easily a material may become electrically polarized in the presence of a voltage *v* across it. This influences, for instance, the energy *U* stored in a capacitor:

$$U_C = \frac{1}{2}CV^2 = \epsilon\left(\frac{A}{2d}\right)v^2 \tag{T3.11}$$

The permittivity of a vacuum or thin gas is $\epsilon_0 \approx 8.854 \times 10^{-12}$ $C^2/(N \times m^2)$, and the permittivities of many materials are defined as multiples ϵ_r of this "fundamental" value:

$$\epsilon = \epsilon_r \epsilon_0 \tag{T3.12}$$

ϵ_r is called the "relative permittivity" of the material. Furthermore, the material in a capacitor cannot sustain unlimited voltage without "breaking down" and passing

TABLE T3.3
Active and Passive Electrical Components and Their Defining Characteristics

Component	Symbol	Characteristics
	Passive	
Resistor		Provides resistance: $R = \rho \frac{L}{A}$ (ohmic resistor)
Capacitor	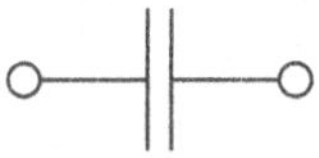	Provides capacitance: $C = \epsilon \frac{A}{d}$ (parallel-plate configuration)
Inductor	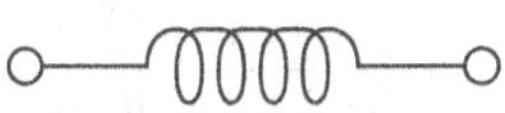	Provides inductance: $L = \mu \frac{N^2 A}{\ell}$ (solenoid configuration)
	Active	
Diode[a]	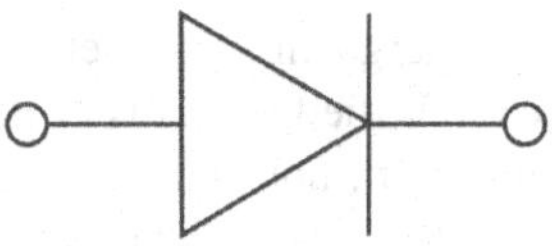	Nonlinear V–i characteristics; functions as a one-way current "valve"
Transistor	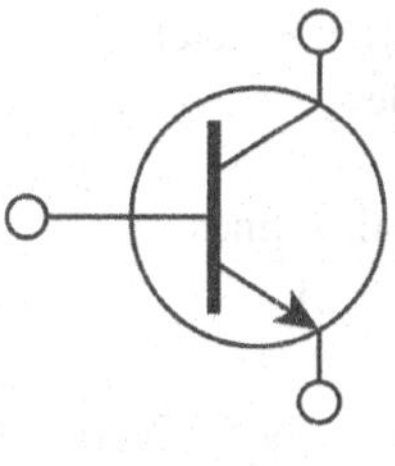	Provides amplification and acts as a controllable current "gate"

[a] Diodes are often categorized as passive, nonlinear components, but their application in shaping currents in dramatic ways makes the "active" label more apt.

current directly between the parallel plates. A material's resistance to breakdown is an electrical property similar to a strength; this property is called the **dielectric strength** and has units [voltage/length].

Example T3.3:

You want to construct a capacitor that stores a charge of 2.2×10^{-11} C at 12 V. If your opposing plates are 1.0 mm^2 in area and the plate spacing is d = 50. μm, what must the relative permittivity ϵ_r of the dielectric be? The capacitor equation is

$$Q = Cv$$

SOLUTION

From **Equation T3.11**, the capacitance is

$$C = \epsilon \frac{A}{d}$$

and $\epsilon = \epsilon_r \epsilon_0$ (from **Equation T3.12**) so

$$Q = \epsilon \frac{A}{d} v = \epsilon_r \epsilon_0 \frac{A}{d} v$$

This gives

$$\epsilon_r = \frac{Qd}{\epsilon_0 A v} = \frac{\left(2.2 \times 10^{-11}\right)\left(50. \times 10^{-6}\right)}{\left(8.85 \times 10^{-6}\right)\left(1.0 \times 10^{-6}\right)(12)} = \mathbf{10}$$

(note that, as a *relative* permittivity, this quantity has no units).

An inductor is a device that stores energy in a magnetic field. Recall that magnetic fields have magnitudes that are quantified in terms of the number of magnetic field lines (or "flux" lines) that form closed, non-intersecting loops connecting the magnetic "**N**" pole to an "**S**" pole in a field source. The number of flux lines that pass through a particular cross section with area A determines the magnetic flux density **magnetic flux density B** with magnitude

$$B = \frac{\#\text{ flux lines}}{A} \tag{T3.13}$$

Here, $[B]$ = [flux lines/area] = Wb/m^2 = (N s)/(C m) = T and indicates the field's ability to produce magnetic force (i.e., the field's "size" or "intensity"). For a simple coil inductor – a "solenoid" with **inductance** L – the stored energy depends on the current i, the coil geometry (area A, length ℓ, and number of windings N):

$$U_L = \frac{1}{2} L i^2 = \mu \left(\frac{N^2 A}{2\ell} \right) i^2 = \frac{B^2 \ell A}{2\mu} \tag{T3.14}$$

The parameter μ is the **permeability** of the material embedded in the coil. When this material is a vacuum (or a thin gas), the permeability is $\mu_0 = 4\pi \times 10^{-7}$ N/A^2 = $1/(c^2\epsilon_0)$. Similarly to the permittivity of materials, the permeability of materials is sometimes written in terms of a "relative permeability" μ_r:

$$\mu = \mu_r \mu_0 \tag{T3.15}$$

Some relative permeabilities are given in **Table T3.4**.

TABLE T3.4
Permeabilities of Selected Materials

Material	μ_r [unitless]
Fe (99.9% pure)	100,000
Fe (99.8%)	5000
Mild carbon steel	100
Ni (99%)	500
Ferritic stainless steels	1200
Co-Fe alloys	18,000
Co (99%)	250
50Ni50Fe	70,000
Co-Ni-Zn-Fe_2O_3 alloys	80
Mn-Zn-Fe_2O_3 alloys	8000
Ti[a]	$\gtrsim 1$
Al[a]	$\gtrsim 1$
Ag[a]	$\lesssim 1$
Cu[a]	$\lesssim 1$

[a] These materials have values of μ that are only *slightly* different than μ_0 (see the descriptions of diamagnetism and paramagnetism, below).

The magnetic behavior of a material falls into one of a few categories according to the availability and organization of atomic magnetic moments within the material. Magnetic moments **m** are vector quantities with magnitude $m = |\mathbf{m}|$ and $[m]$ = [current × length²] = A m². Contributions to the magnetic moment of an atom come from the motion of individual electrons in their orbits (i.e., the orbit forms a "current loop" with magnetic moment $\mathbf{m}_\ell$) and the intrinsic spin of the electrons themselves (i.e., the spin moment with magnitude $m_s = \pm\frac{1}{2}$). The possibilities for magnetism in most materials are

Diamagnetism. If all the $m_s = +\frac{1}{2}$ electrons in the orbitals of an atom are paired with another electron of antiparallel spin ($m_s = -\frac{1}{2}$), then these pairs of intrinsic magnetic moments will "cancel", each other. This means that there will be no net contribution to an atom's overall magnetic moment from electron spins, only from their orbits. Diamagnetic materials are made of atoms such as these, and we typically think of them as being fully "non-magnetic" (though in fact they are repelled by a magnetic field of sufficient size). They are characterized by values of μ slightly *less* than μ_0.

Paramagnetism. *Unpaired* electron spin contributions are possible in orbitals/states that only contain one electron. These residual m_s-related contributions increase the net magnitude of an individual atom's magnetic moment **m**. However, in the absence of any forces that act to *align* individual atom's magnetic moments with those of their neighbors, these **m** orientations are random and the average magnetic moment of large clusters of atoms is miniscule. Externally applied magnetic fields **B** will align the moments in the material, producing an enhancement of the overall magnetic field in the vicinity (see below). Paramagnetic materials are characterized by values of μ slightly *greater* than μ_0 (i.e., the magnetic response of these materials is weak attraction).

Ferromagnetism/Ferrimagnetism/Anti-ferromagnetism. If the atomic magnetic moments in a material become aligned on a permanent basis across a materials structure, then a "permanent magnet" will result. This phenomenon is called ferromagnetism and occurs in materials where a quantum-mechanical force called the "exchange interaction" aligns the moments in a stable arrangement. These "stacked" atomic moments produce an overall significant field with fixed orientation that persists in time. "Ferrimagnetism" is like ferromagnetism, but the overall field is smaller because a minority of the atomic moments stabilize in a countervailing direction to those in the majority. A material with an equal and opposite number of stable moments in its structure is called "anti-ferromagnetic". (Anti-ferromagnets appear to be unremarkable materials at first but can be used in specialized kinds of sensors and computer-memory devices.) Ferromagnetic materials are typically associated with values of $\mu \gg \mu_0$.

Superconductance. Materials that exhibit superconductivity also have extreme magnetic properties. They exhibit a behavior likened to "perfect diamagnetism" and completely exclude or nullify any magnetic fields in their interior; the superconducting state thus produces a kind of "anti-magnet". For this reason, an external magnetic field **B** will repel a superconductor with significant force.

The capabilities of electromagnetic devices like the inductor typically depend on the characteristics of the overall field **B**. An applied field with magnitude B_0 is modified by the presence of materials with significant **magnetization** M in the vicinity. The magnetization M is the magnitude of the net moment m in the material per unit volume:

$$M = \frac{m}{V} \tag{T3.16}$$

The total magnetic field has magnitude

$$B = B_0 + \mu_0 M \tag{T3.17}$$

Equation T3.17 has the advantage of separating the contributions of the applied field from those of the material in an expression of overall magnetic field intensity. This equation is sometimes written in terms of the parameter $H = B_0/\mu_0$, called the **magnetic field strength**:

$$H = \frac{B_0}{\mu_0} = \frac{B}{\mu_0} - M \tag{T3.18}$$

Note that H and M have the same units: $[H] = [M] = \text{A/m}$. For example, from **Equation T3.14**, we obtain

$$U_L = \mu_0^2 \left(\frac{B}{\mu_0}\right)^2 \left(\frac{\ell A}{2\mu}\right) = \mu_0 \frac{(H+M)^2 \ell A}{2\mu_r} \qquad \text{(T3.14')}$$

in which the applied field H and the material contribution M appear as commensurable quantities.

Example T3.4:

Recall that the magnetic field produced by a solenoid coil is

$$H = \frac{N}{\ell} i$$

where N is the number of turns in the solenoid coil, ℓ is the solenoid coil's length, and i is the current. Suppose that, in a given device, $N = 27$, $\ell = 20.$ cm, and the maximum current the coil can sustain is $i_{max} \approx 3$ A. If the coil has a manganese-zinc-ferrite core with $\mu_r = 6\underline{0}0$, what is the largest flux density it can produce?

SOLUTION

Since $B = \mu H = \mu_r \mu_0 H$ is the resulting field's flux density, we obtain

$$B_{max} = \mu_r \mu_0 H_{max} = \mu_r \mu_0 \left(\frac{N}{\ell}\right) i_{max} \approx 6\underline{0}0 \left(4\pi \times 10^{-7}\ \frac{\text{H}}{\text{m}}\right)\left(\frac{27}{0.20\ \text{m}}\right)(3\ \text{A}) = \underline{\mathbf{0.3\ T}}$$

Another important consequence of the *H-M* interaction is the ability of an applied field to change the magnetization of a ferromagnet. This effect is shown in **Figure T3.6**. The ability of a ferromagnetic material to exert a permanent field depends on whether the discrete magnetic regions (or "domains") within the material are aligned or not. When the domains have random orientations, as would be the case of a ferromagnet cooled below its "Curie temperature" T_C[1] in the absence of an applied field, the net magnetization of the material is $M = 0$. As shown in **Figure T3.6(a)**, when a field is applied to the non-magnetized material, the material's magnetization changes to match the intensity and direction of the field. A large H produces a large M in the same orientation as magnetic domains fix their alignment with H. Reversing the direction of H drives domains into the opposite orientation, but this does not happen uniformly; the distribution of domains becomes fractured. In this state, the domains with opposing orientations cancel each other out, and the net magnetization returns to zero. Increasing the field intensity drives the domains toward the fully aligned (but oppositely oriented) state. This "history-dependent" microscopic magnetization dynamic produces "hysteresis" in the *H-M* curve, as shown in **Figure T3.6(b)**. The overall size of the hysteresis loop indicates how easy it is (as an energy cost) to change the magnetization of a material. Materials with small loops are called "soft" ferromagnets, and materials with large loops are called "hard" ferromagnets.

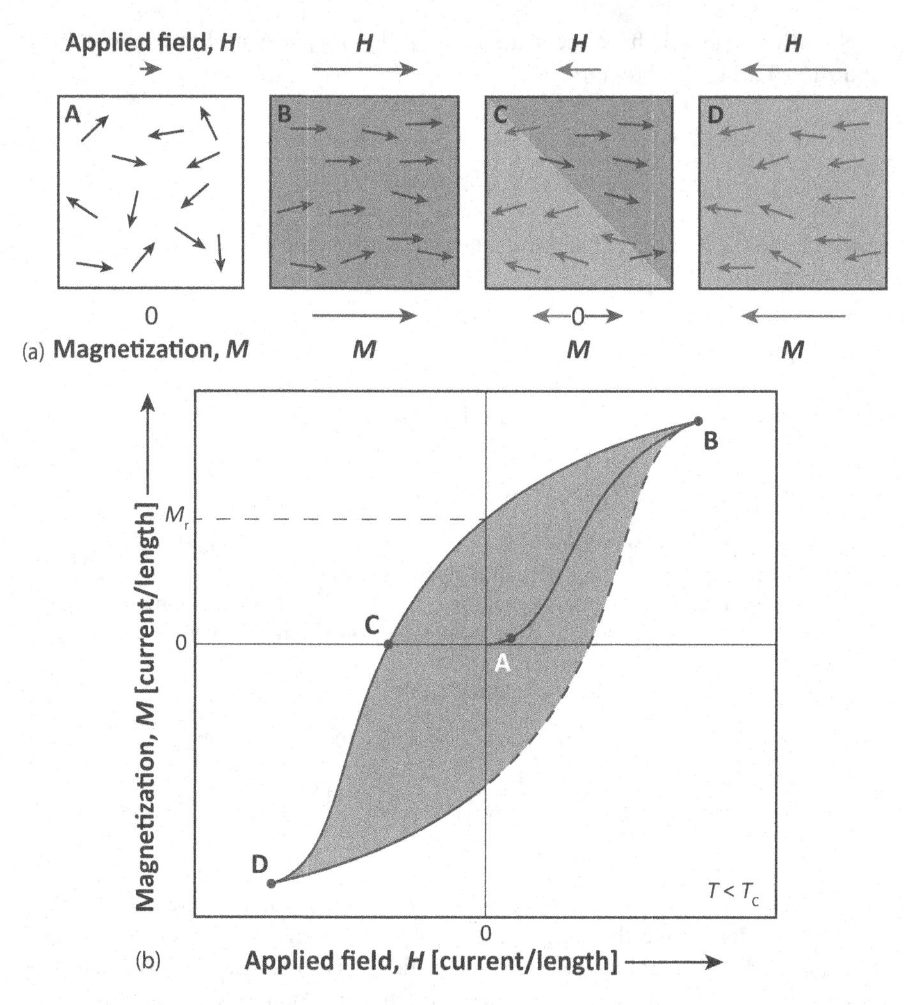

FIGURE T3.6 Magnetization M of a ferromagnetic material in the presence of an applied field. (a) The external field with intensity H drives the alignment of moment-carrying regions of the material. The magnetization grows to reflect the intensity of the field. When the direction of the field is switched, magnetization is driven toward the new orientation. (b) The history dependence of the magnetization means that the magnetization doesn't switch with the applied field in a perfectly reversible way, leading to hysteresis. M_r is the "remnant" magnetization retained by the material when $H = 0$.

CASE STUDY – MATERIAL SELECTION FOR A HIGH-FIELD MAGNET WINDINGS[2]

Generating large magnetic fields (>20 T) is not easy, and large fields are necessary for applications like magnetic resonance imaging (MRI) and precision physics experiments. Even the most powerful permanent magnets can only achieve a field intensity B ~1.3 T. The only workable solution to the problem of creating larger fields seems to be the production of larger currents in solenoids, but producing such currents is associated with another set of materials limitations.

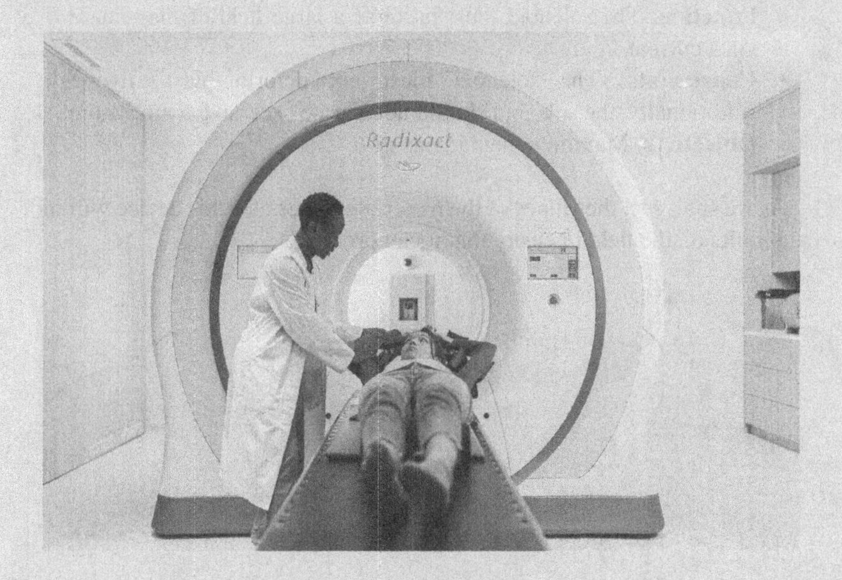

The diagram below shows a solenoid magnet whose field is produced by a current that flows through the (empty) coil. The field intensity B produced by this device is

$$B = \mu_0 \left(\frac{Ni}{\ell} \right) \lambda f(\ell, R, d)$$

The parameter λ captures how efficiently the wire is packed into the coil:

$$\lambda = \frac{\text{Area of wire in cross section}}{\text{Area of cross section}} = \frac{NAw}{\ell d}$$

where A_w is the cross-sectional area of the wire. The function f indicates the dependence of the magnet's field intensity on its geometry (viz., ℓ, R, and d). There are several limitations on the magnetic-field–producing device that require analysis from a materials-design perspective:

1. *Yielding* of the coil material produces permanent distortion in the magnet shape and negatively influences its performance.
2. *Overheating* of the coil by resistive dissipation of electrical power can produce unwanted changes (e.g., oxidation, annealing) to the wire and damage the device and its immediate environment (e.g., fire).

A good design will minimize these effects, leading to a device that is more reliable, has a longer lifespan, and is safer for its operators and any associated equipment. To this end, our design must satisfy the following:

- **Function.** The solenoid must produce a large field of magnitude B via a current i.
- **Constraints.** The solenoid must not deform plastically/yield. Additionally, the solenoid should not accrue excessive temperature.
- **Objectives.** Maximize field intensity B.

As we shall see, the values of the materials indices for this device will indicate limits on the field intensity that it can produce.

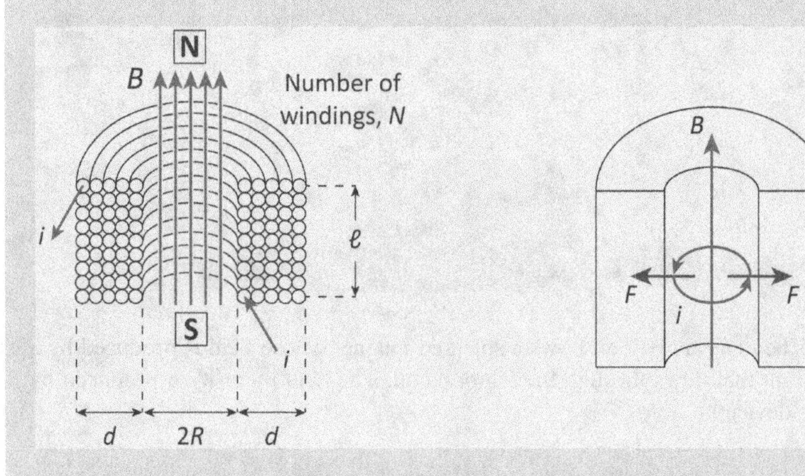

The magnetic field produced by the solenoid exerts a force F on the material in the coil. This force is not transferred to the coil from an external applicator (like in the case of a typical loaded structure) but rather exists throughout the material immersed in the field (i.e., it is a "body force"). F is perpendicular to the direction of current flow *and* the field orientation. This means the force will be transverse to the coil axis and acting outward. This force is equivalent to a force per unit area p that tends to expand the coil diameter like an internal pressure (see the design example of **Topic 1**). We can determine

$$p = \frac{B^2}{2\mu_0 f}$$

and the corresponding stress in the coil would be

$$\sigma = \frac{pR}{d} = \frac{B^2 R}{2\mu_0 f d}$$

Since the solenoid should not operate at a field intensity B_1 that produces a stress that exceeds the yield strength σ_y of the coil material ($\sigma < \sigma_y$), we have

$$B < B_1 = \sqrt{\frac{2\mu_0 fd}{R}} \times \sqrt{\sigma_y}$$

We, therefore, identify one material index here as

$$h_1 = \sqrt{\sigma_y} \quad \text{or} \quad h_1 = \sigma_y$$

The other material index is related to the electrical resistivity ρ_e of the coil. The energy dissipated as a power P via the resistance is

$$P = i^2 R_e$$

where R_e is the electrical resistance and i is the current. During a given field production run lasting t seconds, the energy U dissipated (assuming constant current and resistance) is

$$U = \int P(t)\,dt = \int [i(t)]^2 R_e\,dt \approx i^2 R_e t = \frac{i^2 \rho_e \ell_w t}{A_w} = \frac{i^2 \rho_e V_w t}{A_w^2}$$

where ℓ_w is the length of the wire, and $V_w = \ell_w A_w$ is the volume of wire. If *all* of the energy U is dissipated as heat, then the change in temperature for wire mass m_w is

$$\Delta T = \frac{U}{m_w c_p}$$

where c_p is the specific heat capacity of the material and m_w is the mass of the wire. Now, for a wire with mass density $\rho_w = m_w/V_w$, we get

$$\Delta T = \frac{U}{m_w C_p} \approx \frac{i^2 \rho_e V_w t}{m_w C_p A_w^2} = \frac{i^2 \rho_e t}{\rho_w C_p A_w^2}$$

And since $i = B\ell/\mu_0 N\lambda f$, we have

$$\Delta T = \frac{\rho_e \ell^2 t}{\mu_0^2 N^2 \rho_w C_p A_w^2 \lambda^2 f^2} B^2$$

or

$$B_2 = \frac{\mu_0 N d \lambda^2 f \Delta T}{t} \times \sqrt{\frac{\rho_w C_p}{\rho_e}}$$

This means that, in order to limit the temperature change ΔT to a manageable level, we must have $B < B_2$. The material index h_2 for B_2 is recognizable as

$$h_2 = \sqrt{\frac{\rho_w Cp}{\rho_e}} \quad \text{or} \quad h_2 = \frac{\rho_w Cp}{\rho_e}$$

With these material indices in hand, we can assess what materials are most suitable. Consider the schematic plot of the maximum field magnitude possible B at a given run length t. The "crossover time" t_c indicates the value of t at which the field limit changes from being governed by h_1 to being governed by h_2. That is, at low $t < t_c$, the field is limited by the strength (σ_y), and at high $t > t_c$ the field is limited by the electrical and thermal properties (ρ_w, C_p, ρ_e). Materials that are not suitable for short runs might be suitable for long runs because the indices are mostly independent.

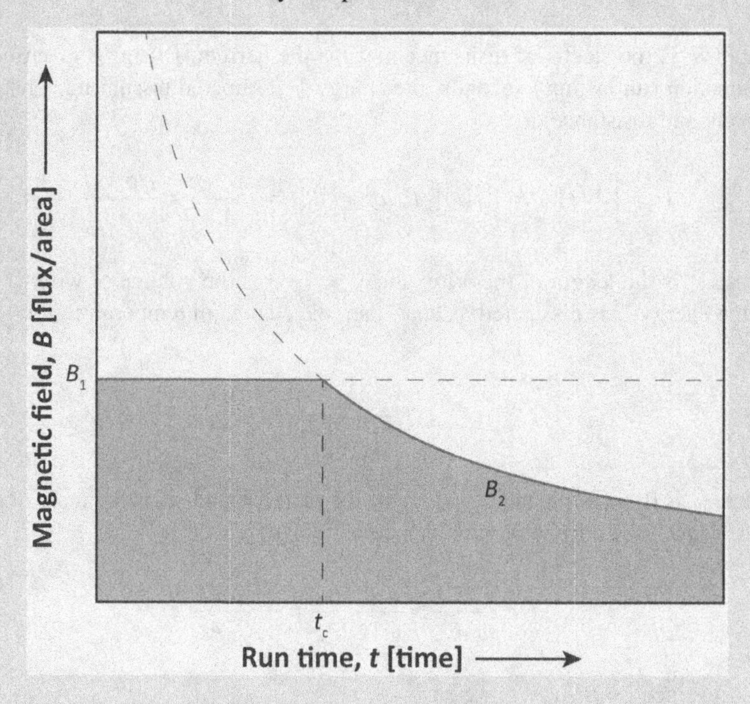

Material	h_1 [MPa]	h_2 [10^6 J/Ω m^4 K]	Notes
High-conductivity Cu	75	2.0	Fairly common; for low-*B*, high-*t* applications
Cu-Nb alloys	780	1.4	High-*B*, high-*t* applications
Cu-Al_2O_3 composite	500	0.95	High-*B*, high-*t* applications
Ag	250	1.5	Expensive; for low-*B*, high-*t* applications
Low-alloy steel	1000	0.14	High-*B*, low-*t* applications

The data in the selection table are ranked by their index h_1. The order changes very little if we instead rank by h_2, so we will treat h_2 as supplemental information that influences the decision very little. With this ranked shortlist in hand, let's consider the candidates individually.

- **Copper.** Copper can be obtained and processed at a relatively low cost to produce highly conductive materials. The low yield strength makes it unsuitable for small-*t* generators that produce intense fields.
- **Cu-Nb alloys.** These materials are much improved compared to pure Cu in terms of yield strength by virtue of their alloy structure. Conductivity suffers slightly, but the material is still well-balanced.
- **Cu-Al_2O_3 composite.** This material has a radically different set of properties than pure Cu, and so it is suitable for different generators. Its strength meant that overall larger fields are possible, but its lack of conductivity indicates poor suitability for long-run fields.
- **Ag.** Silver has superlative conductivity but relatively low strength. It is more expensive than Cu by several times.
- **Low-alloy steel.** This type of high-strength selection is suitable for the production of intense fields that only last a short time.

The picture of magnet design we developed here is considerably simplified. Modern devices utilize sophisticated geometries of nested magnets (sometimes superconducting) to produce different constraints. However, the general design principle illustrated here – one that counterposes the magnet's strength and heat buildup – is conceptually valid.

T3.4 SEMICONDUCTOR DEVICES

Some examples of "active" electronic devices are provided in **Table T3.3**. These devices are capable of conditioning current flows in drastic ways that "passive" components typically cannot. The most important examples are the *pn*-junction **diode** and the bipolar-junction **transistor**. These components are important in power electronics, signal conditioning, digital logic, and many other applications. These devices are constructed from semiconductor materials that have been prepared in a way that modifies their band structure. **Figure T3.7** illustrates the doping of a semiconductor material. At a typical location in a Si lattice, a Si atom is bonded to four of its neighbors via covalent bonds. This utilizes all four of the Si atom's valence electrons. These bonds are stable and do not contribute any free charges that can carry a current (see **Figure T3.4**). Left this way, the Si crystal is an "intrinsic" semiconductor that can only conduct if electrons can be promoted to the conduction band across the bandgap.

If, however, substitutional lattice defects are introduced that disrupt the intrinsic semiconductor's bonding, then a number of charge carriers become available for conduction. The substitution of an atom with valence three will form bonds with three

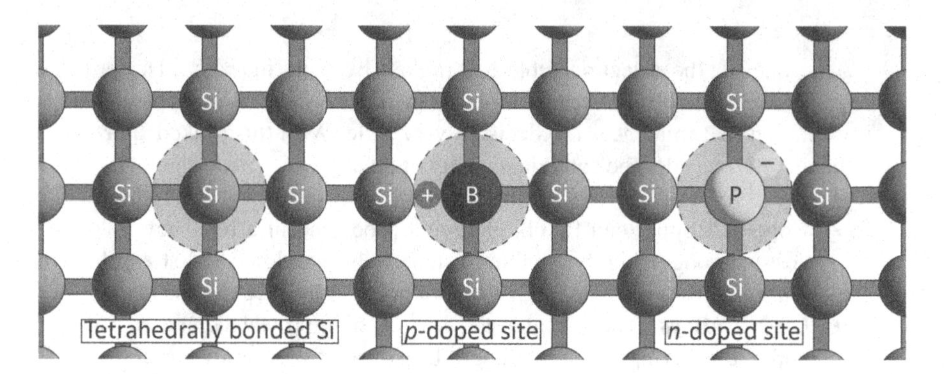

FIGURE T3.7 Types of doping. Normally, Si bonds with four of its neighbors. This configuration is electrically static since all the available electrons are involved in bonding. In *p*-doped Si, the addition of a Group-3 element like B produces a lattice defect associated with "half" of a bond. This electron-hole defect is positively charged compared to a neutral tetrahedrally bonded Si site. Similarly, inserting an element into the lattice with five valence electrons (i.e., Group 5) produces a defect of one unbonded electron.

of its four Si neighbors. This "dangling bond" defect introduces an "acceptor level" into the bandgap of the material. This new state can accept an electron from the filled valence band, leaving behind a positively charged **electron hole** that can carry charge through the crystal in the valence band. A semiconductor that contains these hole-inducing defects is said to be *p* doped. If you introduce an atom of an element with a valence of five rather than three, you produce *n*-type doping. This replacement produces an excess of negatively charged electrons in a "donor level" just below the conduction band. With a little additional energy, this electron can move through the conduction band and pass a current. *p*-type and *n*-type doped semiconductors are called "extrinsic" semiconductors. **Figure T3.8** depicts the modified bandgap diagrams for extrinsic semiconductors.

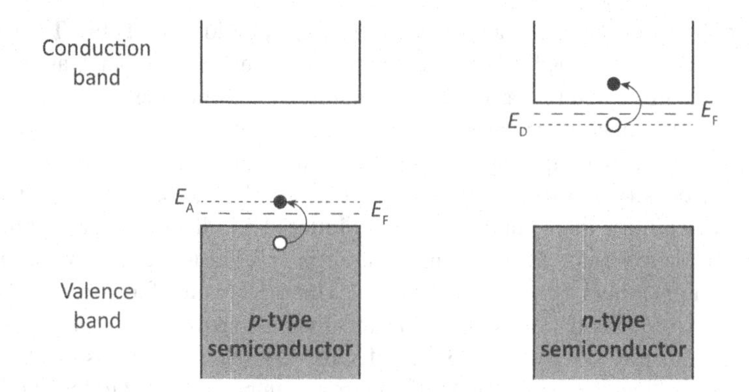

FIGURE T3.8 Bandgap diagrams of doped semiconductors. *p*-type doping produces an acceptor level just above the valence band whose states carriers can occupy. Occupancy of these new states leaves behind conductive holes in the valence band. In *n*-type materials, electrons in the donor level can move into the conduction band with just a little additional energy.

The behavior of a *pn*-junction diode is depicted in **Figure T3.9**. In **Figure T3.9(a)**, an adjustable, reversible power supply is connected across the terminals of the device, along with a load resistance, voltmeter, and ammeter. Though a diode has two terminals like a resistor, it is unlike a resistor in that it is a polar device. One side of the diode is the anode, and the other is the cathode. The operation of the device depends on the *sense* of the voltage applied across it, as indicated by the *v*-*i* curve. If

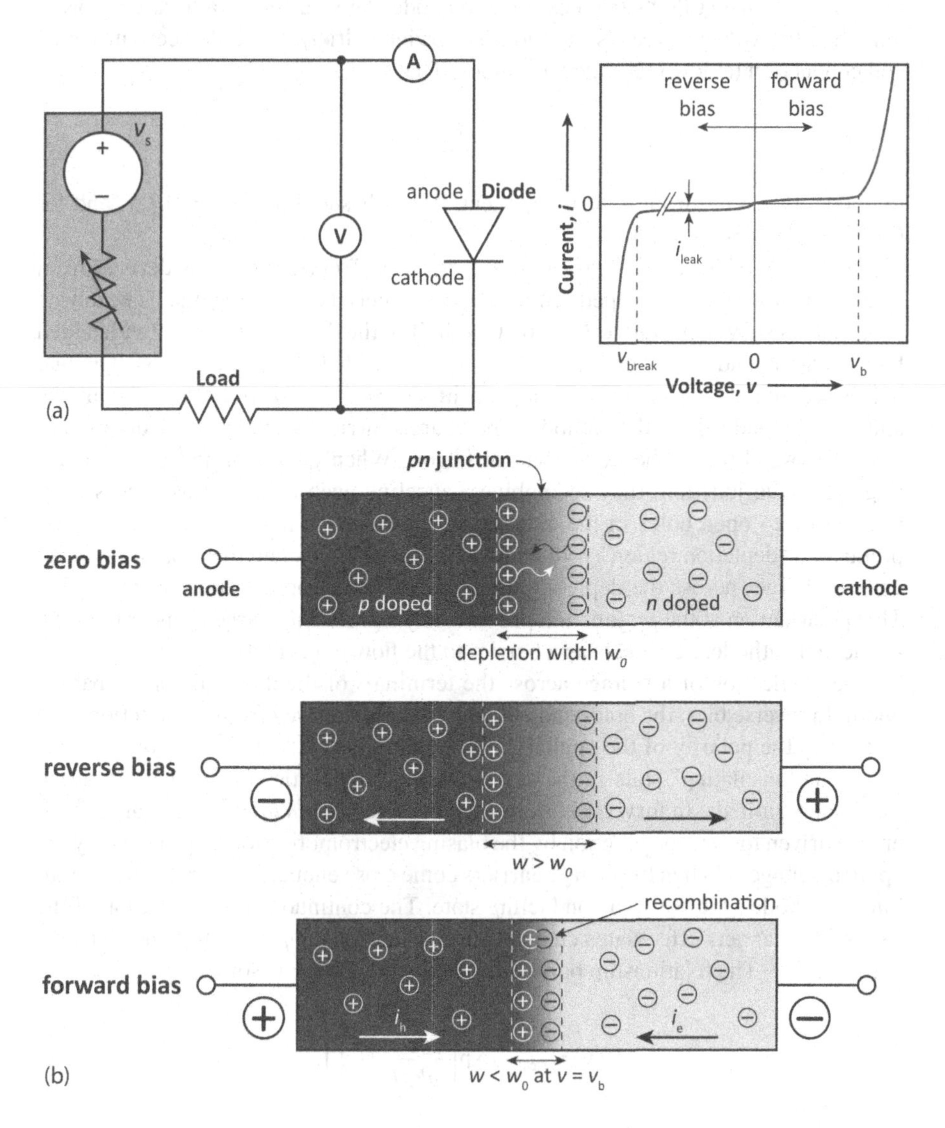

FIGURE T3.9 Operating principles of a *pn*-junction diode. In zero bias, electron holes from the *p*-doped side and electrons from the *n*-doped side are separated by a potential barrier of width w_0 at the junction. In reverse bias, the barrier width increases: $w > w_0$. In these configurations, little current flow is possible. In forward bias, the barrier width decreases so that electrons and electron holes can overlap, producing a current.

the voltage across the device is $v = 0$, there is no net current passing through it; this is a condition of "zero bias". If the voltage is higher at the cathode and lower at the anode, i.e., $v < 0 < v_{\text{break}}$, the device is in "reverse bias". In this state, the current i_{leak} through the device (anode to cathode) is small and typically negligible. (If $v < v_{\text{break}}$, then the device "breaks down" and becomes fully conductive. This is *typically* undesirable from a device-function standpoint.) When the device is placed in "forward bias", $v > 0$, current flows from cathode to anode. This current is a trickle at low v, but when the voltage exceeds the diode's **barrier voltage** v_b the device "turns on" and conduction (at low resistance) develops. That is,

$$v > v_b$$

to obtain a current through the component. For Si-based diodes, $v_b \approx 0.7$ V, and for Ge-based devices, $v_b \approx 0.3$ V.

The ability of the diode to operate as a "one-way" current valve is derived from its construction from two doped semiconductor materials joined together. The device schematic is shown in **Figure T3.9(b)**. One-half of the device is p doped (with added hole carriers), and the other half is n doped (with added electron carriers). The two halves are joined, coming together at a ***pn* junction**. The p-doped side is the anode, and the n-doped side is the cathode. The charge carriers introduced by doping are free to move through the device via diffusion. When holes (+) and electrons (–) migrate to the junction, they recombine, canceling each other out as excess electrons fall into open holes in the electron structure of the material. This migration produces a "depletion region" with width w_0 at the junction, and the separation of (+) and (–) charges across the depletion region gives rise to polarization in the device. This polarization at the *pn* junction prevents any significant charge transfer between anode and cathode, i.e., it acts as a barrier to the flow of current.

The application of a voltage across the terminals of the diode alters the barrier width. In reverse bias, the holes and electrons are pulled away from the junction even further by the polarity of the applied field, *expanding* the depletion region and producing an "insulating" state in the device. In this state, little current flows between the diode terminals. In forward bias, the depletion region *shrinks* as the charge carriers are driven toward the junction by the biasing electromotive forces provided by the applied voltage v. When the charge carriers come close enough to overlap and recombine, the diode switches to a conducting state. The continuous recombination of the (+) and (–) carriers effectuates charge transport across the *pn* junction and thus the entire device. The relationship between current and voltage is sometimes taken as

$$i(v) = i_{\text{leak}}\left[\exp\left(\frac{q_e v}{n k_B T}\right) - 1\right] \tag{T3.19}$$

where $n = 2$ for Si diodes and $n = 1$ for Ge diodes. This valving capability is sometimes called "rectification" when the voltage applied across the diode is an AC voltage that switches polarity; the diode "filters out" the negative-voltage portions of the applied voltage and passes the positive ones.

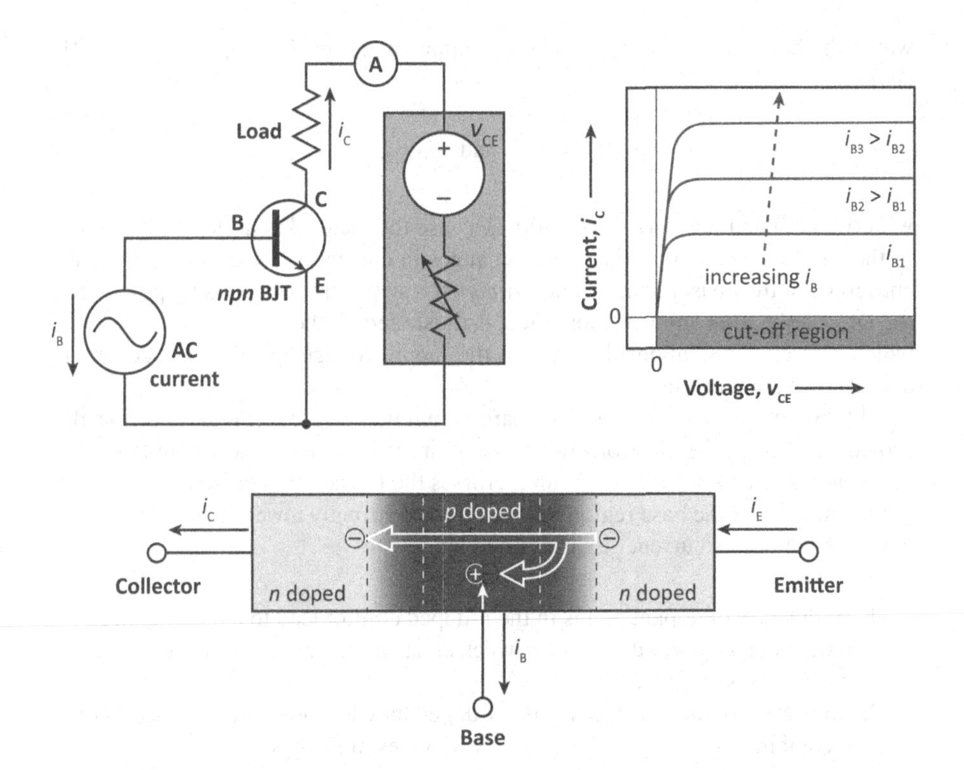

FIGURE T3.10 Operation of bipolar-junction transistor. The power delivered to the load comes from the current between the emitter (E) and the collector (C) of the BJT. This current is controlled or "gated" by the AC current supplied to the base (B) of the device. Small changes in i_B can produce large changes in (or even cessation of) the current $i_C = \beta i_B$. The internal layout of the BJT device resembles that of two *pn*-junction diodes placed "back-to-back". Charges injected into the *p*-doped base region are strongly driven toward the collector in the presence of a biasing current at the base.

A transistor is another active semiconductor-based component that can control the flow of current via an adjustable "valve". The operation of a **bipolar-junction transistor** (BJT) is shown in **Figure T3.10**. The component has three terminals, named the "collector" (C), the "emitter" (E), and the "base" (B). These three terminals are separated by *pn* junctions between the base and the emitter (BE) and between the base and the collector (BC). Current flows from the emitter to the collector when the device is biased by setting the voltages

$$v_E + v_b < v_B < v_C$$

where v_b is the BE junction's barrier voltage in forward bias. (The BC junction is in reverse bias.) In the BJT, the incoming/outgoing currents obey

$$i_C = i_E - i_B \tag{T3.20}$$

where the base current is the minority component: $|i_B| \ll |i_E| < |i_C|$. In typical BJT devices

$$i_C = \alpha i_E \quad \text{and} \quad i_C = \beta i_B \tag{T3.21}$$

with $\alpha \approx 0.99$ and $\beta = \alpha/(1 - \alpha) \approx 100$. Because the factor β is large, small changes in the base current can produce large changes in current at the collector (or halt it entirely). For this reason, one is able to *amplify* a signal at the base using power from the DC supply. This amplification effect is illustrated by the "characteristic curves", which indicate the relationship between the supply voltage v_{CE}, the base current i_B, and the collector current i_C.

The two *pn*-junction diodes that share a common anode at the base of the BJT provide a pathway for electrons to be driven at high voltages toward the load. Any electron that emerges from the emitter, crosses the BE junction barrier, and diffuses to the midpoint of the base region will be directed strongly toward the collector. The development of this current flux is assisted by

1. High levels of dopant atoms in the *n*-doped emitter lead to a larger concentration of charges at the emitter, which enhances the net diffusion across the *p*-doped base region.
2. Low levels of dopant atoms in the *p*-doped base lead to overall low levels of recombination of emitted electrons with holes in the base.
3. A very thin *p*-doped base region provides increased odds of electrons diffusing across the midpoint of the base.

The 95%+ of emitted electrons that are swept across the base and that reach the collector terminal determine the parameter $\alpha = i_C/i_E$. This overall level of current i_C is then adjusted by adjusting the base current i_B (which increases or decreases the BC bias). Since the voltage gain A_v in a typical device is

$$A_v = \frac{v_{CE} - v_C}{v_B - v_E} \tag{T3.22}$$

and the current gain A_i in the device is

$$A_i = \beta \tag{T3.23}$$

The power gain A_P in a typical transistor amplifier configuration is

$$A_P = A_v A_i = \beta \frac{v_{CE} - v_C}{v_B - v_E} \tag{T3.24}$$

Thus, small signals applied at the base can be made large with excellent fidelity up to the intrinsic limits of the BJT.

APPLICATION NOTE – BIPOLAR SWITCHES AND DIGITAL LOGIC

The description of (*npn*) BJTs used as amplifiers required that

$$v_E + v_b < v_B < v_C$$

in order for the device to maintain its bias so current can flow from emitter to collector. If conditions should ever change so that

$$v_B < v_E + v_b$$

the BE-junction side of the transistor will fall out of bias and the current in the transistor will be zero. That is, control of the voltage at the base permits the device to be activated and deactivated completely. In this way, the BJT can be operated as a "switch" that is opened (or closed) by applying base voltages that are less than (greater than) the emitter voltage. The schematic picture is provided below. The voltage applied at the base might vary with time, changing between two levels of ON and OFF that correspond to levels that bring the device into and out of bias. When v_B = ON, $i_C > 0$, and power is delivered to the load. When v_B = OFF, the load is de-energized. Since BJTs are electrical rather than mechanical devices, transistor switches are more reliable, accurate, and rapid than other switches (like relays). They can also be operated at low power since the base draws little current.

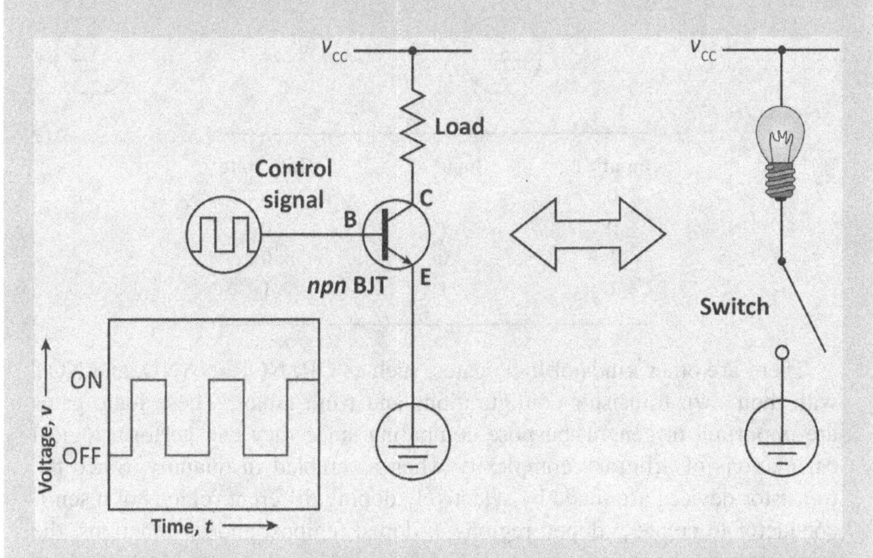

In the example above, a single control signal is used to select the ON/OFF state of the load. In the more sophisticated arrangement shown below, the load is controlled using *two* inputs. At any given time, either of the BJTs may be in or out of bias. We associate the states of the individual BJTs with the logical values 0 = OFF and 1 = ON and call this multiple-input device a **logic gate**. The logic gate shown is called an AND gate since we must have Input #1 = 1 AND Input #2 = 1 (i.e., *both* transistors ON and *both* switches closed) in order for current to flow to the load. If either switch is open, the circuit is not complete. There are four possible states of the BJT switches, and these are collected in the table below, along with the logical state (0 or 1) of the load. (This kind of table is sometimes called the "truth table" for the logic gate and lists every possible state.)

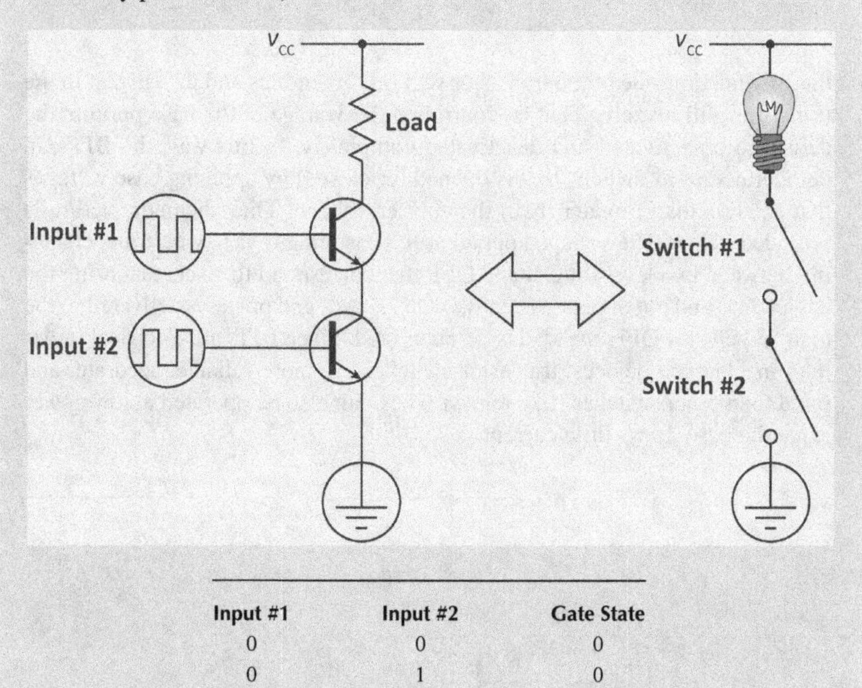

Input #1	Input #2	Gate State
0	0	0
0	1	0
1	0	0
1	1	1

There are other kinds of logic gates, such as OR, NOR, NAND, and NOT with their own transistor configurations and truth tables. These logic gates are important in general-purpose computing since they can perform logical calculations of arbitrary complexity when assembled in quantity. Since the transistor devices are made by selectively doping different regions of a semiconductor to create *p*-doped regions, *n*-doped regions, and *pn* junctions, the logic gates can be made very small and integrated on the same block of material. Sophisticated microprocessors can integrate some billion individual transistors on a compact chip.

T3.5 CLOSING

The electrical and magnetic properties of materials are determined by the collective properties of the material's electrons. Wholesale motion of electrons produces conduction; magnetic moments attached to the electrons contribute to a net magnetic field. The capacity of materials to sustain currents, polarize without passing currents, and magnetize gives rise to devices such as conductors, resistors, capacitors, inductors, and some sensors like thermocouples. These "passive" components play important roles in electrical and electronic systems. Semiconductor devices, such as diodes and transistors, play "active" roles in such systems, and the modern forms of these devices are produced from doped semiconductors brought together to form junctions. Such active components have applications in digital devices and microprocessors.

T3.6 CHAPTER SUMMARY

Key Terms

bandgap
barrier voltage
bipolar-junction transistor
capacitance
capacitor
conduction band
conductivity
current
diamagnetism
dielectric strength
diode
electric permittivity
electromagnetic
electron band

electron hole
electronics
Fermi level
ferromagnetism
inductance
inductor
logic gate
magnetic field strength
magnetic flux density
magnetization
Ohm's law
paramagnetism
***pn* junction**
permeability
resistivity
resistor
superconductance
transistor
valence band
voltage

Important Relationships

$$v = iR \qquad \text{(Ohm's law)}$$

$$R = \rho\frac{\ell}{A} \qquad \text{(resistance)}$$

$$\sigma = \frac{1}{\rho} \qquad \text{(conductivity)}$$

$$E(n_x, n_y, n_z) = \frac{1}{2m_e}\left(\frac{h}{2\ell}\right)^2\left(n_x^2 + n_y^2 + n_z^2\right) \qquad \text{(box energy levels)}$$

$$F(E) = \frac{1}{1 + \exp\left(\frac{E - E_F}{k_B T}\right)} \qquad \text{(Fermi function)}$$

$$J = \sigma|\mathbf{E}| \quad \text{and} \quad \sigma = \mu N e \qquad \text{(electron flux)}$$

$$\rho(T) = \rho 0\left[1 + \alpha(T - T_0)\right] \qquad \text{(resistivity)}$$

$$U_C = \frac{1}{2}CV^2 = \epsilon\left(\frac{A}{2d}\right)v^2 \qquad \text{(capacitor energy)}$$

$$U_L = \frac{1}{2}Li^2 = \mu\left(\frac{N^2 A}{2\ell}\right)i^2 = \frac{B^2 \ell A}{2\mu} \qquad \text{(inductor energy)}$$

$$B = B_0 + \mu_0 M \qquad \text{(flux density w/material)}$$

$$H = \frac{B}{\mu_0} - M \qquad \text{(field strength)}$$

$$U_L = \mu_0\frac{(H + M)^2 \ell A}{2\mu_r} \qquad \text{(inductor energy)}$$

$$i(v) = i_{\text{leak}}\left[\exp\left(\frac{q_e v}{n k_B T}\right) - 1\right] \qquad \text{(diode equation)}$$

$$i_C = \alpha i_E \quad \text{and} \quad i_C = \beta i_B \qquad \text{(transistor relations)}$$

$$A_v = \frac{v_{CE} - v_C}{v_B - v_E} \qquad \text{(voltage amplification)}$$

$$A_i = \beta \qquad \text{(current amplification)}$$

$$A_P = \beta \frac{v_{CE} - v_C}{v_B - v_E} \qquad \text{(power amplification)}$$

T3.7 QUESTIONS AND EXERCISES

Concept Review

CT3.1 What thermocouple type would you consider to be best for measuring the temperature of an Al sample from room temperature up to its melting point? Justify your choice.

CT3.2 Describe the relationship between the decrease in the size of transistors and the increase in computing power of computer chips.

PROBLEMS

PT3.1 Suppose that you have a piece of cylindrical wire with resistance R_1 and another similar piece of wire that has twice the diameter and is made of a material with 1/3 of the resistivity ρ_e of the first. How much longer must the second wire be so that they have the same resistance ($R_2 = R_1$)?

PT3.2 You want to construct a capacitor that stores a charge of 2.2×10^{-11} C at 12 V. If your opposing plates are 1.0 mm^2 in area and the plate spacing is d = 50. μm, what must the relative permittivity ϵ_r of the dielectric be? How much energy can the capacitor store at 12 V?

PT3.3 Suppose you have an inductor that is a cylindrical coil of wire and requires a magnetic material for the core of the device. What would a material index h look like that reflects the maximum field for the minimum cost?

PT3.4 What is the emitter voltage v_E? What is the collector current i_C? Take the base-emitter barrier voltage as 0.5 V and $\alpha = 0.95$.

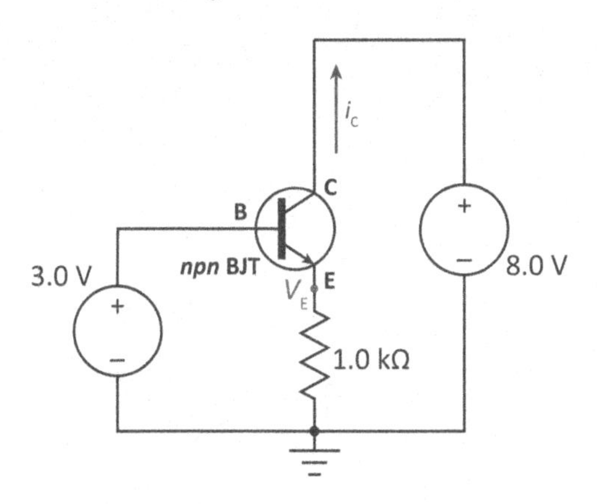

MATLAB Exercises

MT3.1 Write a MATLAB function `fermi(E, T)` that computes the value of the Fermi function at the temperature T and energy E. Take $E_F = 11.5$ eV. If the bandgap is 1.0 eV, what is F at the edge of the conduction band?

MT3.2 Recall the model for the i–v behavior of a diode in reverse and forward bias (minus the breakdown regime):

$$i(v) = i_{\text{leak}}\left[\exp\left(\frac{q_e v}{n k_B T}\right) - 1\right]$$

Plot the $i(v)$ curve for a Si diode at room temperature. Take i_{leak} as 5 μA.

NOTES

1. Above a material's Curie temperature, the magnetization drops to 0.
2. After Michael F. Ashby. *Materials Selection in Mechanical Design*. Fourth edition. Oxford, England, UK: Butterworth-Heinemann, 2011.

Appendix A: Working with Numerical Data Sets in MATLAB

Essential Data-analysis Techniques

A.1 IMPORTING DATA SETS

Data sets typically come in the form of *tables* of values (typically text or numbers) organized in rows and columns. The data files come in many different filetypes: spreadsheets, text-based comma-separated values (".csv"), tab-separated values (".tsv"), or binary files (like the MATLAB® ".mat" file format). Reading the information in the data file and *importing* it into the desired analysis software is a crucial step in working with data. The data from the imported file are parsed and placed into a data structure or table that the analysis software can access, display, and modify using built-in or user-defined methods.

In MATLAB, this process typically involves the `importdata()` function or the `load()` function. The `importdata()` function applies to spreadsheet files, delimited text files, and fixed-width text files, while the `load()` function is for saved MATLAB workspaces (".mat" files). The data are imported into MATLAB in the form of data *structures* whose *fields* are the different data components, e.g., text column headers and arrays of numerical data. Individual fields can be addressed by using a dot(.)-based notation, e.g., `structName.fieldName` gives the contents of the field `fieldName` in the structure `structName`.

Example A.1:

The data file "MATLAB_Import_Example.txt" contains tabular numerical values:

```
Squares and square roots of the
numbers 1-100
n           n^2           n^(1/2)
1             1               1
2             4           1.414213562
3             9           1.732050808
...          ...              ...
```

Import the information in the file by using `importdata()` and display the values. Save the imported data in a ".mat" file.

SOLUTION

The contents of the data file can be stored in a MATLAB variable `myData` by using

```
>> myData = importdata('MATLAB_Import_Example.txt')
myData =
   struct with fields:
          data: [100×3 double]
      textdata: {2×3 cell}
    colheaders: {'n'  'n^2'  'n^(1/2)'}
>>
```

The data are stored in a `struct` with three fields named `data`, `textdata`, and `colheaders`. They can be inspected using the dot-based notation:

```
>> myData.textdata
ans =
  2×3 cell array
    {'Squares and roots of...'} {0×0 double} {0×0 double}
    {'n'} {'n^2'} {'n^(1/2)'}
>> myData.colheaders
          ans =
  1×3 cell array
    {'n'}     {'n^2'}     {'n^(1/2)'}
>> myData.data
          ans =
   1.0e+04 *
    0.0001    0.0001    0.0001
    0.0002    0.0004    0.0001
    0.0003    0.0009    0.0002
    0.0004    0.0016    0.0002
[etc...]
>>
```

(Note that the digits here are truncated in the display *only*; the full values are always present in the `data` array. Use `format long` or a similar command to increase the number of digits displayed.) The numerical values associated with the columns `n`, `n^2`, and `n^(1/2)` are in an array with three columns and 100 rows. The data may be saved in the MATLAB file by using

```
>> save('MATLAB_Import_Example.mat', 'myData')
          >>
```

Note that the MATLAB file preserves the variable name, and this variable name will appear in the MATLAB workspace after loading.

Example A.2:

Import the data in the "MATLAB_Import_Example.mat" file by using the `load()` function.

SOLUTION

The contents of the data file can be stored in a MATLAB variable `myData` by using

```
>> myData = load('MATLAB_Import_Example.mat')
>> myData
myData =
  struct with fields:
          data: [100×3 double]
      textdata: {2×3 cell}
    colheaders: {'n'  'n^2'  'n^(1/2)'}
>>
```

A.2 WORKING WITH DATA ARRAYS

As shown in **Example A.1**, the imported numerical data is an $n \times m$ array. That is, the data has n rows and m columns. We may address any individual value `myVal` in the data field `myData` by taking

```
myVal = myData.data(i, j)
```

where `i` is the row number and `j` is the column number. We may select an entire column `myCol` by taking

```
myCol = myData.data(:, j)
```

where `j` is the column number. Here, the symbol ":" means "all rows", and `myCol` is an $(n \times 1)$ array. Similarly, we may obtain an entire row `myRow` using

```
myRow = myData.data(i, :)
```

where `i` is the row number. (":" in the second position means "all columns".) This is a $(1 \times m)$ array. Multiple columns or rows can be obtained using, for instance

```
mySelection = myData.data(:, [1 3])
```

which produces a $(n \times 2)$ array consisting of columns 1 and 3. Using these array indexing techniques, we may "chop up" the larger array into the subarrays that we require for our analysis. (See **Example A.3** for an application.)

Example A.3:

Plot the data in the `myData.data` array as two series: `n^2 vs. n` and `n^(1/2) vs. n` on the same set of axes.

SOLUTION

The contents of the data file can be stored in a MATLAB variable `myData` by using `load()`. The plot variables `X` (= column 1), `Y` (= column 2), and `Z` (= column 3) can then be set:

```
>> load('MATLAB_Import_Example.mat');
>> X = myData.data(:, 1);
>> Y = myData.data(:, 2);
>> Z = myData.data(:, 3);
>>
```

The plot is then constructed using

```
>> yyaxis left
>> scatter(X, Y)
>> yyaxis right
>> scatter(X, Z)
>>
```

This produces something like

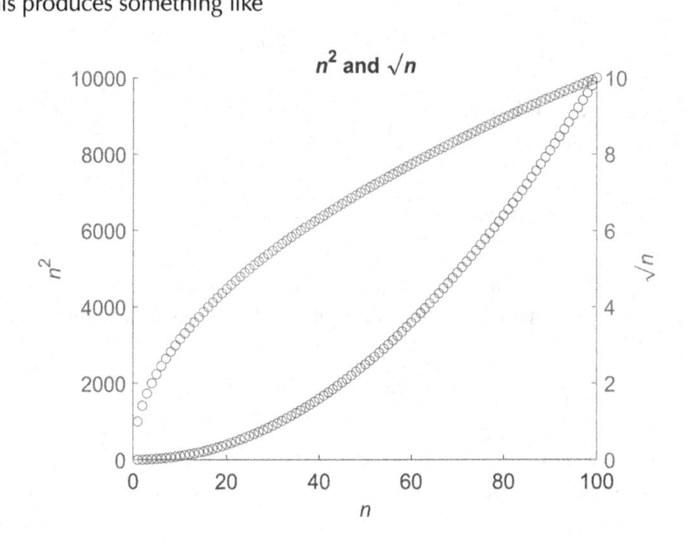

Appendix B: Lattice Positions, Directions, and Planes

Finding Your Way in a Crystal

B.1 CRYSTALLOGRAPHIC COORDINATES, DIRECTIONS, AND PLANES IN COMMON LATTICES

We introduced in **Chapter 1** the idea of lattices and lattice locations, and in **Chapter 3** the idea of crystallographic (or "lattice") coordinates: a position-labeling scheme that is convenient for specifying (and performing calculations with) the geometry of a crystal lattice. Some examples are given in **Figure B.1**. **Figure B.1(a)** shows a unit cell (base-centered monoclinic) with its lattice positions labeled using such crystallographic coordinates. The lattice parameters a, b c, γ, etc. are unspecified in the example but are secondary considerations given the choice of directions and units of the axes. The coordinates within a unit cell have possible values $0 \leq x \leq 1$, $0 \leq y \leq 1$, and $0 \leq z \leq 1$. As we have seen, these coordinates are sufficient for discriminating between all the possible planes within the crystal.

Multiple adjacent unit cells are shown in **Figure B.1(b)** along with several example vectors **u**, **v**, and **w**. Such vectors are called "crystallographic directions" or "lattice directions". As vector in three dimensions, crystallographic directions have three components: $\mathbf{u} = [u_1\ u_2\ u_3]$. Similar to the way we compute the components of a vector in normal Euclidean 3D space, we find the difference between the coordinates in a "tip-minus-tail" fashion. Since the tip of **u** is at 1 0 1, and the tail is at 0 0 0, we obtain

$$\mathbf{u} = [u_1\ u_2\ u_3] = [(1-0)\,(0-0)\,(1-0)] = [1\ 0\ 1]$$

We say that the vector **u** is the "[1 0 1] direction". Like any vector, a crystallographic direction has an orientation/direction and magnitude (though we are not typically interested in the latter), and a given vector may be hung at any location in the lattice without any change to its orientation. Notice how in **Figure B.1(b)** the vector **u** has the same essential properties in both locations.

Similar to the way we found the components of **u**, we can find $\mathbf{v} = [0\ 2\ 1]$ and $\mathbf{w} = [½\ ½\ 1½]$. Conventionally, we only use integers for the crystallographic directions, so we multiply **w** by a scalar constant (= 2) to produce

$$\mathbf{w} = 2 \times \left[\frac{1}{2}\ \frac{1}{2}\ 1\frac{1}{2}\right] = [1\ 1\ 3]$$

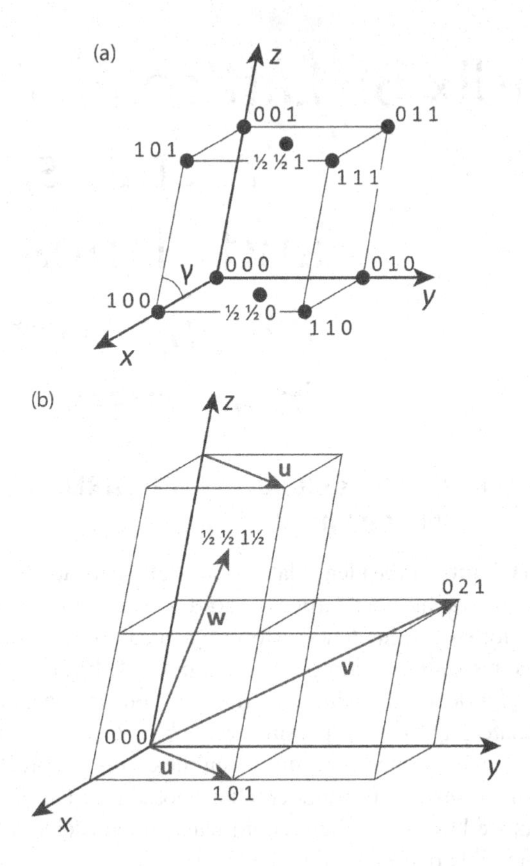

FIGURE B.1 Crystallographic coordinates and directions. (a) All of the crystallographic coordinates of the lattice positions in the base-centered monoclinic lattice. (b) Crystallographic vectors **u**, **v**, and **w** shown in an extended system of unit cells.

We have deliberately chosen the value of the scalar constant (= 2) so that the integers used as the components of **w** are as small as possible. (Recall that multiplication by a scalar changes a vector's magnitude but *not* its orientation.) If a particular component of a vector has a negative value, we use an "overbar" as an indicator. For example, a vector **u** with a component that points in a negative direction **u** = [2 "–1" 1] is written $[2\ \bar{1}\ 1]$.

In the same way that three components determine a vector in a 3D space, three points determine a plane. Recall the rules for identifying the Miller indices of a plane introduced in **Chapter 3**:

1. The points **A**, **B**, and **C** are located where the plane intersects the x, y, and z axes in the unit cell: **A** = x_0 0 0, **B** = 0 y_0 0, and **C** = 0 0 z_0. Where x_0, y_0, and z_0 are the respective intercepts.
2. The indices h, k, ℓ of the plane are the reciprocals of the intercepts x_0, y_0, z_0 and are enclosed in parentheses: $(h\ k\ \ell) = (1/x_0\ \ 1/y_0\ \ 1/z_0)$.
3. Planes that are parallel to (i.e., do not intersect with) an axis are assigned an intercept of "∞".

4. h, k, and ℓ will always be integers. Multiplication of h, k, and ℓ by the same constant c does not change anything about the geometry of the plane.
5. Negative values of h, k, and ℓ may result; those values are identified using an overbar (e.g., "$\bar{2}$" rather than "–2").
6. If the plane passes through the origin 0 0 0 in the standard coordinate system above, shift the coordinate system so that its origin is at a different, more convenient corner.

Some examples are shown in **Figure B.2**. **Figure B.2(a)** shows the (011), (110), and (101) planes and the details of the determination of their indices following the

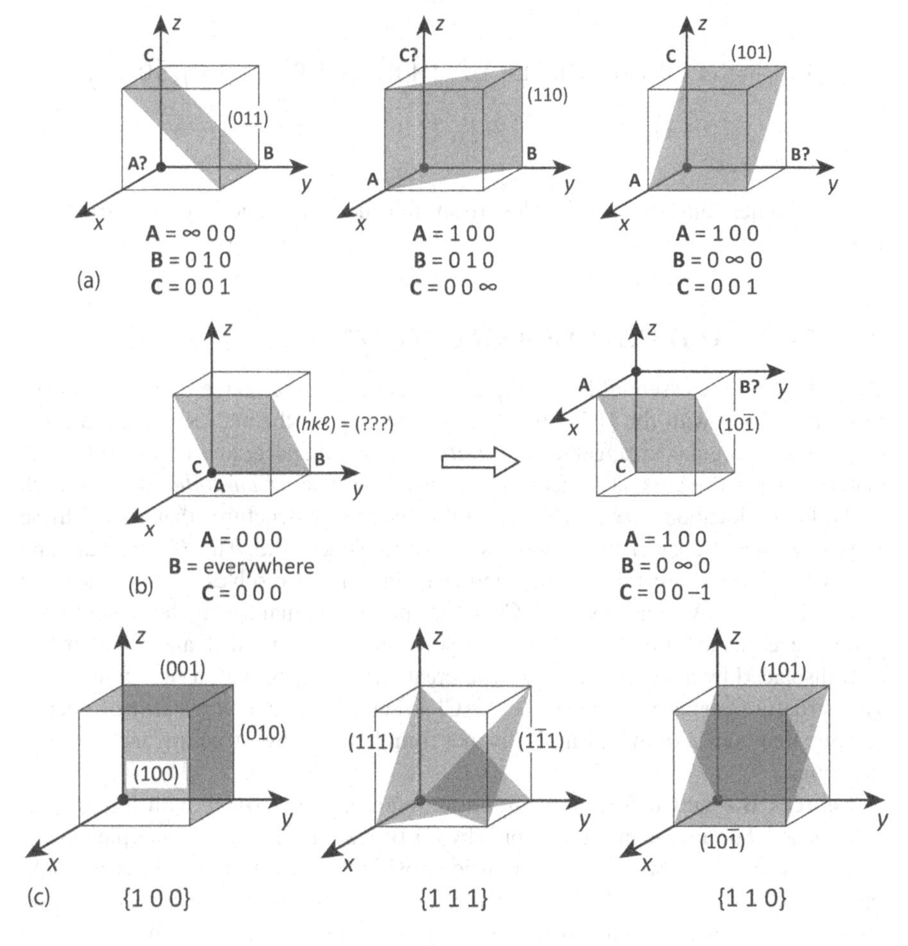

FIGURE B.2 Example calculations of the Miller indices of planes. In (a), identifying the intercepts with the various axes – x_0, y_0, z_0 – is sufficient to determine $h = 1/x_0$, $k = 1/y_0$, and $\ell = 1/z_0$ in each case. In cases where the plane intersects the origin, such as that illustrated in (b), the determination of the Miller indices requires a shift in the origin of the coordinate systems to a different vertex of the unit cell. Some additional examples of planes are provided in (c), grouped by the family that they belong to.

rules above. **Figure B.2(b)** shows a "problematic" case where the rules give nonsensical results for h, k, and ℓ. In this case, the difficulties are resolved by placing the origin from which the intercepts are calculated at a different location. (There is nothing privileged about where the origin is placed in a crystal made of repeated, stacked units.) A shift of the origin to the next lattice point in the $+z$ direction permits the indices to be computed as in the other cases: $(h\ k\ \ell) = (1\ 0\ \bar{1})$. Note that the origin might as easily have been shifted in the $+x$ direction, giving $(h\ k\ \ell) = (\bar{1}\ 0\ 1)$. In this case, both sets of indices are wholly equivalent. In fact, all of the planes illustrated in **Figure B.2** belong to a "family" of (110)-like planes (sometimes abbreviated using curly brackets as the "{110} family"). These planes are all geometrically similar, and the family includes:

$$\{1\,1\,0\} = \{(1\,1\,0),\ (0\,1\,1),\ (1\,0\,1),\ (\bar{1}\,1\,0),\ (1\,\bar{1}\,0),\ (\bar{1}\,\bar{1}\,0),\ (0\,\bar{1}\,1),$$
$$(0\,1\,\bar{1}),\ (0\,\bar{1}\,\bar{1}),\ (\bar{1}\,0\,1),\ (1\,0\,\bar{1}),\ (\bar{1}\,0\,\bar{1})\}$$

Some other important examples from families of planes are illustrated in **Figure B.2(c)**.

B.2 BASES AND EXTENDED STRUCTURES

Table 1.4 lists the essential lattice types – 14 of them – that make up all crystalline materials (along with the HCP structure, see below). Mathematically, these are the only possible regular arrangements of *lattice* points in space, and many crystalline materials have these exact structures. However, by placing *multiple atoms* at each of the lattice locations, we can define unit cells/crystal structures that extend these 14 possibilities. Consider the structures shown in **Figure B.3**. The SC structure and the BCC structure are both lattices and structures in themselves; if you place one identical atom of **A** at each SC or BCC lattice point, you make a crystal. Alternately, if you place one atom or ion of **A** at each SC lattice site along with another atom/ion of **B** displaced by a vector [½ ½ ½], you create a SC lattice with a *two-atom basis*. This structure, sometimes called the "CsCl structure" after the crystalline mineral/ceramic that comes in this form, is distinct from both the SC structure and the BCC structure.

Figures B.4 and **B.5** show some more examples of lattices with a multiple-atom basis. The face-centered–cubic (FCC) Bravais lattice can be supplemented with an atom/ion of **A** + **B** at each lattice position; this can make either a "rocksalt" structure or a "zincblende" structure. The rocksalt structure is associated not just with NaCl but with oxides like MgO and FeO. The zincblende structure is that of the mineral ZnS or the ceramic SiC. When the **A** and **B** atoms are the same type – e.g., as in Si or crystalline C/diamond, the zincblende structure is called "diamond cubic". The cubic "perovskite" structure, "fluorite" structure, and "cristobalite" structure are in **Figure B.5**. Perovskite is a common structure for oxides of type metal **A** + metal **B** + 3O (such as $CaTiO_3$), and the fluorite

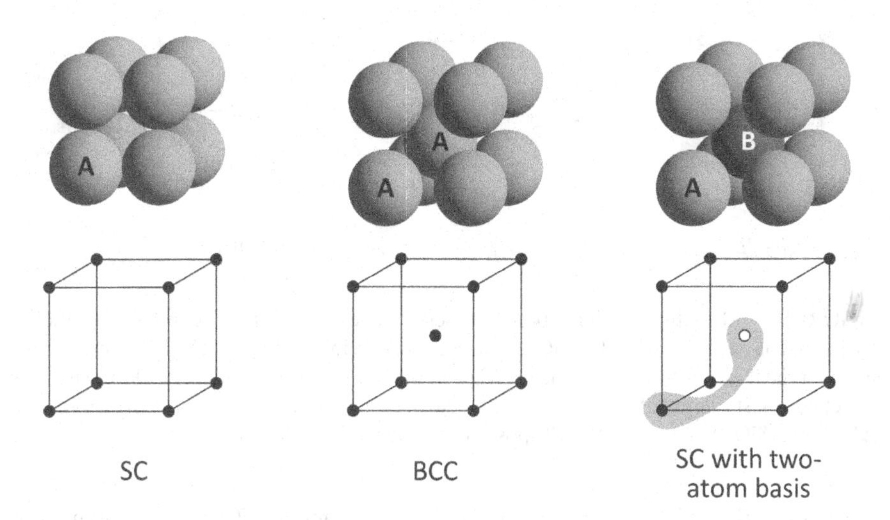

FIGURE B.3 Comparison of three different crystal structures based on two lattices. The SC structure has one atom of **A** in its unit cell, and the BCC structure has two atoms of **A** in its unit cell. The CsCl structure is derived from the SC lattice but possesses one atom of **A** and one atom of **B** in its unit cell. An **A** + **B** basis unit is placed at each corner of the SC lattice to make the CsCl structure.

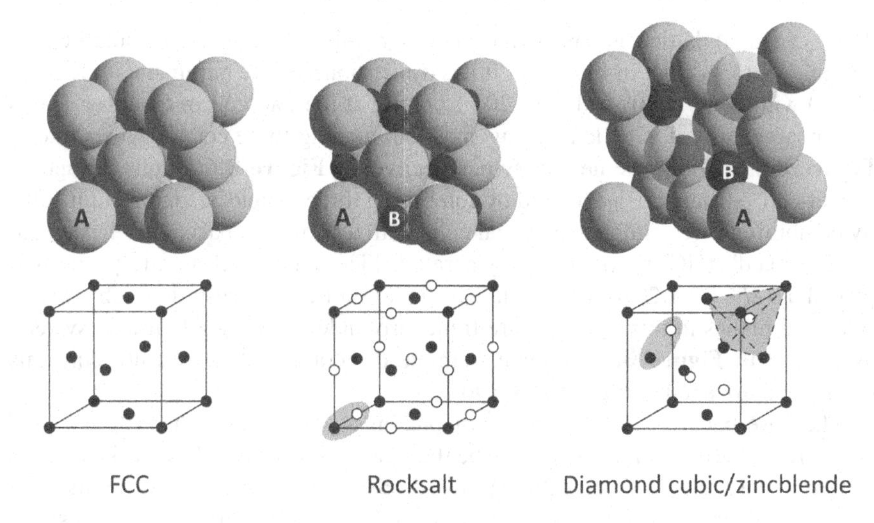

FIGURE B.4 Structures with two-atom bases derived from the FCC lattice. The FCC structure has four atoms of **A** in its unit cell. The rocksalt structure is derived from the FCC structure/lattice but contains four atoms/ions of **A** and four atoms/ions of **B** in its unit cell, as shown. (At each FCC lattice position, place an **A** + **B** basis.) In a diamond cubic/zincblende unit cell, the FCC lattice is supplemented with additional atoms at each of four tetrahedrally coordinated locations in the cell. When the **A** and **B** atoms are the same (e.g., C or Si), the structure is diamond cubic; when the atoms are of different types (e.g., Zn + S or Si + C), the structure is that of zincblende.

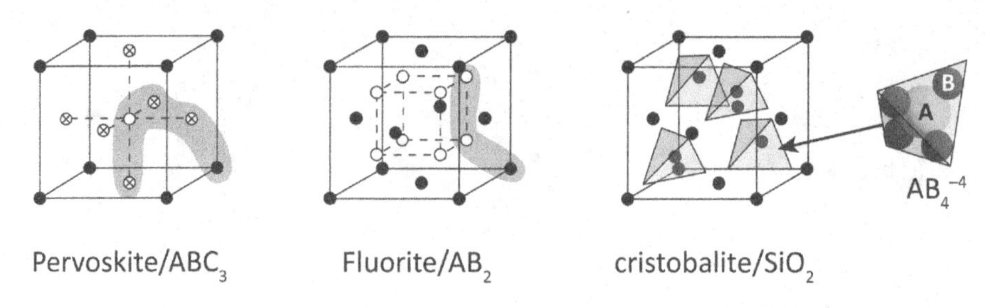

FIGURE B.5 Structures with multi-atom bases. The perovskite structure is based on the SC lattice and has a basis of **A** + **B** + 3**C** at each SC location in its unit cell. The FCC lattice is the basis for the fluorite structure; at each FCC position, there is an **A** + 2**B** basis. The cristobalite structure (most commonly associated with quartz at high temperatures) has two tetrahedral $\mathbf{AB_4}^{-4}$ (i.e., SiO_4^{-4}) ions at each FCC position, so the basis is 2**A** + 8**B**.

structure is that of oxides of type metal **A** + 2O (such as UO_2 and TeO_2) in addition to minerals like CaF_2. Cristobalite is the high-temperature form of crystalline SiO_2 or "quartz".

B.3 CRYSTALLOGRAPHIC COORDINATES AND DIRECTIONS IN THE HEXAGONAL SYSTEM

The hexagonal lattice is important since a number of important metallic and ceramic systems exist in this state. It has dimensions/lattice parameters $a \times a \times c$ ($a \neq c$) with $\alpha = \gamma = 90°$ and $\beta = 120°$. Looking at the unit cell in **Chapter 1**, it is not apparent why it is called "hexagonal", but placing three cells together like in **Figure B.6** provides the necessary perspective. In **Figure B.6(a)**, one hexagonal "super-cell" structure is generated by combining three simple hexagonal unit cells; by combining three simple hexagonal cells with a two-atom basis, the "hexagonal close packed" (HCP) structure is generated. [The additional atomic position is placed at (2/3 1/3 1/2) in the simple hexagonal unit cell.] Crystallographic directions and planes in this structure are frequently indexed using a four-axis system, as defined in **Figure B.6(a)**. Some example directions and planes utilizing this convention are shown in **Figure B.6(b)**.

The close-packed planar configuration of atoms occurs in both the FCC and HCP structures. These structures may even be thought of as stacks of close-packed planes instead of parallelepiped unit cells. The question may occur to you: if both structures are stacks of close-packed planes, then why do they represent different crystal structures? The answer to the question is that the two structures have slightly different stacking arrangements, as shown in **Figure B.7**. **Figure B.7(a)** shows the respective close-packed planes in the two structures: {111}-type planes in the case of FCC and (0001) and (0002) planes in HCP. As shown in **Figure B.7(b)**, the stacking order differs between the two structures. The misalignment between the close-packed planes in their stacks makes for different possible orderings. In the case of FCC, three

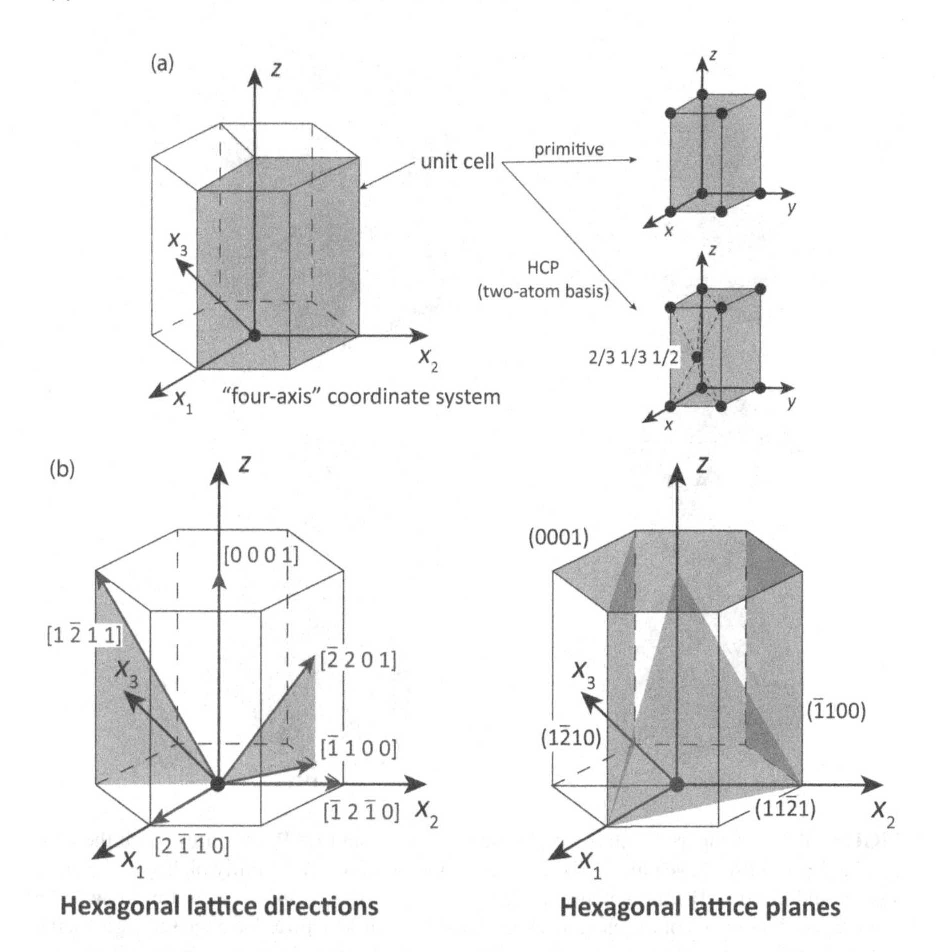

FIGURE B.6 The layout of the hexagonal (and hexagonal close-packed) lattice. The unit cell is nominally a six-sided parallelepiped in which directions and planes can be defined with three components/indices but is sometimes represented more conveniently with eight sides (= three unit cells combined), as shown in (a). In the eight-sided representation, four axes are used in the description of directions and planes. The hexagonal close-packed structure is based on the primitive unit cell but incorporates a two-atom basis. In (b), some examples of directions and planes based on the four-axis system are provided.

distinct layer alignments, "A", "B", and "C", are repeated in A-B-C-A order. In HCP structures, two distinct layer alignments are repeated.

The HCP structure is present in pure metals such as titanium, zirconium, and magnesium and determines a lot about their mechanical performance (see **Chapter 3**). HCP-like organizations are present in some other materials as well, such as alumina (Al_2O_3). The structure of alumina is illustrated in **Figure B.8**. The primary structure is comprised of O^{-2} ions arranged in an HCP-like arrangement. This arrangement incorporates six O^{-2} ions per unit cell, and the inclusion of four Al^{+3} ions completes the compound in the unit cell.

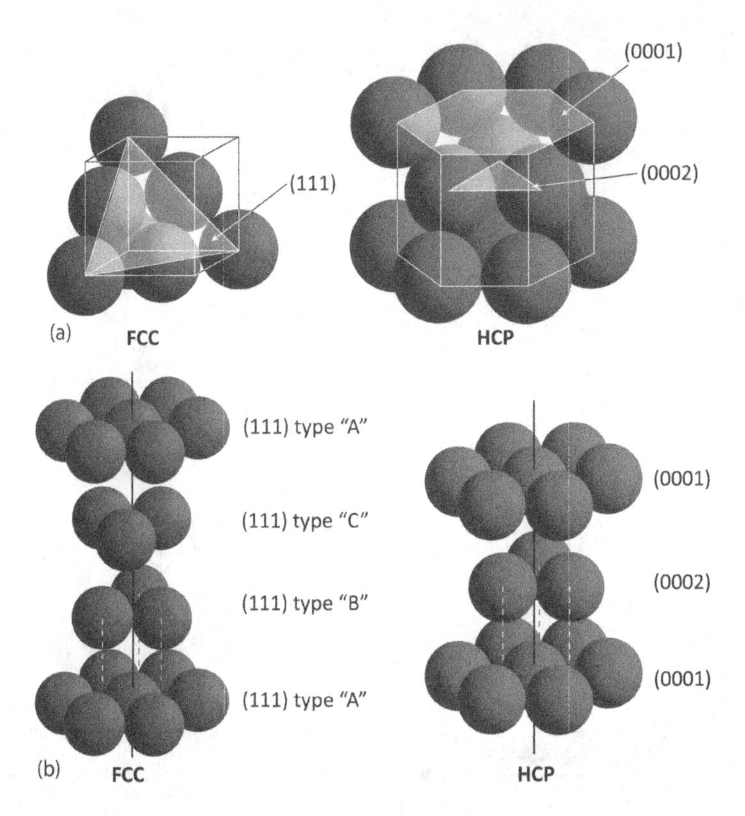

FIGURE B.7 Comparison between close-packed FCC and HCP structures. Both the FCC and HCP structures have close-packed planes; these are the {111} family planes in the cubic system and the {0001} family planes in the hexagonal system. (Some atoms are cut away for the sake of clarity.) All of the close-packed planes in both structures have the same geometry, but the difference between the two systems is in the stacking order of the planes.

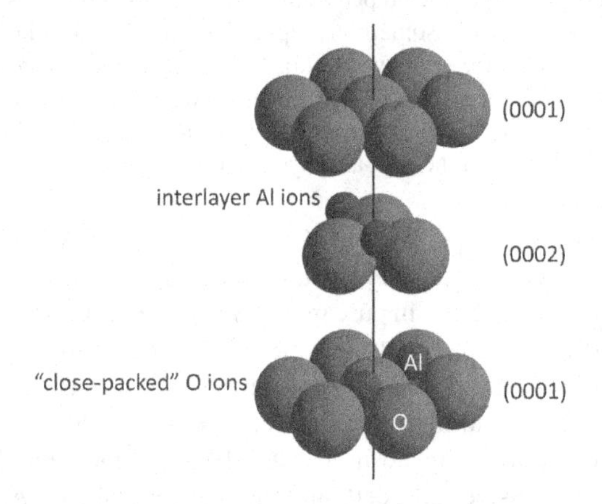

FIGURE B.8 The structure of alumina. Close-packed planes of oxygen ions are stacked in an HCP configuration. Al ions occupy positions in between otherwise close-packed layers.

Index

F

G

H

I

K

L

M

N

O

P

Q

R

S

Printed in the United States
by Baker & Taylor Publisher Services